The Biogeography of Host–Parasite Interactions

EDITED BY

Serge Morand
University of Montpellier II, France

Boris R. Krasnov
Ben-Gurion University of the Negev, Israel

OXFORD
UNIVERSITY PRESS

OXFORD
UNIVERSITY PRESS

Great Clarendon Street, Oxford OX2 6DP

Oxford University Press is a department of the University of Oxford.
It furthers the University's objective of excellence in research, scholarship,
and education by publishing worldwide in

Oxford New York

Auckland Cape Town Dar es Salaam Hong Kong Karachi
Kuala Lumpur Madrid Melbourne Mexico City Nairobi
New Delhi Shanghai Taipei Toronto

With offices in

Argentina Austria Brazil Chile Czech Republic France Greece
Guatemala Hungary Italy Japan Poland Portugal Singapore
South Korea Switzerland Thailand Turkey Ukraine Vietnam

Oxford is a registered trade mark of Oxford University Press
in the UK and in certain other countries

Published in the United States
by Oxford University Press Inc., New York

British Library Cataloguing in Publication Data
Data available

Library of Congress Control Number: 2010929118

Typeset by SPI Publisher Services, Pondicherry, India
Printed in Great Britain
on acid-free paper by
CPI Antony Rowe, Chippenham, Wiltshire

ISBN 978–0–19–956134–6 (Hbk.)
     978–0–19–956135–3 (Pbk.)

1 3 5 7 9 10 8 6 4 2

Foreword

Robert E. Ricklefs

The globalization of human activity, along with extensive habitat modification, has exposed humans to new pathogens, many of which have caused disease on epidemic proportions. This phenomenon of emerging diseases is hardly new. Over the ages, humans and their domesticated animals and plants have been plagued by pests and pathogens. The essential elements for these emerging and recurring diseases have been population crowding—monoculture in the case of crops and domesticated animals—and the access provided by travel and by agricultural development resulting in exposure to new habitats with their disease-bearing hosts and vectors. Although we are most acutely aware of emerging diseases in our own population, all species harbour parasites of various kinds and are potential hosts for new pathogens. Indeed, the distribution of parasites with respect to host taxa and geography reveals a history of mobility along both axes. The study of emerging diseases in natural populations can provide insight into the human condition, of course, as well as inform us more generally about the factors influencing the biogeography and community relationships of organisms. Indeed, the success of some invasive species has been linked either to the absence of disease organisms in the new range, releasing the invader from pathogen control, or exotic pathogens brought as allies in the ecological battle against native species.

Emerging diseases have geographic, ecological, and evolutionary components. Switching to a new host occurs at a particular place and within an appropriate ecological context, between two hosts that are compatible with a particular parasite. This compatibility might reflect a shared evolutionary history. In some cases, however, pathogens have jumped between distantly related species, and so the evolutionary background can be more complex, difficult to understand, and difficult to predict. A major issue is whether host switching requires some initiating change in environment, habitat, or host distribution, as opposed to happening spontaneously as the result of chance or some random genetic change in the parasite or host population.

Ecologists emphasize the role of the external environment in driving shifts in distribution, sometimes with conflicting effects on host organisms, parasites, and vectors. In this way, environmental change, including the climate change that the Earth is experiencing at present, can rearrange host and parasite distributions and create opportunities for host switching and new emerging diseases. This type of ecological dynamic poses formidable challenges to biologists attempting to unravel the history of host–parasite relationships. A variety of biogeographic and co-phylogenetic analyses have been brought to bear on this problem. In general, the greater the degree of host specificity, the closer the tie between host and parasite history and the more straightforward is the interpretation of contemporary distributions. All too often, switching between distantly related hosts and long-distance dispersal between regions complicate the picture and obscure the signal of the underlying processes. Given time, and the vast number of host–parasite interactions on Earth, extremely unlikely and rare events will happen with close to certainty, placing a long-lasting stamp on patterns of host–parasite distribution.

Although ecology and ecological change—the external environment—play major roles in patterns of host and parasite distribution, co-evolutionary dynamics that are intrinsic to the host–parasite interaction also contribute, and might even outweigh the influence of the external environment.

Although the coevolution of virulence and resistance can achieve a stable steady state, the delicate balance is continually challenged by the appearance of new mutations in either player, and by random changes in the environment of the coevolutionary relationship, including alternative hosts and competing parasites—what can be thought of as environmental noise, as opposed to pervasive environmental change. It is possible that such 'intrinsic' changes in a system might alter the host–parasite relationship so dramatically that the host initiates a phase of population and geographic expansion—sometimes called taxon pulses—or of population and ecological contraction. Evidence of such dynamics is pervasive and well-appreciated, but the possible (likely?) role of host–parasite coevolutionary relationships has not been firmly established.

Evolutionary biologists are concerned with the degree to which parasites have played a role in the evolution and diversification of their hosts. Certainly hosts evolve defences against parasites, at some cost, but is that the extent of the parasite influence? The introduction of parasites to evolutionary naïve hosts—malaria and pox virus in the Hawaiian avifauna comes to mind—shows that disease organisms can exert powerful selective forces, and change the abundance and distribution of hosts. Other kinds of influence are possible with potential long-term consequences for diversification of hosts and parasites. For example, a parasite endemic to one host population might be pathogenic in a sister population and prevent these populations from existing in sympatry. This appears to be the case with some introduced species, such as the North American grey squirrel, which is currently displacing the red squirrel in the United Kingdom, apparently with the help of a viral parasite for which it is an effective host. Failure of sister populations to achieve secondary sympatry ultimately could reduce the rate of diversification of a host clade and perhaps cause diversification to be self-limiting. This is speculation at this point, but the possibility exposes our lack of knowledge about these critically important host–parasite relationships.

As in the case of several well-studied mutualisms, host–parasite relationships have spatial and temporal dimensions. When dispersal of host and pathogen individuals through a host population is slow, the outcomes of coevolutionary relationships can be localized within the host population and lead to variations in host abundance—presumably even its absence—throughout its range, with consequences for other interacting species and perhaps even ecosystem processes. The establishment of such a mosaic of coevolution, as John Thompson has called it, depends on the relative rates of spread and coevolutionary change in the system, which might vary predictably among different types of organisms. As readers of this book will note, the geography of disease is beginning to receive the attention it deserves from parasitologists, ecologists, biogeographers, and evolutionary biologists. Parasites influence all aspects of the diversity and distributions of their hosts, and they will undoubtedly determine, to a large degree, the way in which the Earth's biota responds to the rapid changes in the environment caused by human activities.

Contents

Contributors

Nadir Alvarez Institute of Biology, University of Neuchâtel, Emile-Argand 11, CP 158, CH-2009 Neuchâtel, Switzerland

Jon S. Beadell Department of Ecology and Evolutionary Biology, Yale University, 21 Sachem St., ESC 150, 208107, New Haven, CT 06520, USA

Frédéric Bordes Institute of Evolutionary Sciences (Institut des Sciences de l'Évolution), Centre National de la Recherche Scientifique (CNRS), Université de Montpellier 2, F-34095 Montpellier, France

Daniel R. Brooks Department of Ecology and Evolutionary Biology, University of Toronto, 25 Harbord Street, Toronto, ON M5S 3G5, Canada

Nathalie Charbonnel Centre for Population Biology and Management (Centre de Biologie et de Gestion et des Populations), Institut National de Recherche Agronomique (INRA), Campus International de Baillarguet, CS 30016, F-34988 Montferrier-sur-lez, France

Jean-François Cosson Centre for Population Biology and Management (Centre de Biologie et de Gestion et des Populations), Institut National de Recherche Agronomique (INRA), Campus International de Baillarguet, CS 30016, F-34988 Montferrier-sur-lez, France

Yves Desdevises Université Pierre et Marie Curie, Observatoire Océanologique de Banyuls, Laboratoire Arago, Avenue du Fontaulé, F-66651 Banyuls-sur-Mer, France

Julie Deter Departement Environment, Microbiology, et Phycotoxins, Laboratoire de Microbiologie, French Research Institute for Exploitation of the Sea (IFREMER), Z.I. de La Pointe du Diable, B.P. 70, F-29280 Plouzané, France

Katharina Dittmar Department of Biological Sciences, University of Buffalo, 109 Cooke Hall, Buffalo, NY 14260, USA

László Z. Garamszegi Department of Biology, University of Antwerp, Campus Drie Eiken, Universiteitsplein 1, B-2610 Wilrijk, Belgium

Joëlle Goüy de Bellocq Evolutionary Biology Group, University of Antwerp, Groenenborgerlaan 171, B-2020 Antwerpen, Belgium

Vincent Herbreteau Cemagref, Territories, Environment, Remote Sensing, and Spatial Information Joint Research Unit (UMR TETIS), Maison de la Télédétection, 500 rue J-F Breton, F-34093 Montpellier, France

Eric P. Hoberg US National Parasite Collection, ARS, US Department of Agriculture, Animal Parasitic Diseases Laboratory, BARC East 1180, 10300 Baltimore Avenue, Beltsville, MD 20705, USA

Lenny Hogerwerf Laboratoire de Lutte Biologique et Ecologie Spatiale, Université Libre de Bruxelles, Avenue Franklin Roosevelt, 50 CP 160/12, B-1050 Brussels, Belgium

Mariah E. Hopkins Department of Anthropology, University of Texas, Austin, 1 University Station C3200, Austin, TX 78727-0303, USA

Martine Hossaert-McKey Centre for Functional and Evolutionary Ecology (Centre d'Ecologie Fonctionnelle et Evolutive), Centre National de la Recherche Scientifique (CNRS), 1919 route de Mende, F-34293 Montpellier, France

Jean-Pierre Hugot Origin, Structure and Evolution of Biodiversity, Centre National de la Recherche Scientifique (CNRS), Museum National d'Histoire Naturelle (MNHN), 55, rue Buffon, F-75231 Paris cedex 05, France

Emmanuelle Jousselin Centre for Population Biology and Management (Centre de Biologie et de Gestion et des Populations), Institut National de Recherche Agronomique (INRA), Campus International de Baillarguet, CS 30016, F-34988 Montferrier-sur-lez, France

Finn Kjellberg Centre for Functional and Evolutionary Ecology (Centre d'Ecologie Fonctionnelle et Evolutive), Centre National de la Recherche Scientifique (CNRS), 1919 route de Mende, F-34293 Montpellier, France

Boris R. Krasnov Mitrani Department of Desert Ecology, Swiss Institute for Dryland Environmental Research, Jacob Blaustein Institutes for Desert Research, Ben-Gurion University of the Negev, Sede-Boqer Campus, 84990 Midreshet Ben-Gurion, Israel

Armand M. Kuris Department of Ecology, Evolution, and Marine Biology and Marine Science Institute, University of California, Santa Barbara, CA 93106, USA

Kevin D. Lafferty Western Ecological Research Centre, US Geological Survey, c/o Marine Science Institute, University of California, Santa Barbara, CA 93106, USA

Kevin D. Matson Animal Ecology Group, Centre for Ecological and Evolutionary Studies, University of Groningen, Kerklaan 30, NL-9750AA Haren, The Netherlands

Doyle McKey Centre for Functional and Evolutionary Ecology (Centre d'Ecologie Fonctionnelle et Evolutive), Centre National de la Recherche Scientifique (CNRS), 1919 route de Mende, F-34293 Montpellier, France

Anders P. Møller Laboratoire d'Ecologie, Systématique et Evolution, Centre National de la Recherche Scientifique (CNRS), Université Paris-Sud, Bâtiment 362, F-91405 Orsay, France and Centre for Advanced Study, Drammensveien 78, NO-0271 Oslo, Norway

Serge Morand Institute of Evolutionary Sciences (Institut des Sciences de l'Évolution), Centre National de la Recherche Scientifique (CNRS), Université de Montpellier 2, 34095 Montpellier, France and Animal and integrated management of risks, Agricultural Research Centre for International Development (CIRAD), TA C-22/E Campus International de Baillarguet, F-34398 Montpellier cedex 05, France

Caroline Nieberding Evolutionary Biology Group, BDIV Research Centre, Université Catholique de Louvain, Carnoy building, Croix du Sud, 4–5, B-1348 Louvain-la-Neuve, Belgium

Charles L. Nunn Department of Anthropology, Harvard University, Peabody Museum, 11 Divinity Ave., Cambridge, MA 02138, USA

Susan L. Perkins Sackler Institute for Comparative Genomics and Division of Invertebrate Zoology, American Museum of Natural History, Central Park West at 79th Street, New York, NY 10024, USA

Pascale Perrin Institute of Evolutionary Sciences (Institut des Sciences de l'Évolution), Centre National de la Recherche Scientifique (CNRS), Université de Montpellier 2, F-34095 Montpellier, France

Benoît Pisanu Department of Ecology and Management of Biodiversity, Museum National d'Histoire Naturelle (MNHN), Centre National de la Recherche Scientifique (CNRS), 61 rue Buffon, CP 53, F-75231 Paris cedex 05, France

Robert Poulin Department of Zoology, University of Otago, P.O. Box 56, 9054 Dunedin, New Zealand

Robert E. Ricklefs Department of Biology, University of Missouri, 8001 Natural Bridge Road, St. Louis, MO 63121-4449, USA

Stéphane de la Rocque Agricultural Research Centre for International Development, (CIRAD), Campus de Baillarguet, F-34398 Montpellier, France and Food and Agriculture Organisation of the United Nations, Viale delle Terme di Caracalla, I-00153 Rome, Italy

Klaus Rohde Zoology University of New England, Armidale NSW 2351, Australia

Jan Slingenbergh Animal Health Service, Food and Agriculture Organization of the United Nations, Viale delle Terme di Caracalla, I-00153 Rome, Italy

Mark E. Torchin Smithsonian Tropical Research Institute, Apartado 0843-03092 Balboa, Ancon, Republic of Panama

Eric Waltari Tufts Cummings School of Veterinary Medicine, 200 Westboro Road, North Grafton, MA 01536, USA

Introduction

Serge Morand and Boris R. Krasnov

Studies of host–parasite relationships are strongly associated with the understanding that hosts and their parasites interact over both relatively long evolutionary and relatively short ecological time. This highlights the importance of spatial phenomena that contribute to the observed patterns, processes, and temporal dynamics of host–parasite interactions.

The idea behind this book is indebted to two most influential contributions. First, Daniel Brooks and Deborah McLennan in their book *Nature of Diversity* (Brooks and McLennan 2002) have shown the importance of history, depicted by phylogenetics, as a background that is necessary for understanding processes and contingencies that may explain the macroevolutionary patterns of geographical variation in organism diversity. Second, John Thompson in his book *Geographic Mosaic of Evolution* (Thompson 2005) substantiated the spatial context of the coevolutionary processes of interactions (such as parasitism and symbiosis) that have led to the concept of hot and cold spots of coevolution.

Indeed, recently the study of host–parasite relationships has started to catch the attention of a growing number of researchers (both ecologists and parasitologists) due to a central perspective in ecology that considers dynamical processes in a spatial context. Between the two approaches of Thompson (2005) and Brooks and McLennan (2002), there is a huge gap for empirical contributions from various domains including: community ecology, phylogeography, landscape ecology, evolutionary ecology, immuno-ecology, behavioural ecology, and health ecology.

Furthermore, investigations of host–parasite relationships in a spatial context have special importance due to the modification of the global epidemiological environment. Human activities on Earth are dramatically modifying the ecosystems through ongoing global changes (climate, biotic invasion, landscape modification) that affect the biology of hosts and their parasites, which may be displaced within and outside their geographical ranges (de la Rocque *et al.* 2008). In other words, the threats of emerging infectious diseases are spatially contextualized (Jones *et al.* 2008).

The first important aim of this book is to provide an overview of recent advances in the investigations of host–parasite relationships in a spatial (and historical) context. Major advances in concepts, methods, and tools contribute to the renewed views of host–parasite interactions. The contributions presented here do not mask the inevitable gaps in our knowledge, but they all try to give new perspectives of research. The second important aim of this book is to illustrate how the sciences of ecology and evolution contribute significantly to the applied fields of conservation and health.

The present book is organized in five parts, namely (1) historical biogeography; (2) ecological biogeography and macroecology; (3) geography of interactive populations; (4) invasion, insularity, and interactions; and (5) applied biogeography.

The first part of this book starts with the statement that hosts and their parasites have co-interacted in a historical framework, which is revealed by co-phylogenetic studies (Page 2003). Diversification involves both coevolution and colonization, which explain complex patterns of host–parasite associations.

Starting from the hypothesis that most observed host–parasite associations can be explained by historical interaction of ecological fitting, Eric Hoberg and Daniel Brooks, in the first chapter, show how major episodes of environmental change have driven both persistence and diversification of host–parasite systems. Major environmental changes create opportunities for host switching during periods of geographic expansion and for coevolution and cospeciation during periods of geographic isolation.

In the second chapter, Katharina Dittmar reviews the knowledge concerning the palaeogeography of parasites, and the exploration of the evolution of parasitism as an ecological process tied to ever-changing ecosystems. This chapter, reviewing taxonomic and genetic evidence gathered from fossil and archaeological data, gives a comprehensive treatment of past parasite migrations and host interactions.

In the next chapter, Nadir Alvarez and coauthors illustrate how studying the historical biogeography and phylogeography of obligate specific mutualisms can tell us much about these coevolutionary processes. In evolutionary time, associated pairs of mutualists are expected to respond quite differently to environmental changes that drive population expansion and contraction, range shifts, and population differentiation. The main questions asked are (a) how do obligate specific mutualisms persist in the face of such environmental change? and (b) do different demographic and evolutionary responses of mutualist pairs to environmental change affect the stability of mutualisms, or drive their diversification?

Earth ecosystems are human-dominated, and humans are affected by numerous parasites and pathogens. Pascale Perrin and coauthors, in Chapter 4, show that these pathogens are not randomly distributed on Earth and that most of them have been captured by humans during their historical migrations. They review the animal and geographic origins of parasites, which may explain the actual pattern of parasite and pathogen distribution. This chapter exemplifies how parasites and the diseases they can cause have imposed strong selection on human genetics (particularly on the genes of immune defence) and human behaviour.

The last chapter of the first part starts from the statement that co-phylogeographic patterns are used to investigate the nature of interactions, but have rarely been used for predictions. Caroline Nieberding and coauthors propose a methodological framework to predict co-phylogeographic pat-terns from various potential traits (dispersal ability, geographical range etc.) and vice versa. This framework could be used to select among competing co-diversification scenarios and to better understand how diverse species interactions lead to different evolutionary outcome.

The second part of the book takes a different viewpoint by focusing on the environmental factors that may contribute to the observed geographical patterns.

In Chapter 6, Klaus Rohde investigates the links between parasite diversity and environmental gradients in a marine environment. The marine environment is less heterogeneous than terrestrial and fresh water habitats, and more suitable for evaluating the importance and effects of gradients such as gradients in temperature, habitat heterogeneity, area, and productivity. Klaus Rohde show how recent new theories, such as the metabolic theory and some recent theoretical studies using ecosystem models, can help in understanding diversity gradients.

Frédéric Bordes and coauthors, in Chapter 7, examine the parasite diversity of terrestrial mammals. Focusing on latitudinal gradients of diversity, they show how and explore why parasite diversity does not follow the latitudinal rule and may depend on intrinsic properties of parasites.

In the next chapter, Boris Krasnov and Robert Poulin consider the relative importance of species-specific stability and spatial variation in two traits of parasites: their abundance and host specificity. They consider patterns of geographic variation in these characters and investigate evolutionary and ecological causes of these patterns. Extrinsic and intrinsic factors shape the patterns of parasite assemblages, which display a range of biogeographical patterns.

Following this, Rober Poulin and Boris Krasnov (Chapter 9) investigate the patterns in local assemblages and their taxonomic identity in a spatial context. They show how similarity in the species composition of different assemblages may vary with increasing distance between them, and explore the likely mechanisms that underpin these patterns.

All the preceding contributions at some point draw attention to the lack of knowledge due to unequal sampling for parasites across space and time. Mariah Hopkins and Charles Nunn (Chapter 10) emphasize that continental and global analyses require methods to control for sampling effort, requiring approaches to first quantify heterogeneity in sampling. They present newly developed approaches based on 'gap analysis', which provides a means to identify geographical variation in our knowledge of parasites. They review these methods and their application in studies of parasites in primates.

In Part 3, the biogeography of interactions is explored using more mechanistic approaches using the natural history of the organism.

In Chapter 11, Susan Perkins and Eric Waltari describe ecological niche modelling (ENM) and its use for parasite distribution. ENM predicts species geographic distributions based on the attributes of environmental conditions at locations of known occurrence. ENM techniques have begun to be used to predict parasite distributions, although additional dimensions are added to their host use. The authors evaluate the power of niche modelling of parasite distributions by examining predicted parasite distributions in relation to host distributions, absence data, and parasite ecology.

Serge Morand and coauthors (Chapter 12) emphasize that evolutionary and ecological theories are helpful to build several hypotheses on geographical variability in the investment in immune defences. By confronting these hypotheses with empirical data, they show how the interplay between life-history traits and immunity may depend on parasite diversity and how this may explain gradients in immune defence. In particular, they illustrate how some of components of the immunity (MHC, cell-mediated responses) may be influenced by parasite diversity and assemblages.

In Chapter 13, Julie Deter and coauthors introduce the aims of landscape epidemiology, initially the search for predictors of disease risks and the study of abiotic and biotic factors that may influence the distribution of pathogens. As spatial heterogeneity and evolutionary processes strongly influence the outcome of host-pathogen interactions, they show how the use of metapopulation theory and community ecology contribute to the understanding of occurrence of adaptation/maladaptation and the evolution of virulence, whereas life-history theory

and immuno-ecology may explain the dynamics of immuno-genetics of host resistance/susceptibility.

In Part 4, three contributions show the irrigation of parasitism in invasion ecology and conservation ecology, and the importance of insularity.

In Chapter 14, Mark Torchin and coauthors review the biogeography of host and parasite invasions. By emphasizing that parasites can be lost, transferred, and gained during invasion processes, they thoroughly explore the role of parasites in biological invasions and evaluate how geography, environment, and life-histories influence the invasion of parasites and their impacts on their hosts. The concept of parasite release may give support to the hypothesis that parasite loss may ultimately favour host diversification and speciation.

Anders Møller and László Garamszegi investigate the proximal cause of the success of host invasion in Chapter 15. Successful invasion requires an ability to disperse, become established, and finally expand, and each of these three components is affected by parasitism and host defences. They review evidence that the ability of hosts to resist parasites contributes to the successful invasion of islands and continents, which may have future implications.

Islands are laboratories for investigation evolution at work. In Chapter 16, Kevin Matson and Jon Beadell review the possible paths that may lead to immunological differentiation and increased disease susceptibility in insular birds. After summarizing case studies highlighting the interest in immunity and parasitism in insular birds, they assess the impacts of different diseases on other insular birds.

Understanding complex links between environment and disease is the central topic of the emerging discipline of health ecology (Part 5).

The emergence of infectious diseases takes place in a spatial context, as exemplified in Chapter 17 by Jan Slingenbergh and Stéphane de la Rocque. Taking avian influenza viruses as an example, they emphasize that a change in host niche availability may lead to enhanced transmission and produce an ecological invasion process at issue in the geographical spread of disease in poultry, the shifting avian host range, and the adjustment of the disease features in the new host species.

Finally, Vincent Herbreteau in the last chapter (Chapter 18) advocates that the geography of heath should meet health ecology. In an explanatory approach to diseases, health geography has contributed to a spatial understanding of local and global dynamics by multiplying scales and broadening the analysis of causative factors. The ecology of health has emerged recently as a new domain that specifically investigates the ecology of diseases. Here, illustrations are given for how the field of health ecology has developed through the advanced methods shared with health geography.

We hope that this book will be stimulating and that scientists and students will find in it illustrations and encouragement to integrate host–parasite interactions in biogeography and vice versa.

References

Brooks, D.R. and McLennan, D.A. (2002). *The Nature of Diversity: An Evolutionary Voyage of Discovery*. University of Chicago Press, Chicago.

de la Rocque, S., Morand, S., and Hendrix, G. (2008). *Climate Change and Pathogens*. Revue Scientifique et Technique, World Animal Health (OIE) 27.

Jones, K.E., Patel, N.G., Levy, M.A., Storeygard, A., Balk, D., Gittleman, J.L., and Daszak, P. (2008). Global trends in emerging infectious diseases. *Nature*, **451**, 990–93.

Page, R.D.M. (2003). *Tangled Trees: Phylogeny, Cospeciation, and Coevolution*. University of Chicago Press, Chicago.

Thompson, J.N. (2005). *The Geographical Mosaic of Coevolution*. University of Chicago Press, Chicago.

PART I

Historical Biogeography

Beyond vicariance: integrating taxon pulses, ecological fitting, and oscillation in evolution and historical biogeography

Eric P. Hoberg and Daniel R. Brooks

1.1 Introduction: challenging the vicariance paradigm

Faunal or biotic structure in complex symbiont or host–parasite systems represents the interplay of geographical, ecological and host-evolutionary associations and processes on a temporal and spatial continuum. Geographic distributions for complex biotas are structured by taxon pulses (TP) as the primary determinants of episodic expansion, geographic colonization, and isolation on varying temporal (evolutionary to ecological time) and spatial scales (local, regional to global) (Erwin 1985; Halas *et al.* 2005; Lim 2008). Ecological associations emerge from TP and are explained in the context of ecological fitting (EF) defining the potential for events of host colonization and extent for initial host range (Janzen 1985; Brooks and McLennan 2002; Brooks *et al.* 2006; Agosta and Klemens 2008). Actual or realized host ranges are influenced strongly by trophic segregation, dynamics of foraging guilds, and geographic isolation (allopatry) (Brooks and Ferrao 2005; Hoberg and Brooks 2008). A linkage of macro- and microevolutionary processes for cospeciation and coadaptation (summarized in Brooks and McLennan 2002), respectively, is further played out under the dynamic for Oscillation (OS) which constitutes a narrowing of host range or ecological associations on a temporal continuum (Janz *et al.*, 2006; Janz and Nylin 2007).

Macroevolutionary structure emanating from an interaction of TP and EF represents the historical backbone from which local to regional faunas emerge (Hoberg and Brooks 2008). Increasingly, empirical evidence favours geographical and ecological drivers (TP + EF) as primary mediators of structure, with microevolutionary determinants involved in OS (coadaptation) emerging secondarily (e.g. Nieberding *et al.* 2008). Represented in this view of the biosphere is a unifying hypothesis or model for exploring processes that have driven biogeography and diversification for host–parasite systems in evolutionary through ecological time (Hoberg and Brooks 2008).

1.2 Vicariance and the dynamics of TPs

Vicariant speciation has represented the paradigm over the past 30 years for exploring patterns and processes that have served to structure complex biotas in space and time (Nelson and Platnick 1981). Conceptually, this defined a biosphere where formation of geographical barriers resulted in the origins of sister-species whose distributions reflected a history of geological or biotic discontinuity, isolation, and allopatric speciation. A history for faunal associations or biotas was revealed from general and synchronic biogeographical patterns denoting isolation and diversification across multiple clades within a particular spatial/temporal context reflecting local, regional, or global events. As outlined by Halas *et al.* (2005) diversification under vicariance reflects large scale processes:

(1) correlated with extensive geological, geographical, or climatological events resulting in the formation of barriers;
(2) where sympatric speciation (=lineage duplication) occurs between bouts of vicariance and is largely consistent with general area relationships;
(3) where speciation events involving dispersal across barriers are rare, idiosyncratic, clade-specific, and inconsistent with general area relationships; and
(4) where extinctions (=lineage sorting) result in the absence of particular clades in areas of endemism defined by vicariance.

In this world, congruence in area relationships across clades was confounded by processes for dispersal (post speciation), peripheral isolates speciation (allopatric speciation by dispersal), and extinction, but dispersal was invoked only to explain the departures from a general pattern.

Vicariance (VC) and its coevolutionary counterpart represented by maximum cospeciation have been the interrelated models for describing the structure and history of complex host–parasite assemblages (e.g. Brooks and McLennan 1993; Page 2003; Brooks and Ferrao 2005). An expanding empirical framework indicating primary roles for dispersal and processes of geographical and host colonization challenges the explanatory power and generality for vicariance/cospeciation (Hoberg and Klassen 2002; Brooks and McLennan 2002; van Veller and Brooks 2001; van Veller *et al.* 2003; Hoberg and Brooks 2008). This view is consistent with recognition that both vicariance and dispersal are important as mediators of diversity and biotic structure (Brooks 2005), however, TP, or episodes of correlated range contraction and expansion usually over extended timeframes, provide a more comprehensive model for historical biogeography (Erwin 1985; Halas *et al.* 2005; Brooks and Ferrao 2005; Lim 2008).

The dynamics for TP differ from VC in fundamental ways (Halas *et al.* 2005; Folinsbee and Brooks 2007; Lim 2008) which can be identified (Fig. 1.1).

(1) Diversification is driven by or related to biotic expansion, such that general biogeographical patterns should be associated with dispersal and not solely related to vicariance. General patterns in TP result from breakdown of barriers or ecological isolation.
(2) Episodes of biotic expansion under TP result in complex mosaics, reticulated histories, and disparate ages, origins, and sources for specific components of biotas. Geographical heterogeneity operating at varying temporal and spatial scales during the expansion phase may result in differential effects for diversity among populations, individual species or species-assemblages established through the interplay of micro- and macroevolutionary processes.
(3) Absence of clades from a regional setting or area may be indicative of failure to disperse or successfully colonize during a particular episode of general expansion, rather than to secondary extinction. It is apparent that the dynamics of both TP and VC have been operating across temporal and spatial scales extending from deep evolutionary to shallow ecological time (integrating historical biogeography and phylogeography) involving faunas, species, and populations (Lieberman 2000; Hoberg and Brooks 2008; Koehler *et al.* 2009). Further, Hoberg and Brooks (2008), emerging in part from historical biogeographic studies by Hoberg (1992, 1995), introduced an exploration of episodic or cyclical processes as drivers for TP and as determinants of distribution and diversification for complex host–pathogen systems across all scales in earth history.

Host–parasite associations examined in the context of TP address both general and unique patterns in diversification and biogeography (Erwin 1985; Halas *et al.* 2005; Brooks and Ferrao 2005). General patterns are:

(1) wholesale isolation of a biota by the formation of a barrier (a vicariance event), leading to two adjacent, sister biotas (pairs of sister species in many clades); and
(2) wholesale expansion of a biota following the breakdown of the original barrier.

Episodes of vicariance produce stable and isolated/endemic ecological associations and alternate with episodes of biotic expansion, during which members of ecological associations living in adjacent geographic areas may come into primary or secondary contact. Consequences for ecological

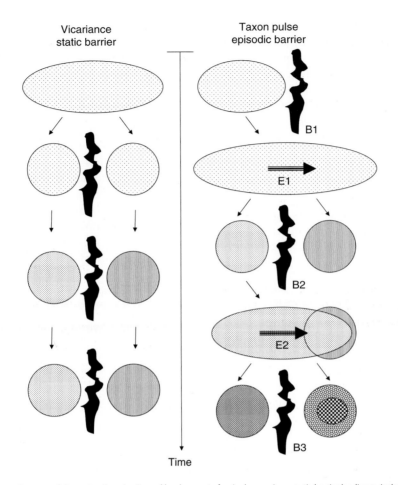

Figure 1.1   A schematic approach in contrasting vicariance (development of a single or unique static barrier leading to isolation) with taxon pulses (episodic development and breakdown of geographic barriers) and the downstream consequences for biotic structure. Shown is the simple scenario or mode for TP based on cycles or episodes for alternation in development and breakdown of barriers (B1–B3) with phases for asymmetrical expansion (E1–E2). Episodic geographic colonization and isolation result in complex mosaics for faunas, ecosystems, species, and populations as determined by spatial and temporal scale.

perturbation driving episodes of expansion and contact between evolutionarily divergent biotas are interactive and may be manifested as establishment, competition, accommodation, or extinction for associated lineages of hosts and parasites (Hoberg and Brooks 2008). Given that resulting faunal associations are complex, analyses of co-occuring clades is requisite in exploring diversification by TP, as such permit identification of reticulated area relationships (Halas *et al.* 2005; Wojcicki and Brooks 2005; Brooks and Ferrao 2005; Folinsbee and Brooks 2007; Lim 2008).

1.3   TPs and the structure of host–parasite faunas

Episodic processes and TP have operated as determinants of parasite–host diversity across all scales in earth history extending into deep evolutionary time and have been important drivers for geographic distribution and faunal radiation (Hoberg and Brooks 2008). A diverse array of potential examples, not restricted to particular host or parasite taxa, have been demonstrated across marine, aquatic, and terrestrial systems (e.g. see Brooks and

McLennan 1993; Hoberg and Klassen 2002; Hoberg and Brooks 2008). We briefly explore biogeographic and faunal connections for two iconic regions, Eurasia and Beringia, where the dynamic for episodes of biotic expansion (TP) interacting with guild associations, trophic structure, and mechanisms for ecological isolation (EF) are particularly evident.

1.4 Eurasia—establishing links across the continents

Processes of biotic expansion linking Eurasia and Africa by TP and host switching consistent with EF were drivers for distribution and diversification among phylogenetically disparate assemblages of hosts and parasites:

(1) among antelopes, associated Bovidae, and gastrointestinal nematodes (*Haemonchus*, Ostertagiinae) (Hoberg *et al.* 2004, 2008a);

(2) among Anthropoid primates, hookworms, and pinworms (*Oesophagostomum* and *Enterobius*) (Brooks and Ferrao, 2005);

(3) among Carnivora, tapeworms, and nematodes (*Taenia* and *Trichinella*) (Brooks and Hoberg 2006; Zarlenga *et al.* 2006); and

(4) among Carnivora, Hominoidea, and tapeworms (*Taenia*) (Folinsbee and Brooks, 2007; Hoberg *et al.*, 2001).

Episodes of TP extend across the late Tertiary to Pleistocene and involve differential timing for expansion events from Eurasia into Africa (or out of Africa), variable duration for periods of occupation among respective mammalian groups, and further interactions with environmental variation (Fig. 1.2).

PACT analyses ('phylogenetic analysis for comparing trees'—see Wojcicki and Brooks 2005) among nematodes in primates, and among mammalian host groups, explored historical biogeography in Africa and Eurasia (Brooks and Ferrao 2005; Folinsbee and Brooks 2007). In systems involving pinworms and hookworms in anthropoids a substantial history of cospeciation (about 70 per cent) was demonstrated, with remaining associations having resulted from events of host colonization

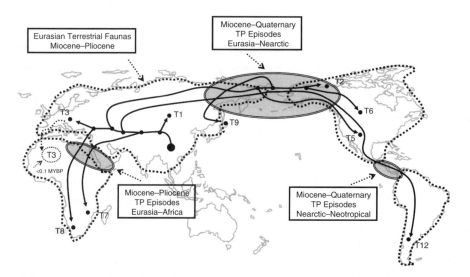

Figure 1.2 Taxon pulses out of Eurasia showing a history for episodes of biotic expansion extending across the late Tertiary into the Quaternary. Shown here is the phylogeny for the subclade of encapsulated species of *Trichinella* mapped onto a projection of contemporary geography, emphasizing the pervasive history and timing for TP and events of geographic colonization in diversification (based on Zarlenga *et al.* 2006; Pozio *et al.* 2009). Taxa are designated as follows: T1 (*T. spiralis*), T2 (*T. nativa*), T3 (*T. britovi*), T5 (*T. murrelli*), T7 (*T. nelsoni*), and 4 currently unnamed species represented by T6, T8, T9, and T12. Further indicated are the major terrestrial pathways for expansion and TP linking Eurasia with Africa and Eurasia with the Nearctic/Neotropical regions since the Miocene.

consistent with ecological fitting (shifts among primate host species). A history for speciation in *Oesophagostumum* and *Enterobius* among the anthropoid primates involved alternating episodes of isolation and subsequent movements between Africa and Eurasia consistent with TP. Among events of host switching, most were correlated with episodes of biotic expansion for primates linked to cycles of global climate change since the Miocene (Brooks and Ferrao 2005).

Episodes for TP, with geographic and host colonization since the Miocene, also served to structure gastrointestinal nematode faunas among African and Eurasian ungulates, and set the stage for further expansion and diversification into the Nearctic (Hoberg *et al.* 2004, 2008a; Hoberg 2005). Among the trichostrongyles, Haemonchinae and Ostertagiinae are sister groups with putative origins in Eurasia. Radiation for a component of this fauna was centred in Africa, coincidental with sequential expansion, arrival, and occupation by a diverse assemblage of artiodactyls over the Miocene and Pliocene (Hoberg *et al.*, 2004). Diversification for *Haemonchus* was driven by retro-colonization by nematodes from antelopes to Bovinae, Camelidae, and Giraffidae following patterns of range expansion and faunal associations linked to habitat structure which established the potential for host switching.

The contemporary distribution for *Trichinella* nematodes provides further evidence for the dynamic of pervasive host switching and episodic expansion out of Eurasia (Zarlenga *et al.* 2006, Pozio *et al.* 2009). These nematodes of carnivorans, suids, and rodents radiated initially in Eurasia and demonstrate a history of independent biotic expansion events into Africa (2 Miocene/Pliocene; 1 Quaternary), into the Neotropics through the Nearctic (1 Miocene) and the Nearctic (2 Pliocene/ Quaternary) (Fig. 1.2). As with other faunal elements linking Eurasia and Africa, and Eurasia and the Western Hemisphere, expansion during the Miocene and Pliocene and events that led to the radiation of the Nearctic fauna in the Quaternary were facilitated by episodic development of land connections and dispersal corridors which led to a breakdown in geographic isolation. Further, for *Trichinella* which circulates through carnivory and scavenging, processes for EF

through guild associations promoted colonization events among hyaenids, felids, canids, ursids, and mustelids in respective biogeographic or regional settings (Zarlenga *et al.*, 2006). Collectively, these patterns demonstrated for an array of phylogenetically and historically disparate host–parasite systems arising in Eurasia, the role of episodic expansion, geographic colonization, and isolation.

1.5 Beringia—connections across the roof of the world

The generality for TPs as drivers or determinants of faunal structure in shallow evolutionary and ecological time is further revealed in the Beringian region, or the nexus linking Palearctic and Nearctic biotas at high latitudes from the Kolyma River in Siberia to the Mackenzie River in the Northwest Territories (Hoberg and Brooks 2008; Koehler *et al.* 2009). Initially this nexus for Eurasian and Nearctic faunas was permanently emergent through much of the Tertiary, serving as a primary pathway for interchange of terrestrial faunas during times of relatively equitable climate. Tectonics and global cooling through the late Tertiary altered this dynamic. As a crossroads for the northern continents, Beringia strongly influenced patterns of distribution and speciation, alternately serving as a barrier or pathway for expansion of marine and terrestrial faunas and as a centre for diversification over the past 4–5 Myr (Hopkins 1959; Sher 1999). The episodic nature of interstadial–stadial cycles, numbering 20 events during the late Pliocene/Quaternary, together with patterns of biotic expansion and isolation at inter- and intracontinental scales, and formation/dissolution of intermittent refugia have had pervasive effects on the history (speciation, extinction) and distribution of a complex mosaic of host–parasite systems from the Holarctic to the Neotropics (e.g. Rausch 1994; Hoberg *et al.* 2003; Hoberg 2005; Cook *et al.* 2005; Dragoo *et al.* 2006; Zarlenga *et al.* 2006; Pozio *et al.* 2009) (Fig. 1.3).

Phylogeography of large nematodes (*Soboliphyme baturini*) in martens (*Martes caurina*, Pacific marten; and *M. americana*, American marten) and other mustelid hosts illustrates a critical role for TP in the temporally shallow history of the broader Beringian fauna (Koehler *et al.* 2009). This assemblage is

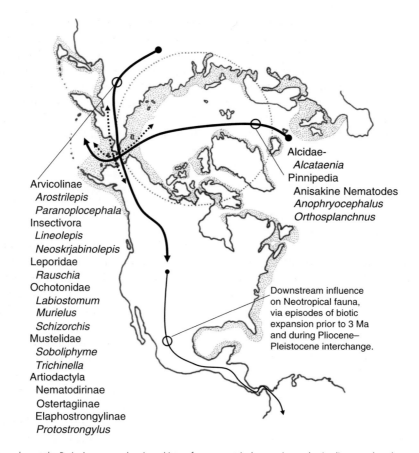

Arvicolinae
Arostrilepis
Paranoplocephala
Insectivora
Lineolepis
Neoskrjabinolepis
Leporidae
Rauschia
Ochotonidae
Labiostomum
Murielus
Schizorchis
Mustelidae
Soboliphyme
Trichinella
Artiodactyla
Nematodirinae
Ostertagiinae
Elaphostrongylinae
Protostrongylus

Alcidae-
Alcataenia
Pinnipedia
Anisakine Nematodes
Anophryocephalus
Orthosplanchnus

Downstream influence
on Neotropical fauna,
via episodes of biotic
expansion prior to 3 Ma
and during Pliocene–
Pleistocene interchange.

Figure 1.3 Taxon pulses at the Beringian nexus, showing a history for asymmetrical expansion and episodic events largely over the late Pliocene and Pleistocene as primary determinants of structure in terrestrial and marine faunas. Development of impermeable barriers to dispersal in marine and terrestrial systems alternated respectively with stadial and interstadial cycles; shown (stippled) is the extent of exposed continental shelf during glacial maxima. Cyclical breakdown of these barriers led to waves of episodic expansion for an oceanic fauna, including seabirds, marine mammals, and their parasites, primarily from the Atlantic through the Arctic Basin to the Pacific (e.g. Hoberg 1992; Hoberg 1995; Hoberg and Adams 2000; Hoberg and Klassen 2002). In parallel, the development of a diverse terrestrial assemblage of mammalian hosts and parasites resulted from biotic expansion from the Palearctic into the Nearctic (e.g. Rausch 1994; Haukisalmi *et al.* 2001, 2006; Cook *et al.* 2005; Hoberg 2005; Zarlenga *et al.* 2006; Waltari *et al.* 2007; Galbreath 2009; Koehler *et al.* 2009). The diversity of hosts and parasites depicted here is not exhaustive and constitutes a representation of substantially more complex faunas in both terrestrial and marine environments involving an array of macro- and microparasites. Schematic representation is based on conclusions for historical biogeography in marine and terrestrial systems.

representative of the mosaic process for establishment of the Beringian/Nearctic fauna through expansion involving a diverse array of soricomorphs, arvicoline rodents, lagomorphs, carnivorans, ungulates, and their associated parasites (Rausch 1994; Waltari *et al.* 2007; Hoberg and Brooks 2008).

Mustelids and *Soboliphyme* form a geographically widespread assemblage across the Holarctic

and amphiberingian region involving a single parasite species and multiple hosts, primarily represented by species of *Martes* (Koehler *et al.* 2007). Reticulate host associations are apparent, there is minimal evidence for cospeciation and distributions among mustelids have largely been determined by the process of ecological fitting and colonization (Brooks *et al.* 2006; Agosta and Klemens 2008). Phylogeographic analyses exploring sequence data from multi-gene systems demonstrated a relatively shallow history with minimal diversification (speciation), but with substantial population structure for parasites (in north-eastern Eurasia, interior Alaska, and coastal south-eastern Alaska) that is related to geographical and physical events rather than to hosts and host phylogeny (Koehler *et al.* 2009).

Subsequent to an origin in Eurasia, geographical expansion and colonization by *S. baturini* extended eastward across Beringia into Alaska with the ancestor of *M. caurina* + *M. americana* prior to 1 MYBP during an early Pleistocene stadial cycle (Stone *et al.* 2002; Koepfli *et al.* 2008). As of the late Pleistocene, coincidental with the early maximum of the terminal Wisconsin glacial cycle 65–122 KYA, these two species of martens in North America were distributed south of the Cordillera and Laurentide in western (*M. caurina*) and eastern (*M. americana*) refugia (Stone *et al.*, 2002). Although host speciation occurred over this million years of divergence, similar patterns for diversification of *Soboliphyme* are not recognized, and the associations of *S. baturini* with *M. caurina* and *M. americana* represent a complex temporal and spatial mosaic.

Phylogenetic and phylogeographic structure for *Soboliphyme* (partitioned in Siberian, Interior Alaska and coastal clans) is consistent with geographic colonization from Eurasia across Beringia and an initial north to south expansion with sequential isolation in insular (south-east Alaskan archipelago) and coastal refugial zones extending along the western shore of North America. Relatively extensive genetic diversity in coastal and insular populations of *S. baturini* is consistent with an extended period of geographic occupation and persistence of the assemblage in putative ice-free refugial zones for hosts and parasites leading up to the terminal Wisconsin (Koehler *et al.* 2009); the host–parasite assemblage may have entered the temperate zone of continental North America relatively early (mid-Pleistocene) with *Soboliphyme* distributed from the Pacific Coast and perhaps into the Great Basin.

Later, through subsequent stadial–interstadial cycles, martens were distributed across temperate latitudes of North America south of the former Cordilleran and Laurentide ice (Stone *et al.* 2002). Differentiation for martens occurred in eastern and western refugia and genetic divergence was cumulative across multiple stadial–interstadial cycles; parasites, in contrast, exhibited a failure to speciate. The parasite was likely persistent in populations of *M. caurina* through the late Pleistocene. During the Holocene, populations of *S. baturini* appear to have been influenced by waves of expansion eastward to Wyoming and northward into the coastal zone (Koehler *et al.* 2009); parasite populations in the coastal zone are consequently represented by a mosaic of older endemics (with extensive and local genetic variation) and more recently derived colonizers (where genetic diversity is not strongly partitioned geographically). It appears, however, that *S. baturini* was lost in populations of *M. americana* during isolation in eastern continental refugia south of the Laurentide in the late Pleistocene; consistent with missing the boat (see Hoberg 2005). Thus, the contemporary and currently limited occurrence of *S. baturini* among populations of American martens (and absence from central and eastern North America) is attributed to two or more independent events for secondary host colonization in the Holocene. Geographical colonization during the post-Pleistocene in interior Alaska and south-east Alaska was asynchronous, resulting from American martens tracking deglaciation of the Laurentide and Cordilleran ice, leading to secondary contact with *M. caurina* other mustelids and *S. baturini*. A recent history is reflected in patterns of low genetic variation (Koehler *et al.* 2009). A dynamic history for episodic dispersal events and isolation controlled by inter- and intracontinental barriers during alternating stadial/interstadial cycles is recognized in this system. Multiple expansion events are also identified in the history for species of *Trichinella* nematodes, played out across Beringia, in an initially broader spectrum of carnivoran hosts during time frames extending from the Miocene across the Pliocene through the Quaternary (Zarlenga *et al.* 2006; Pozio *et al.* 2009).

Beringian parasite assemblages, both marine and terrestrial, represent elegant exemplars for exploring the intricacies of geographic colonization in shallow evolutionary time (Rausch 1994; Hoberg 1992, 1995; Hoberg *et al.* 2003; Wickström *et al.* 2003; Cook *et al.* 2005; Haukisalmi *et al.* 2001, 2006; Zarlenga *et al.* 2006; Waltari *et al.* 2007; Koehler *et al.*

2009; Galbreath 2009). Episodic faunal expansion and geographic colonization has occurred at different modes and tempos extending from the late Tertiary through the Quaternary in response to fluctuations in climate, and shifts in environmental and ecological structure (e.g. Rausch 1994; Hoberg 1992, 1995, 2005; Lister 2004; Waltari *et al.* 2007; Koehler *et al.* 2009). Biotic expansion in terrestrial systems has been predominately asymmetrical, involving eastward dispersal from Eurasia into the Nearctic as evidenced by diverse assemblages of mammals (soricomorphs to ungulates) and their associated micro- and macroparasite faunas (Waltari *et al.* 2007). Ecologically and phylogenetically disparate terrestrial faunas including nematodes in lagomorphs and artiodactyls (Hoberg 2005), those inhabiting carnivores (Zarlenga *et al.* 2006; Koehler *et al.* 2009), the cestodes infecting arvicoline rodents and Ochotonidae (Haukisalmi *et al.* 2001, 2006; Wickström *et al.* 2003; Cook *et al.* 2005; Galbreath 2009) all exhibit general patterns of episodic biotic expansion and TP between the Palearctic and Nearctic at specific times during the late Tertiary and Quaternary. Adjacent Beringian marine biotas exhibit complementary patterns linking the Atlantic, Arctic, and North Pacific basins within similar temporal limits (e.g. Hoberg 1992, 1995; Hoberg and Adams 2000; Hoberg *et al.* 2003). Beringia further represents an important empirical system where outcomes for natural events can serve as a model for understanding factors that either limit or facilitate the introduction of potentially invasive species under anthropogenic control, and the responses of biotic systems to ongoing environmental perturbation linked to accelerated climate change (Cook *et al.* 2005; Hoberg 2005; Brooks and Hoberg 2006, 2007; Hoberg *et al.* 2008b).

1.6 A generality for TPs

TPs set the stage for diversification driven by episodes of environmental disruption and biotic expansion leading to a breakdown in ecological structure and mechanisms for ecological isolation within and between populations, species, faunas, and regions (Hoberg and Brooks 2008). TP provides a more robust explanation than vicariance for the history and structure of complex biogeographical

and faunal associations (Halas *et al.* 2005; Folinsbee and Brooks 2007; Lim 2008). The dynamic for TP is a foundation for understanding the fundamental patterns and distribution of diversity, but in complex systems involving symbionts and host–pathogen associations is insufficient and additional considerations are necessary. Increased levels of complexity which characterize host–parasite assemblages and the intimate evolutionary and ecological linkages (parasite species and populations nested within host species and populations, more broadly nested within guilds, ecosystems, and faunas) contrast with ecological and faunal structure among free-living organisms (e.g. Hudson *et al.* 2006; Dobson *et al.* 2008). For example, episodic processes of ecological disruption (breakdown of an ecological or physical barrier) define the initiation of an expansion phase in TP that may be the precursor for pervasive geographical and host colonization (Hoberg and Brooks 2008). A cascade of events across macroevolutionary to microevolutionary scales ensues to determine spatial and temporal mosaics for parasites relative to hosts and geography (Thompson 2005; Hoberg and Brooks 2008). Secondary and interacting processes involve EF (Janzen 1985; Brooks and McLennan 2002; Brooks *et al.* 2006; Agosta and Klemens 2008) determining the potential and realized host range (arena of opportunity and potential for colonization) and OS (Janz and Nylin 2007) which influences actual host range in an evolutionary continuum leading to specialization (outcomes in evolutionary to ecological time) following the initiation of an association among symbionts or hosts and parasites (Brooks and Ferrao 2005). Thus, TP sets the geographic stage for host colonization, which appears to be maximized during the periods of faunal expansion and minimized during periods of faunal isolation (Hoberg and Brooks 2008).

Under the dynamic of TP, parasite lineages are predicted to show alternating geographic patterns of expansion and isolation, just like free-living species. For parasites, geographic expansion may lead to a variety of host associations where TP, EF, and OS interact as hierarchical components of a larger process. The TP establishes the 'opportunity' or defines the arena for associations over time (alternating episodes for expansion and vicariance).

The expansion phase of TP and EF define the potential for processes of colonization that may initially result in broader host range. Subsequently, OS comes into play with the downstream narrowing of host range or associations over time following the initial events of expansion and host colonization. Under the simultaneous interaction of TP and OS, four outcomes are possible as outlined by Hoberg and Brooks (2008):

(1) coevolutionary affinities retained—hosts and parasites exhibit the same expansion—there is no host switching and any trophic change on the part of the parasites is mirrored by trophic change in all hosts in the life-cycle;
(2) parasite colonization—parasites disperse and switch hosts;
(3) host colonization—hosts disperse and acquire new parasites; and
(4) host and parasite colonization—hosts and parasites disperse and host switching occurs in both directions.

Further, clade-specific biotic expansion may produce clade-specific cases of host switching and cyclical or temporally recurrent biotic expansion may be a driver for episodic host switching. These examples and concepts emphasize episodic processes and TPs involving disparate assemblages of hosts and parasites and a series of discrete events in space and time. Further is the idea of historical backbones from which processes at phylogeographic scales are emergent; conceptually this suggests that we need a deeper scale to understand the foundations for processes emergent at fine spatial and temporal scales.

The dynamic for TP should be predicted in association with adjacent regions or faunas where there has been episodic development (and breakdown) of largely impermeable barriers to dispersal; barriers or 'dispersal corridor' may be variable in duration and as such may influence the subsequent timing and extent of biotic expansion and patterns of diversification. Further, TP should also be linked to cyclical episodic processes associated with climatic or oceanographic variability that have driven patterns of repeated glaciation, structural shifts in habitat, or marine transgression/regression. Perhaps such is best exemplified by stadial–interstadial cycles in

the Beringian region over the Pliocene and Quaternary (discussed here), or as observed for intermittent marine transgression in the Amazonian basin since the Miocene (Webb 1995). The alternating or periodic origin of barriers and subsequent breakdown with faunal expansion are also evident as factors influencing the structure of terrestrial faunas in Africa, Eurasia, the Nearctic, and the Neotropical regions (Zarlenga *et al.* 2006; Folinsbee and Brooks 2007; Hoberg and Brooks 2008; Lim 2008), and serve to demonstrate the generality of this phenomenon in evolutionary time. This contrasts with instances in earth history where relatively permanent, large scale barriers resulting in vicariance have emerged from 'isolated events' such as the sundering of Pangea or, later, Gondwana and Laurasia (Brown and Lomolino 1998; Benton 2009).

Explanatory power for TP and EF extends to complex faunal mosaics at the population level and in phylogeography. Recurrent climatological/physical events can drive the structure of faunal associations in shallow evolutionary and near ecological time (Thompson 2005; Dragoo *et al.* 2006; Hoberg and Brooks 2008; Koehler *et al.* 2009; Galbreath 2009). Cyclical expansion leading to faunal interchange can drive development of recurrent secondary to tertiary contact events resulting in highly complex and reticulate histories emanating from repeated episodes or opportunities for host/geographic colonization. The structure of resulting mosaics may be ephemeral to irreversible in space and time. Symmetry or asymmetry of expansion events resulting from episodic establishment and breakdown of geographic barriers (e.g. consider events at the Beringian nexus) further may influence the outcomes at new or recurrent ecotones (Hoberg 2005; Waltari *et al.* 2007). Thus, episodic events and TP strongly contrast with single, unique and isolated events leading to the establishment of barriers and to vicariant patterns.

1.7 Host switching and emerging infectious disease

Host switches are most often driven by biotic expansion, geographic dispersal, and colonization in a regime of ecological perturbation. In such instances, host switches occur at the intersection of geographic dispersal and EF based on the observation that no matter how ecologically specialized an association between species at any particular place and time, the traits (resources) characterizing the association may be phylogenetically conservative (Janzen 1985; Brooks and Ferrao 2005). Historical host switching, or acquisition of novel and naïve hosts in the context of an evolutionarily conservative ecological setting (EF), is analogous to processes linked to contemporary emergence of infectious agents or disease (EID) (Brooks and Hoberg 2006). An equivalence or uniformity for processes in space and time suggests that history (e.g. coevolution and biogeography) can be applied to a broadened understanding of the EID crisis providing a basis for prediction and a shift from a reactive to proactive approach (Brooks and Ferrao 2005).

Lessons learned from the study of historical diversity and diversification serve as analogues for defining determinants in shallow time or in contemporary systems, an issue particularly significant in effectively predicting and responding to emergent or invasive parasites and pathogens (Brooks and Ferrao 2005; Brooks and Hoberg 2006; Brooks *et al.* 2006; Miura *et al.* 2006). Phylogenetic conservatism, the changing arena for apparency of specificity, ecological fitting, and the potential for host colonization are critical in limiting or facilitating introduction, establishment, and emergence following processes of passive or natural dispersal, or in situations involving anthropogenic translocation (e.g. Torchin *et al.* 2003; Hoberg 2005). Parasites in introduced species may rapidly colonize resident hosts, even if the parasite appears to be highly host-specific in its native range, and introduced hosts may acquire resident parasites (Brooks and Ferrao 2005). Furthermore, parasites, either introduced or native, may persist in a colonized host after the original host goes extinct.

Complexity in the biosphere is described by a mosaic of specialized but evolutionarily conservative ecologies into which a substantial diversity of pathogens may fit, depending on circumstances (Brooks and Ferrao 2005). Through history, those circumstances have involved episodes or events of ecological perturbation linked to alterations in geography or climate. Emergence often implies that host

switching has been coupled with geographic movement. In these instances, either novel hosts have expanded into the area of origin of a given pathogen, or the pathogen has expanded beyond its original range into sympatry with susceptible and naïve hosts.

Conceptually we can establish a general evolutionary/ecological model that can be applied to the current crisis for invasive species and emerging infectious diseases on global, regional, and local scales (Brooks and Ferrao 2005; Brooks and Hoberg 2006). Further, modes of host range expansion/dispersal will play a role in determining the outcomes of interactions among endemic and introduced faunas under current regimes of global climate change and anthropogenic forcing (Hoberg 2005; Brooks and Hoberg 2007; Hoberg *et al.* 2008b). We emphasize the need to integrate history and historical processes, linked to empirical data which emerge from detailed taxonomic and natural history inventories, as underlying factors serving to influence the fundamental mechanisms and determinants of ecological structure in complex systems (Brooks and Hoberg 2000, 2006; Hoberg and Klassen 2002; Brooks and Ferrao 2005; Wolfe *et al.* 2007).

1.8 Conclusions

Processes that serve to structure faunas are equivalent or the same irrespective of the scale under consideration. Theoretically and empirically, this continuum has been established through a large and extensive body of research addressing biogeography and diversification in complex host–parasite systems (e.g. Brooks and McLennan 1993, Brooks and McLennan 2002; Hoberg and Klassen 2002; Thompson 2005; Brooks and Hoberg 2008). TP, EF, and OH interact and influence populations, species, and faunal associations on both deep and shallow time frames. Drivers for populations, lineages, species, and faunas across local, regional, or global distributions in evolutionary to ecological time have operated universally as determinants of diversity and biotic structure in complex host–parasite systems (e.g. Hoberg and Brooks 2008). Recovering or revealing history for biogeography of complex associations across macroevolutionary to microevolutionary scales has relied on direct comparisons of phylogenetic trees for hosts, parasites, and areas. Assuming the original input trees to be robust, discovery-based or *a posteriori* methods provide a powerful protocol for exploring evolutionary and biogeographic history for associated clades in a regional context (e.g. van Veller *et al.* 2003; Wojcicki and Brooks 2005; Folinsbee and Brooks 2007; Lim 2008).

Biogeographic and evolutionary analyses are not immediately dependant on molecular-based data. Molecular data provide one advantage in estimation of divergence times which can be used to calibrate nodes and branches in trees, allowing further correlation with episodes or events such as those that lead to expansion or isolation (e.g. Zarlenga *et al.* 2006; Lim 2008). Molecular-based methods, however, do not supplant the basic conceptual foundations and extensive insights for processes in historical biogeography which have been developed over the past 40 years (e.g. Hoberg and Klassen 2002); although fine-scale analyses of populations clearly allow for testing of hypotheses that were not previously accessible (e.g. Avise 2000; Criscione *et al.* 2005; Huyse *et al.* 2005; Dragoo *et al.* 2006; Nieberding and Olivieri 2006; Whiteman *et al.* 2007; Koehler *et al.* 2008; Nieberding *et al.* 2008).

A general model for the evolution of parasite biotas emerges from the combination of TP, EF, and OS to explain biogeographical distributions, host colonization, and the evolution of host range (Janz *et al.* 2006; Brooks *et al.* 2006; Agosta and Klemens 2008; Hoberg and Brooks 2008). The TP establishes the context for geographic distributions (Erwin 1985; Halas *et al.* 2005; Folinsbee and Brooks 2007; Lim 2008). Ecological fitting defines the arena of opportunity and potential for events of host colonization (Brooks *et al.* 2006; Agosta and Klemens 2008), whereas oscillation describes or predicts the outcomes of such events in the continuum of evolutionary to ecological time (Janz and Nylin 2007). Ecological fitting and oscillation interact, where colonization events that initially result in broader host range may be followed by a narrowing of associations with isolation on or in a particular subset of hosts within the original assemblage. Oscillation, further, should result in increasing specialization that is scale dependant in evolutionary to ecological time. In a macroevolutionary sense this involves the narrowing of host range for a symbiont and groups of related symbionts or the narrowing of ecological context for a pathogen with respect to a host or host group following development of an initial association (Janz and Nylin 2007). In a microevolutionary sense this is a function of increasing adaptation (or potential for reciprocal coadaptation) between host and parasite lineages or populations manifested as specificity and narrowing of associations on increasingly fine geographic scales. Microevolutionary or phylogeographic mosaics, ephemeral in space and time (in the sense of Thompson 2005) may be emergent from a deeper macroevolutionary and historical landscape; it is such interactions at the scale of populations of hosts and pathogens which may be particularly important as determinants of emerging diseases (e.g. Dragoo *et al.* 2006). Direct implications for understanding the often ephemeral and explosive emergence of disease and the observations that pathogens describe broader distributions than their associated disease syndromes are apparent (Audy 1958; Brooks and Hoberg 2006). The universal nature of oscillation is thus evident and further strengthens the linkage to the dynamic of TP and EF irrespective of spatial and temporal considerations. Collectively, in this view of the biosphere, a spectrum of episodic processes has been in play across the expanse of earth history and has had a substantial influence on faunal structure including complex host–parasite systems (e.g. Hoberg and Klassen 2002; Hoberg and Brooks 2008).

Acknowledgments

We thank Joe Cook at University of New Mexico and Anson Koehler now at the University of Otago for insights and discussion about Beringia, mustelids, and *Soboliphyme*. Research by EPH was in part supported by the Beringian Coevolution Project (with J.A. Cook at University of New Mexico), an interdisciplinary programme funded by the National Science Foundation (US) (DEB 0196095 and 0415668), exploring evolution and historical biogeography of complex host–parasite systems across the roof of the world. DRB acknowledges support from the Natural Sciences and Engineering Research Council (NSERC) of Canada through an Individual Discovery Grant.

References

Agosta, S.J. and Klemens, J.A. (2008). Ecological fitting by phenotypically flexible genotypes: implications for species associations, community assembly and evolution. *Ecology Letters*, **11**, 1–12.

Audy, J.R. (1958). Localization of disease with special reference to the zoonoses. *Transactions of the Royal Society Tropical Medicine and Hygiene*, **52**, 309–28.

Avise, J.C. (2000). *Phylogeography: The History and Formation of Species*. Harvard University Press, Cambridge.

Benton, M.J. (2009). The Red Queen and the Court Jester: species diversity and the role of biotic and abiotic factors through time. *Science*, **323**, 728–32.

Brooks, D.R. (2005). Historical biogeography in the age of complexity: expansion and integration. *Revista Mexicana de Biodiversidad*, **76**, 79–94.

Brooks, D.R. and Ferrao, A. (2005). The historical biogeography of coevolution: emerging infectious diseases are evolutionary accidents waiting to happen. *Journal of Biogeography*, **32**, 1291–99.

Brooks, D.R. and Hoberg, E.P. (2000). Triage for the biosphere: the need and rationale for taxonomic inventories and phylogenetic studies of parasites. *Comparative Parasitology*, **67**, 1–25.

Brooks, D.R. and Hoberg, E.P. (2006). Systematics and emerging infectious diseases: from management to solution. *Journal of Parasitology*, **92**, 426–29.

Brooks, D.R. and Hoberg, E.P. (2007). How will global climate change affect parasite-host assemblages? *Trends in Parasitology*, **23**, 571–74.

Brooks, D.R., León-Règagnon, V., McLennan, D.A., and Zelmer, D. (2006). Ecological fitting as a determinant of the community structure of platyhelminth parasites of anurans. *Ecology*, **87**, S76–S85.

Brooks, D.R. and McLennan, D.A. (1993). *Parascript: Parasites and the Language Of Evolution*. Smithsonian Institution Press, Washington, D.C.

Brooks, D.R. and McLennan, D.A. (2002). *The Nature of Diversity: An Evolutionary Voyage of Discovery*. University of Chicago Press, Chicago.

Brown, J.H. and Lomolino, M.V. (1998). *Biogeography*. Sinauer Associates, Sunderland.

Cook, J.A., Hoberg, E.P., Koehler, A. *et al.* (2005). Beringia: intercontinental exchange and diversification of high latitude mammals and their parasites during the Pliocene and quaternary. *Mammal Study*, **30**, S33–S44.

Criscione, C.D., Poulin, R., and Blouin, M.S. (2005). Molecular ecology of parasites: elucidating ecological and microevolutionary processes. *Molecular Ecology*, **14**, 2247–57.

Dobson, A., Lafferty, K.D., Kuris, A.M., Hechinger, R.F., and Jetz, W. (2008). Homage to Linnaeus: How many parasites? How many hosts? *Proceedings of the National Academy of Sciences of the USA*, **105**, 11482–89.

Dragoo, J.W., Lackey, J.A., Moore, K.E., Lessa, E.P., Cook, J.A., and Yates, T.L. (2006). Phylogeography of the deer mouse (*Peromyscus maniculatus*) provides a predictive framework for research on hantaviruses. *Journal of General Virology*, **87**, 1997–2003.

Erwin, T.L. (1985). The taxon pulse: a general pattern of lineage radiation and extinction among carabid beetles. In G.E. Ball, ed. *Taxonomy, Phylogeny and Biogeography of Beetles and Ants*, pp. 437–72. W. Junk, Dordrecht.

Folinsbee, K.E. and Brooks, D.R. (2007). Early hominid biogeography: pulses of dispersal and differentiation. *Journal of Biogeography*, **34**, 383–97.

Galbreath, K.E. (2009). *Of Pikas and Parasites: Historical Biogeography of an Alpine Host-Parasite Assemblage*. PhD Thesis, Cornell University, Ithaca, New York.

Halas, D., Zamparo, D., and Brooks, D.R. (2005). A historical biogeographical protocol for studying diversification by taxon pulses. *Journal of Biogeography*, **32**, 249–60.

Haukisalmi, V., Henttonen, H., and Hardman, L. (2006). Taxonomy, diversity and zoogeography of *Paranoplocephala* spp. (Cestoda: Anoplocephalidae) in voles and lemmings of Beringia, with a description of three new species. *Biological Journal of the Linnean Society*, **89**, 277–99.

Haukisalmi, V., Wickström, L.M., Hantula J., and Henttonen, H. (2001). Taxonomy, genetic differentiation and Holarctic biogeography of *Paranoplocephala* spp. (Cestoda: Anoplocephalidae) in collared lemmings (*Dicrostonyx*; Arvicolinae). *Biological Journal of the Linnean Society*, **74**, 171–96.

Hoberg, E.P. (1992). Congruent and synchronic patterns in biogeography and speciation among seabirds, pinnipeds and cestodes. *Journal of Parasitology*, **78**, 601–15.

Hoberg, E.P. (1995). Historical biogeography and modes of speciation across high-latitude seas of the Holarctic: concepts for host-parasite coevolution among the Phocini (Phocidae) and Tetrabothriidae. *Canadian Journal of Zoology*, **73**, 45–57.

Hoberg, E.P. (2005). Coevolution and biogeography among Nematodirinae (Nematoda: Trichostrongylina), Lagomorpha and Artiodactyla (Mammalia): exploring determinants of history and structure for the northern faunas across the Holarctic. *Journal of Parasitology*, **91**, 358–69.

Hoberg, EP., Abrams, A., and Ezenwa, V.O. (2008a). An exploration of diversity among the Ostertagiinae (Nematoda: Trichostrongyloidea) in ungulates from sub-Saharan Africa with a proposal for a new genus. *Journal of Parasitology*, **94**, 230–51.

Hoberg, E.P. and Adams, A. (2000). Phylogeny, history and biodiversity: understanding faunal structure and biogeography in the marine realm. *Bulletin of the Scandinavian Society of Parasitology*, **10**, 19–37.

Hoberg, E.P., Alkire, N.L., de Queiroz, A., and Jones, A. (2001). Out of Africa: origins of *Taenia* tapeworms in humans. *Proceedings of the Royal Society B: Biological Sciences*, **268**, 781–87

Hoberg, E.P. and Brooks, D.R. (2008). A macroevolutionary mosaic: episodic host-switching, geographic colonization and diversification in complex host-parasite systems. *Journal of Biogeography*, **35**, 1533–50.

Hoberg, E.P., Galbreath, K., Kutz, S., and Cook, J. (2003). Arctic biodiversity: from discovery to faunal baselines-revealing the history of a dynamic ecosystem. *Journal of Parasitology*, **89**, S84–S95.

Hoberg, E.P., and G.J. Klassen (2002). Revealing the faunal tapestry: coevolution and historical biogeography of hosts and parasites in marine systems. *Parasitology*, **124**, S3–S22.

Hoberg, E.P., Lichtenfels, J.R., and Gibbons, L. (2004). Phylogeny for species of *Haemonchus* (Nematoda: Trichostrongyloidea): considerations of their evolutionary history and global biogeography among Camelidae and Pecora (Artiodactyla). *Journal of Parasitology*, **90**, 1085–1102.

Hoberg, E.P., Polley, L.R., Jenkins, E.M., and Kutz, S.J. (2008b). Pathogens of domestic and free-ranging ungulates: global climate change in temperate to boreal latitudes of North America. *Office International des Épizooties Revue Scientifique et Technique*, **27**, 511–28.

Hopkins, D.M. (1959). Cenozoic history of the Bering land bridge. *Science*, **129**, 1519–28.

Hudson, P.J., Dobson, A.P., and Lafferty, K.D. (2006). Is a healthy ecosystem one that is rich in parasites? *Trends in Ecology and Evolution*, **21**, 381–85.

Huyse, T., Poulin, R., and Theron, A. (2005). Speciation in parasites: a population genetics approach. *Trends in Parasitology*, **21**, 469–75.

Janz, N. and Nylin, S. (2007). The oscillation hypothesis of host-plant range and speciation. In K.J. Tilman, ed. *Specialization, Speciation, and Radiation: The Evolutionary Biology of Herbivorous Insects*, pp. 203–15. University of California Press, Berkeley.

Janz, N., Nylin, S., and Wahlberg, N. (2006). Diversity begets diversity: host expansions and the diversification of plant-feeding insects. *BMC Evolutionary Biology*, **6**, 4.

Janzen, D.H. (1985). On ecological fitting. *Oikos*, **45**, 308–10.

Koehler, A.V.A., Hoberg, E.P., Dokuchaev, N.E., and Cook, J.A. (2007). Geographic and host range of the nematode *Soboliphyme baturini* across Beringia. *Journal of Parasitology*, **93**, 1070–83.

Koehler, A.V.A., Hoberg, E.P., Dokuchaev, N.E., Tranbenkova, N.A., Whitman, J.S., Nagorsen, D.W., and Cook J.A. (2009). Phylogeography of a Holarctic nematode, *Soboliphyme baturini*, among mustelids: climate change, episodic colonization, and diversification in a complex host-parasite system. *Biological Journal of the Linnean Society*, **96**, 651–63.

Koepfli, K.-P., Deere, K.A., Slater, G.J., Begg, C., Begg, K., Grassman, L., Lucherini, M., Veron, G., and Wayne, R.K. (2008). Multigene phylogeny of the Mustelidae: Resolving relationships, tempo and biogeographic history of a mammalian adaptive radiation. *BMC Biology*, **6**, doi10.1186/1741-7007-6-10.

Lieberman, B.S. (2000). *Paleobiogeography*. Plenum/Kluwer Academic, New York.

Lim, B.K. (2008). Historical biogeography of New World emballonurid bats (tribe Diclidurini): taxon pulse diversification. *Journal of Biogeography*, **35**, 1385–1401.

Lister, A. (2004). The impact of Quaternary ice ages on mammalian evolution. *Philosophical Transactions of the Royal Society of London B*, **359**, 221–41.

Miura, O., Torchin, M.E., Kuris, A.M., Hechinger, R.F. and S. Chiba. (2006). Introduced cryptic species of parasites exhibit different invasion pathways. *Proceedings of the National Academy of Sciences of the USA*, **103**, 19818–23.

Nelson, G. and Platnick, N.I. (eds.) (1981). *Systematics and Biogeography: Cladistics and Vicariance*. Columbia University Press, New York.

Nieberding, C.M., Durette-Desset, M.-C., Vanderpooten, A., Casanova, J.C., Ribas, A., Deffontaine, V., Feliu, C., Morand, S., Libois, R., and Michaux, J.R. (2008). Geography and host biogeography matter for understanding phylogeography of a parasite. *Molecular Phylogenetics and Evolution*, **47**, 538–54.

Nieberding, C.M. and Olivieri, I. (2006). Parasites: proxies for host genealogy and ecology? *Trends in Ecology and Evolution*, **22**, 156–65.

Page, R.D.M. (ed.). 2003. *Tangled Trees: Phylogeny, Cospeciation and Coevolution*. University of Chicago Press, Chicago.

Pozio, E., LaRosa, G., Hoberg, E.P., and Zarlenga, D.S. (2009). Molecular taxonomy, phylogeny and biogeography of nematodes belonging to the *Trichinella* genus. *Infection, Genetics and Evolution*, **9**, 606–16.

Rausch, R.L. (1994). Transberingian dispersal of cestodes in mammals. *International Journal for Parasitology*, **24**, 1203–12.

Sher, A. (1999). Traffic lights at the Beringian crossroads. *Nature*, **397**, 103–4.

Stone, K.D., Flynn, R.W., and Cook, J.A. (2002). Postglacial colonization of northwestern North America by the forest-associated American marten (Martes americana, Mammalia: Carnivora: Mustelidae). *Molecular Ecology*, **11**, 2049–63.

Thompson, J.N. (2005). *The Geographical Mosaic of Coevolution*. University of Chicago Press, Chicago.

Torchin, M.E., Lafferty, K.D., Dobson, A.P., McKenzie, V.J., and Kuris, A.N. (2003). Intriduced species and their missing parasites. *Nature*, **412**, 628–29.

Van Veller, M.G.P. and Brooks, D.R. (2001). When simplicity is not parsimonious: a priori and a posteriori approaches in historical biogeography. *Journal of Biogeography*, **28**, 1–12.

Van Veller, M.G.P., Brooks, D.R., and Zandee, M. (2003). Cladistic and phylogenetic biogeography: the art and science of discovery. *Journal of Biogeography*, **30**, 319–29.

Waltari, E., Hoberg, E.P., Lessa, E.P., and Cook J.A. (2007). Eastward ho: phylogeographic perspectives on colonization of hosts and parasites across the Beringian nexus. *Journal of Biogeography*, **34**, 561–74.

Webb, S.D. (1995). Biological implications of the Middle Miocene Amazon seaway. *Science*, **269**, 361–62.

Whiteman, N.K., Kimball, R.T., and Parker, P.G. (2007). Co-phylogeography and comparative population genetics of the threatened Galapagos hawk and three ectoparasite species: ecology shapes population histories within parasite communities. *Molecular Ecology*, **16**, 4759–73.

Wickström, L., Haukisalmi, V., Varis, S., Hantula, J., Federov, V.B., and Henttonen, H. (2003). Phylogeography of circumpolar *Paranoplocephala arctica* species complex (Cestoda: Anoplocephalidae) parasitizing collared lemmings. *Molecular Ecology*, **12**, 3359–71.

Wolfe, N.D., Panosian Dunavan, C., and Diamond, J. (2007). Origins of major human infectious diseases. *Nature*, **447**, 279–83.

Wojcicki, M. and Brooks, D.R. (2005). PACT: an efficient and powerful algorithm for generating area cladograms. *Journal of Biogeography*, **32**, 755–74.

Zarlenga, D.S., Rosenthal, B.M., La Rosa, G., Pozio, E., and Hoberg, E.P. (2006). Post Miocene expansion, colonization, and host switching drove speciation among extant nematodes of the archaic genus *Trichinella*. *Proceedings of the National Academy of Sciences of the USA*, **103**, 7354–59.

CHAPTER 2

Palaeogeography of parasites

Katharina Dittmar

2.1 Introduction

Palaeogeography of organisms involves the study of patterns and distributions through time, which—by extension—reflect ecological preferences. Like any organism, parasites live in an ever-changing environment. This has prompted them to disperse and extend, or alter their range. In the wake of this, they encountered new hosts, speciated, or went extinct.

Why does a parasite change its range? Intrinsically, the parasitic lifestyle involves the process of niche specialization to or on a particular host or host group. Although the intensity of parasite–host association may vary on spatial and temporal scales, an important cause for parasite dispersal is a range change of their host(s), which prompts the parasite's active or passive dispersal. Thus, if one wants to understand past and extant geographical distributions of parasites, one has to consider host evolution and distribution patterns. Hosts and parasites, however, are also influenced by interactions with the greater environment, which thus becomes another important aspect of understanding parasite dispersal. In the following chapter I will review the body of evidence for palaeo- and historical records of parasites, and explore this data in the context of informing ecology, and dispersal. Although 'palaeo' (Greek: *palaiós*) refers to 'ancient' or 'very old', I will also treat subfossil and historical findings in this chapter, thus presenting a broader range of records.

Since parasites pervade all known domains of life, and make up a considerable part of its biodiversity, it follows that they have been extremely successful throughout evolutionary history. This lofty statement is corroborated by scarce, yet conclusive evidence for their existence in times past.

The long-term preservation of any organism on this planet is prone to entirely fortuitous circumstances, and we cannot hope to find all past life that once existed. For parasites this problem is compounded even further, making for their gross underrepresentation in the fossil record (Dittmar 2009). One reason is their penchant for hidden niches on a host, and their often soft-bodied forms, which are not conducive to preservation. Another randomizing factor to parasite preservation is introduced by the fact that host–parasite relationships are variable, with a few individuals in a population harbouring most of the parasites. Furthermore, a stark bias exists towards the preservation of parasites of terrestrial hosts. Apart from these obvious issues, scientists also perpetuate disparities. Fossils or archaeological materials are mostly collected to study phenomena other than parasitism. Therefore it takes the broad-minded palaeontologist, archaeologist, or parasitologist to connect the dots and launch into an investigation outside their main fields of expertise. Some researchers took on this challenge, and before long, the field of palaeoparasitology was born (Ruffer 1910; Callen and Cameron 1960; Reinhard *et al.* 1988; Araújo and Ferreira 2000). Yet within this field, researchers mostly pay attention to animal parasites in the archaeological record, creating a bias towards a lesser appreciation of fossil animal and plant parasite records.

2.2 Body fossils

The bulk of the available fossil parasite data comes from invertebrate-rich amber deposits of the Cretaceous (145.5–65.5 MYA), and the Oligocene (33.9–23 MYA) (Poinar and Poinar 2007). Examples include amber encased bloodsucking ectoparasites, such as fleas (Whiting *et al.* 2008), mites (Witalinski 2000), and ticks (de la Fuente 2003; Poinar and Brown 2003). Geographically, these records are scattered between present-day Latvia, the Dominican Republic, and The Union of Myanmar (Burma). A well-preserved fossil bird louse (*Megamenopon* sp.) has been recorded from Eocene oil shale in Germany (ca. 55.8–33.9 MYA, Wappler *et al.* 2004), yet shale is an extremely rare preservation medium for parasite fossils. All three previously mentioned epochs underwent major climatic and organismal changes, providing new ecological niches for colonization by parasites. Briefly, the Cretaceous saw the birth of the South Atlantic and Indian Oceans. Flowering plants and insects were on the rise, and early mammals played a minor role in a terrestrial fauna that still was largely dominated by archosaurs. Most important in the Eocene are the rise of the Modern mammals, an epoch marked by the Palaeocene–Eocene Thermal Maximum, which set of subsequent cooling trends cumulating in the Pleistocene glaciations. During the Oligocene, continents drifted towards their current positions, angiosperms continued to expand, and terrestrial and marine fauna became fairly modern.

A more recent parasite find, which may be regarded as a subfossil record, relates to the human head louse, whose presence in Brazil was confirmed by the recovery of a nit glued to a human hair from a cave sediment sample dated at ca. 10,000 years (Araújo *et al.* 2000). Sediment and pollen analysis of the area suggests moister conditions at that time, and multiple occupied shelters in the vicinity of this find point to a thriving human population, providing ample opportunities to sustain a healthy head lice population. All of the above mentioned samples have been found disassociated from their actual host, and thus only their striking resemblance to extant parasites leads us to hypothesize about their supposed similar lifestyle in the past. Due to the lack of host records, we can only make an educated guess about their preferred habitat, based on a multitude of proxy information. For example, several flea fossils from Dominican Oligocene amber (35–30 MYA) have been identified as members of the extant family Rhopalopsyllidae (Poinar 1995; Lewis and Grimaldi 1997). Today, Rhopalopsyllidae mostly occur on Cricetidae and Hystricomorpha. Given the

fossil record of the latter on Hispaniola, we can hypothesize an Oligocene association to these rodents. The same arguments can be made for the amber preserved soft tick *Ornithodoros antiquus* (Argasidae) from Hispaniola, which shares pertinent characters with two extant rodent-specific subgenera (Poinar and Brown 2003). Based on this information their occurrence on a tree-roosting rodent host 30–40 MYA is hypothesized.

2.3 Parasite preservation with the host

The preservation of host and parasite in one specimen is the rarest of all. Such a scenario provides valuable information as to past host associations. Most of the known samples pertain to archaeological specimens, such as human and animal mummies, which yielded equally mummified lice, nits, fleas, helminthes, or protozoans (Bouchet *et al.* 2003; Dittmar *et al.* 2003; Raoult *et al.* 2008). A subfossil tick, indistinguishable from modern *Dermacentor reticulatus*, was found in the auditory canal of a Pliocene woolly rhinoceros (Witalinski 2000).

Only few fossil finds are available, and they stem exclusively from invertebrate hosts. Entomopathogenic mermithid nematodes are known from Dominican and Mexican amber, establishing their presence in the Mid-Tertiary (Poinar 1984). Parasitic fungi have been observed on the cuticles of a mosquito (Diptera: Culicidae), and a fungus gnat (Diptera: Mycetophilidae) from Dominican amber. Further identification of the fungi is not feasible, and a lingering doubt remains as to their potential post mortem colonization of these specimens. Other evidence includes trypanosomatid flagellates in an Early Cretaceous biting midge (Diptera: Ceratopogonidae) from Burmese amber (Poinar and Poinar 2005). Extant biting midges are often infested with trypanosomatids of the genus *Herpetomonas*, yet their relationship to these fossil specimens is difficult to assess from morphology alone. Recently, an acrocerid planidium (first instar larva, Diptera), has been found in Baltic amber, associated to a whirligig mite (Kerr and Winterton 2008), giving new evidence to the previously unthought-of host–parasite relationships between Diptera and Acari. However, the mere presence of a parasite in an organism doesn't automatically imply true parasite–host relationships. Ample extant evidence suggests that parasites may be accidentally ingested and transported without actually infecting their transporter.

2.4 Parasite trace evidence

The most frequent signs of past parasitism are from indirect and direct trace evidence. One cannot launch into a study of indirect trace evidence for past parasites without asking an important question: How do I recognize parasitic interaction on a specimen? Indirect trace evidence may provide information as to the presence of an interaction, yet extrapolations regarding its quality are severely limited. In its most extreme form parasitism is to be understood as an interspecies relationship with an exclusive, unidirectional benefit to the parasite, accompanied by a disadvantage, or harm to the host. Thus, if we recognize an apparent (outward) sign of harm on a fossil or archaeological host specimen, we are safe to consider a parasite among one of the inflicting causes, especially if the observed pathologies are commonly associated with extant parasitic activity. Convincing records of early parasitic relationships in metazoans pertain to pathological evidence from vermiform animals on specimens of Ordovician graptolites (Conway Morris 1981) and Lower Cambrian brachiopods (Bassett *et al.* 2004). At 520 MYA the latter record would be the oldest known evidence for a metazoan parasitic symbiosis. In graptolites it was noted that parasite attachment is not randomly distributed, yet was preferentially located in the mid-dorsal body sections, indicating niche preference. Furthermore, parasitic outgrowths were never found on graptolites whose rhabdosomes were not straight (Bates and Loydell 2000), pointing to possible host specificity.

Yet, parasitism presents but one outcome of a vast continuum of symbiotic relationships, and other interactions within that spectrum may also manifest phenotypically. For instance, the interpretation of the variety of swellings and gall-like structures from Cambrian trilobites as evidence for endoparasitic activity is speculative at best (Conway Morris 1981). Likewise, caution should be taken in the interpretation of signs of bioimmuration (encrusting)

on host fossils, as they may stem from epibiotic activities rather than parasitic ones (Franzen 1974). Numerous shell fossils show circular bore holes, as early as the Proterozoic and Cambrian (Bengtson and Zhao 1992; Conway Morris and Bengtson 1994). The holes have been largely attributed to the activity of platyceratid gastropods (Baumiller 1990), yet it is often unclear whether they were parasites or true predators. Only few fossils show attachment scars, which would indicate a prolonged contact with the driller, thus making a parasitic relationship more likely (Kelley and Hansen 2003). Other evidence comes from gall inducing insects. Proof of the presence of these parasitoids on plants reaches back 300 MYA. Recently, exceptionally well-preserved Pleistocene fossil gall specimens from the Netherlands were used to reconstruct their multitrophic associations. By placement of one fossil into an extant clade of host plant alternation, the presence of an oak species (*Quercus cerris*) could be deduced. This oak is currently unknown from fossil data in that area, yet the presence of a particular parasite added information to the potentially available host landscape (Stone *et al.* 2008). Similar research has used Cretaceous plant fossils from Israel to understand the evolution of endophytic parasite communities in the basal angiosperm radiations (Krassilov 2008).

Occasionally, skeletal, or mummified, archaeological specimens show pathological signs concordant with parasite infection, such as anaemia, osteoporosis, and nutritional abnormities (Leslie and Levell 2006; Reinhard and Bryant 2008). Often however, the actual parasite causing these signs cannot be found anymore, or preservation efforts prohibit destructive sampling. Researchers thus proceed to search for other evidence betraying the presence of a parasite in life, such as parasite specific antigens. Miller *et al.* (1992) were able to trace evidence of *Schistosoma* infection to 5,000-year-old Egyptian mummies by ELISA studies, a technique that was later also used to confirm the presence of *Trypanosoma cruzi* in South American mummies. Other examples for the success of such approaches come from human faecal remains as far ago as 5,300 years, which reacted positive for the parasite *Entamoeba histolytica* in Argentina, the USA, France, Belgium, and Switzerland, speaking to the continued worldwide distribution of this parasite. While caution has to be exercised regarding potential false positive results and cross-reactions, the sensitivity of molecule-based methods to detect parasites is much higher than conventional diagnostics of parasite developmental stages.

Direct trace evidence mainly stems from faecal matter, which is often preserved in the fossil and archaeological record (Reinhard 1988; Bouchet *et al.* 2003; Paabo *et al.* 2004). Coprolites, animal middens, or latrine sediments fall into this category. Parasitological evidence from these sources mainly concerns specimens of endoparasite eggs, and occasionally their larvae or adults. Only rarely are ectoparasites reported from digestive matter (Fugassa *et al.* 2007; Johnson *et al.* 2008). Finding of developmental parasitic stages in faecal matter clearly indicates their presence in a host. To diagnose whether this is a sign of true infection hinges on correct identification of the parasite. This may be fraught with difficulties. First, it is challenging to match the faecal material to the right source, although characteristic shapes and contents may help (Chame 2003). Second, taphonomic processes are known to alter appearance of biological material, thus decreasing an often already sparse diagnostic character set. To date, the successful diagnosis of parasite eggs from fossil and ancient faecal materials spans all three major classes of helminths, namely nematodes, cestodes, and trematodes, as well as protozoans (Aspöck *et al.* 1996; Bouchet *et al.* 2002; Dittmar 2009). Results from these efforts refine extant distribution records, and let researchers hypothesize about past host and parasite distributions, as well as their subsequent dispersal. A comparative approach between subfossil and extant data has been successfully applied to elucidate pre- and post Beringian distribution patterns of hook-, thread-, and whipworm infestations in humans (Hawdon and Johnston 1996; Sianto *et al.* 2005; Montenegro *et al.* 2006; Araújo *et al.* 2008). Current evidence points to the arctic environment as barrier to the survival and distribution of some parasites.

The majority of palaeoparasitological research is centred on a human perspective. Yet, human parasites and those of other animals are linked at multiple levels. First, multiple parasites with a preference for non-human hosts may infest humans, as

evidenced by extant zoonoses, such as trichinellosis (endoparasite), or fleas (ectoparasites). Furthermore, animal parasites may accidentally be present in human samples, without causing disease. The latter scenario is easy to misinterpret, since often parasite eggs can only be diagnosed to a generic level, because they lack taxonomically defining characteristics. Palaeoparasitological evidence corroborating past zoonoses has been recorded many times. For instance, Ferreira *et al.* (1984) confirmed the presence of *Diphylobothrium pacificum* in coprolites of 4,000-year-old Chilean human mummies. *D. pacificum* larvae use shellfish and fish as intermediate hosts. Therefore, this finding not only speaks of the dietary habits of the ancient human populations, but also provides a timeline for this custom, as extant populations still are infested with this parasite. Another example related to a similar marine diet comes from the coprolites of a human mummy from south-east Brazil (600–1,200 BP) (Sianto *et al.* 2005).

Another currently completely understudied line of palaeoparasitological evidence is faecal matter accumulated in protected places (i.e. packrat middens, bat, or bird guano) (Dittmar 2009). Most of the research on this material has concentrated on dietary remains (Poinar *et al.* 2003). However, because these materials tend to accumulate over long timeframes, they may reveal important information about changes in parasite fauna in the context of dietary habits (Fugassa *et al.* 2007).

2.5 Molecular fossils

Recent technical developments have made it possible to use sequence-based ancient DNA methods to elucidate geographical patterns of parasite dispersal over temporal scales. Naturally, given the already scarce material and the unique challenges associated to DNA preservation, only a few records of ancient or historic parasite DNA exist. Studies on 9,000-year-old Chilean mummies revealed the presence of protozoan *Trypanosoma cruzi* kinetoplastid minicircle DNA, and confirmed the presence of a pathogen without outward macropathological evidence (Guhl *et al.* 1999; Madden *et al.* 2001; Aufderheide *et al.* 2004). Similarly, Zink *et al.* (2006) showed *Leishmania* sp. infestation in Egyptian mummy tissue. Parasitic helminth DNA (i.e. *Trichuris* sp., *Ascaris* sp., *Enterobius* sp.) has been sequenced from coprolites by using specific primers to target a particular gene of interest directly (Loreille *et al.* 2001; Iñiguez *et al.* 2003a; Leles *et al.* 2008), or by random oligonucleotide techniques (Iñiguez *et al.* 2003b). Ectoparasite archaic DNA data exist from the human flea (*Pulex irritans*) and head lice (*Pediculus humanus capitis*), and was obtained directly from preserved specimens (Dittmar *et al.* 2003; Raoult *et al.* 2008). Both studies lend support to the idea that these parasites were not introduced with the latest European contact, thus further dispelling lingering notions of a pre-contact parasite free New World. In the case of the head lice study, the analysis also showed that the archaic lice belong to a widespread genotype that originated in Africa, and includes head and body lice. While not proving the pre-European contact presence of body lice in the Americas, it gives evidence to this possibility (Raoult *et al.* 2008). Another striking example for tracing parasite dispersal by means of molecular data pertaining to archaeological material comes from a study of 4,500–7,000-year-old human remains from the Peruacu Valley in Brazil. Lima *et al.* (2008) found molecular evidence for the presence of *Trypanosoma cruzi* I, a genotype of the Chagas Disease-causing pathogen. This genotype is currently absent in the area and, linked to current data on deforestation and microclimate shifts, points to changing epidemiological profiles.

Parasitic invasions can also be found in another arena, essentially making every organism on this planet a living treasure trove of fossils. With the recent explosion of genome-wide analyses genomic scars from past parasitic invasions became visible in many genomes of eukaryotes. One apparent source is autonomous, mobile parasitic (selfish) DNA sequences, known as transposable elements (TEs). Just as with any other parasite, their evolutionary dynamics are complex, involving selection at the host level, transposition regulation, and genetic drift (Le Rouzic and Capy 2005). They are also known to decrease fitness by inserting themselves into coding DNA sequences. Studies showed them to be responsible for 50 per cent of all deleterious mutations in Drosophila. Host switching (horizontal transfer) between species has been shown repeatedly, one

example being the *copia* LTR retrotransposon transfer between the flies *Drosophila melanogaster* and *D. willistoni*. Although separated by ca. 50 million of years of evolution, the TE transfer is recent. From the available data, researchers deduct that the transfer to *D. willistoni* in the Americas most likely happened 100–200 years ago, when the *melanogaster* fly became cosmopolitan in distribution and thus sympatric with neotropical *Drosophila* (King Jordan *et al.* 1999). Other evidence for life's constant exposure to parasites comes from ancient (and recent) remnants of endogenous retroviruses (ERVs) lodged in the genomes of birds, reptiles, amphibians, fish, and mammals. While most of these unwanted residents have been silenced over large evolutionary timeframes, and are rendered inactive, recent studies on an endogenizing koala virus showed its ability to produce an infectious virus (Tarlington *et al.* 2006). This virus entered the germline only ca. 200 years ago, thus pinpointing the event to the geographic area of current koala distribution—Australia. Organisms harbouring intact ERVs, may serve as reservoirs, enable transmission within and across species boundaries, and thus escape host defences.

2.6 Conclusions

While it is likely that over future years new (traditional) fossil evidence for parasites, or their activity, will be discovered, it becomes clear that we only have limited opportunities to directly sample the parasitic past, and thus will only glimpse a snapshot of their history. Because of the strong preservational bias, fossil findings only record isolated occurrences at specific sites (Fig. 2.1). Therefore the real question is: Can single fossil and/or archaeological records really produce crucial information to understand the dynamic process of dispersal?

My answer would be 'YES'! Fossils may provide multiple important information points for parasitological research. The location of the find is only one of them, and it not only speaks to the presence of the parasite in this area, but per association also suggests the occurrence of its host, at the same spatial plane. If compared to extant data, evidence of migrations may be inferred. Many parasites may function as a vector for disease-causing pathogens and, therefore, this reasoning can be extended to the

Figure 2.1   World map of palaeoparasitological evidence (including archaeological records). Triangles mark body evidence from arthropods, circles mark helminths, or protozoan finds, stars mark direct or indirect trace evidence. Dark grey colour denotes fossils, whereas light grey colour stands for historic (archaeological) samples. The different sizes of the shapes are proportional to the number of records. Locations are approximate.

potential presence and/or migration of the pathogen as well. Naturally, in an earth-time-continuum location is relative, and depending on the age of the find, tectonic movements have to be taken into account to evaluate the approximate location at the time of death. Another clue can be derived from the age of the geological stratum or archaeological site of the find, which immediately puts a time point to the evolutionary trajectory of this parasite and, likewise, its host or pathogen. In the context of this trajectory, morphological change can be observed in comparison to extant relatives. A fossil may thus provide important information in understanding adaptive morphological successions. Interestingly, most parasite fossils appear to have changed little. This phenomenon may (among other factors) be related to the often-invoked rapid radiation of parasite taxa, leaving little time for transitional phenotypes. Occasionally however, a fossil parasite cannot be conclusively matched to an extant parasite. In these cases, the extrapolation of their lifestyle is entirely based on certain morphological characteristics of extant taxa that are known to betray parasitism. Examples may be piercing and sucking mouthparts, suggesting a diet based on

animal or plant fluids, or grasping appendages and ctenidia designed to hold on to a host, or circumvent host-grooming efforts. As such, Rasnytsin (2002) classifies a Mesozoic compression fossil—*Strashila incredibilis*—as well as two other records (*Saurophtirus longipes* and *Tawinia australis*) as prefleas. Based on the position of their legs and apparently strong claws he furthermore hypothesizes their similarity to extant bat wing parasites, and postulates their occurrence on pterosaurs, as well as the subsequent shift of fleas from pterosaurs to mammals. While there is clear genetic evidence of a flea radiation with early mammals, Rasnytin's assessment remains controversial (Whiting *et al.* 2008), and further evidence is needed to substantiate this extrapolation.

Parasites play key roles in species interactions, and their temporal and geographic dynamics are important in shaping coevolutionary processes (Thompson 2005). It is possible to estimate these dynamics of parasitic dispersal by sampling the present patterns of genetic diversity, geographical distribution, or ecological parameters, and putting them into an evolutionary perspective. Central to this are the disciplines of historical biogeography and phylogeography (Brooks 1985; Arbogast and Kenagy 2001; Zink 2002). Multiple examples of this will be brought forward in subsequent chapters of this book. Yet, inclusion of past parasite records in extant data can provide for a more realistic scenario for the temporal scales of events. As such they can be used as minimum and maximum calibration points of divergence time estimation for both hosts and parasites. Furthermore, geographical records can help clarify ambiguous estimations of their ancestral distribution areas, by putting a concrete value on nodes of a phylogenetic tree.

However, phylogenetic studies provide only one view of the phenomenon of parasite dynamics. Equally important to understanding parasite evolution on long and short scales are ecological considerations. Climate change is arguably the most pressing ecological (and societal) issue of this century. Undoubtedly our planet is undergoing dramatic ecological changes, including habitat loss and fragmentation, species loss, invasions, and homogenization. Yet, the determination of change and causality in ecosystems is difficult, both philosophically and practically, and these difficulties increase with the scale and complexity of ecosystems. Parasitic evidence from the past provides an additional connecting point among members of an ecosystem long gone. Trace evidence, especially in the form of coprolites or other faecal matter is likely to preserve dietary remnants of plants, insects, or vertebrates, putting the parasite into the context of an ecosystem. Additionally, they can stand in as a proxy for ecological parameters. An example for this is the find of *Dicrocoelium* eggs in an Iron Age site in South Africa (Dittmar and Steyn 2004). While absent from the area today, our knowledge about the extant biology of this parasite immediately implies the presence of an intermediate invertebrate host, and a local climate more conducive to its development than the present dry conditions. In this context, combining past and extant parasitological and climate distributions could lead to a better long-term predictive power regarding the parasitological consequences of climate change on local and global scales.

Despite increasing records of fossil and archaeological data on parasites, it is still impossible to determine the true parasite load at the time of death. Therefore, if such data are available, they can only be taken as an approximation of prevalence, and epidemiological simulations have to be interpreted accordingly.

Finally, an important message reverberates through this last paragraph: evidence from the past is most powerful if compared to, and combined with, extant data. It is to be hoped that future efforts will be made to better capitalize on our knowledge of the past.

References

Araújo, A. and Ferreira, L.F. (2000). Paleoparasitology and the antiquity of human host-parasite relationships. *Memorias do Instituto Oswaldo Cruz*, **95**, 89–93.

Araújo, A., Ferreira L.F., Guidon, N., Maues da Serra Freire, N., Reinhard, K.J., and Dittmar, K. (2000). Ten thousand years of head lice infection. *Parasitology Today*, **16**, 269.

Araújo, A., Reinhard, K.J., Ferreira L. F., and S. L. Gardner. (2008). Parasites as probes for prehistoric human migrations. *Trends in Parasitology*, **24**, 112–15.

Arbogast, B.S. and Kenagy, G.J. (2001). Comparative phylogeography as an integrative approach to historical biogeography. *Journal of Biogeography*, **28**, 819–25.

Aspöck, H., Auer, H., and Picher, O. (1996). *Trichuris trichiura* eggs in the neolithic glacier-mummy from the Alps. *Parasitology Today*, **12**, 255–56.

Aufderheide, A. C., Salo, W., Madden, M. *et al.* (2004). A 9,000-year record of Chagas' disease. *Proceedings of the National Academy of Sciences of the USA*, **101**, 2034–39.

Bassett, M.G., Popov, L.E., and Holmer, L.E. (2004). The oldest-known metazoan parasite? *Journal of Paleontology*, **78**, 1214–16.

Bates, D.E.B. and Loydell, D.K. (2000). Parasitism on graptoloid graptolites. *Paleontology*, **43**, 1143–51.

Baumiller, T.K. (1990). Non-predatory drilling of Mississippian crinoids by platycerid gastropods. Paleontology, **33**, 743–48.

Bengtson, S. and Zhao, Y. (1992). Predatorial borings in late Precambrian mineralized exosceletons. *Science*, **257**, 367–69.

Bouchet, F., Guidon, N., Dittmar, K. *et al.* (2003). Parasite remains in archaeological sites, *Memorias do Instituto Oswaldo Cruz*, **98**, 47–52.

Brooks, D.R. (1985). Historical ecology: a new approach to studying the evolution of ecological associations. *Annals of the Missouri Botanical Gardens*, **72**, 660–80.

Callen, E.O. and Cameron, T.W.M. (1960). A prehistoric diet revealed in coprolites. *New Scientist*, **8**, 35–40.

Chame, M. (2003). Terrestrial mammal feces: a morphometric summary and description. *Memorias do Instituto Oswaldo Cruz*, **98** (Suppl. I), 71–94.

Conway Morris, S. (1981). Parasites and the fossil record. *Parasitology*, **82**, 489–509.

Conway Morris, S. and Bengtson, S. (1994). Cambrian predators: possible evidence from boreholes. *Journal of Paleontology*, **68**, 1–23.

de la Fuente, J. (2003). The fossil record and the origin of ticks (Acari: Parasitiformes: Ixodida). *Experimental and Applied Acarology*, **29**, 331–44.

Dittmar, K. (2009). Old parasites for a New World: the future of palaeoparasitological research. *Journal of Parasitology*, **95**, 215–21.

Dittmar, K. and Steyn, M. (2004). Paleoparasitological analysis of coprolites from K2, an Iron Age archaeological site in South Africa: the first finding of *Dicrocoelium* sp. eggs. *Journal of Parasitology*, **90**, 171–13.

Dittmar, K., Mamat, U., Whiting, M., Goldmann, T., Reinhard, K., and Guillen, S. (2003). Techniques of DNA-studies on prehispanic ectoparasites (*Pulex* sp., Pulicidae, Siphonaptera) from animal mummies of the Chiribaya Culture, Southern Peru. *Memorias do Instituto Oswaldo Cruz*, **98**, 53–59.

Ferreira, L.F., Araújo, A., Confalonieri, U., and Nuñez, L. (1984). The finding of *Diphyllobothrium pacificum* in human coprolites (4,100-1,950 BC) from Northern Chile. *Memorias do Instituto Oswaldo Cruz*, **79**, 175–80.

Franzen, C. (1974). Epizoans on Silurian-Devonian crinoids. *Lethaia*, **7**, 287–301.

Fugassa, M.H., Sardella, N.H., and Denegri, G.M. (2007). Paleoparasitological analysis of a raptor pellet from Southern Patagonia, *Journal of Parasitology*, **93**, 421–22.

Guhl, F., Jaramillo, C., Vallejo, G.A. *et al*. (1999). Isolation of *Trypanosoma cruzi* DNA in 4,000-year-old mummified human tissue from northern Chile. *American Journal of Physical Anthropology*, **108**, 401–12.

Hawdon, J. M. and Johnston, S.A. (1996). Hookworms in the Americas: an alternative to trans-Pacific contact. *Parasitology Today*, **12**, 72–4.

Johnson, K. L., Reinhard, K. J., Sianto L., Araújo, A., Garnder, S. L., and Janovy, J. (2008). A tick from a Prehistoric Arizona coprolite. *Journal of Parasitology*, **94**, 296–97.

Iñiguez, A. M., Reinhard, K. J., Araújo, A., Ferreira, L. F., and Vicente, A.C.P. (2003a). *Enterobius vermicularis*: ancient DNA from North and South American human coprolites. *Memorias do Instituto Oswaldo Cruz*, **98**, 67–69.

Iñiguez, A. M., Araújo, A., Ferreira, L. F., and Vicente, A.C.P. (2003b). Analysis of ancient DNA from coprolites: a perspective with random amplified polymorphic DNA-polymerase chain reaction approach. *Memorias do Instituto Oswaldo Cruz*, **98**, 63–65.

Kelley, P.H. and Hansen, T.A. (2003). The fossil record of drilling predation on bivalves and gastropods. In P.H. Kelley, M. Kowalewski and T.A. Hansen, eds. *Predator-Prey Interactions in the Fossil Record*, pp. 113–33. Kluwer Academic Publishers, New York.

Kerr, P.H. and Winterton, S.L. (2008). Do parasitic flies attack mites: evidence in Baltic amber. *Biological Journal of the Linnean Society*, **93**, 9–13.

King Jordan, I., Matyunina, L.V., and McDonald, J.F. (1999). Evidence for the recent horizontal transfer of long terminal repeat retrotransposon. *Proceedings of the National Academy of Sciences of the USA*, **96**, 12621–25.

Krassilov, V. (2008). Mine and Gall predation as top down regulation in the plant-insect systems from the Cretaceous of Negev, Israel. *Palaeogeography, Palaeoclimatology, and Palaeoecology*, **261**, 261–69.

le Rouzic, A. and Capy, P. (2005). The first steps of transposable elements invasion. *Genetics*, **169**, 1033–43.

Leles, D., Araújo, A., Ferreira, L.F., Paulo Vicente, A. C., and Iñiguez, A.M. (2008). Molecular paleoparasitological diagnosis of *Ascaris* sp. from coprolites: new scenery of ascariasis in pre-Colombian South America times. *Memorias do Instituto Oswaldo Cruz*, **103**, 106–08.

Leslie, K.S. and Levell, N.J. (2006). Cutaneous findings in mummies from the British Museum. *International Journal of Dermatology*, **45**, 618–21.

Lewis, R.E. and Grimaldi, D. (1997). A pulicid flea in Miocene amber from the Dominican Republic (Insecta: Siphonaptera: Pulicidae). *American Museum Novitates*, **3205**, 1–9.

Lima, S. L., Iñiguez, A. M, and Otsuki, K. *et al*. (2008). Chagas disease in ancient hunter-gatherer populations. *Emerging Infectious Diseases*, **14**, 1001–02.

Loreille, O., Roumat, E., Verneau, O., Bouchet, F., and Hänni, C. (2001). Ancient DNA from *Ascaris*: extraction amplification and sequences from eggs collected in coprolites. *International Journal of Parasitology*, **31**, 1101–06.

Madden, M., Salo, W.L., Streitz, J. *et al*. (2001). Hybridization screening of very short PCR products for paleoepidemiological studies of Chagas' disease. *Biotechniques*, **30**, 102–04.

Miller, R.L., Armelagos, G.J., Ikram, S., De Jonge, N., Krijger, F.W., and Deelder, A.M. (1992). Palaeoepidemiology of schistosoma infection in mummies. *British Medical Journal*, **304**, 555–56.

Montenegro, A., Araújo, A., Eby, M., Ferreira, L.F., Hetherington, R., and Weaver, A.J. (2006). Parasites, paleoclimate and the peopling of the Americas: using the hookworm to time the Clovis migration. *Current Anthropology*, **47**, 193–200.

Pääbo, S., Poinar, H., and Serre, D. et al. (2004). Genetic analyses from ancient DNA. *Annual Review of Genetics*, **38**, 645–79.

Poinar, G.O. (1984). Fossil evidence of nematode parasitism. *Revue Nematologique*, **7**, 201–03.

Poinar, G.O. (1995). Fleas (Insecta: Siphonaptera) in Dominican amber. *Medical Science Research*, **23**, 789.

Poinar G. and Brown, A.E. (2003). A new genus of hard ticks in Cretaceous Burmese amber (Acari: Ixodida: Ixodidae). *Systematic Parasitology*, **54**, 199–205.

Poinar, G.O. and Poinar, R. (2005). Fossil evidence of insect pathogens. *Journal of Invertebrate Pathology*, **89**, 243–50.

Poinar, G.O. and Poinar, R. (2007). *What Bugged the Dinosaurs? Insects, Disease, and Death in the Cretaceous*. Princeton University Press, Princeton.

Poinar, H.N., Kuch, M., McDonald, G., Martin, P., and Pääbo, S. (2003). Nuclear gene sequences from a late Pleistocene sloth coprolite. *Current Biology*, **12**, 1150–52.

Raoult, D., Reed, D.L., and Dittmar, K. et al. (2008). Molecular identification of lice from pre-Columbian mummies. *Journal of Infectious Diseases*, **197**, 535–43.

Rasnitsyn, A.P. (2002). Order Pulicida Billbergh, 1820. The fleas. In A.P. Rasnitsyn and D.L.J., eds. *History of Insects*, pp. 240–42. Kluwer Academic Publishers, Dordrecht.

Reinhard, K.J. and Bryant, V.M. (2008). Pathoecology and the future of coprolite studies in bioarchaeology. In A.W.M. Stodder, ed. *Reanalysis andReinterpretation in Southwestern Bioarchaeology*, pp. 205–24. Arizona State University Press, Tempe.

Reinhard, K.J., Confalonieri, U.E., Herrmann, B., Ferreira, L.F., and Araújo, A. (1988). Recovery of parasite remains from coprolites and latrines: aspects of paleoparasitological technique. *Homo*, **37**, 217–39.

Ruffer, M.A. (1910). Note on the presence of *Bilharzia haematobia* in Egyptian mummies of the Twentieth dynasty (1250-1000 BC). *British Medical Journal*, **1**, 16.

Sianto, L., Reinhard, K. J., Chame, M. et al. (2005). The finding of *Echinostoma* spp. (Trematoda: Digenea) and hookworm eggs in coprolites collected from a Brazilian mummified body dated of 600-1,200 years before present. *Journal of Parasitology*, **91**, 972–75.

Stone, G.N., van der Ham, R.W., and Brewer, J.G. (2008). Fossil oak galls preserve ancient multitrophic interactions. *Proceedings of the Royal Society of London B*, **275**, 2213–19.

Tarlington, R.E., Meers, J., and Young, P.R. (2006). Retroviral invasion of the koala genome. *Nature*, **442**, 79–81.

Thompson, J.N. (2005). Coevolution: the geographic mosaic of coevloutionary arms races. *Current Biology*, **15**, 992–94.

Wappler, T., Smith, W.S., and Dagleish, R.C. (2004). Scratching an ancient itch: an Eocene bird louse fossil. *Proceedings of the Royal Society of London B*, **271**, 255–8.

Whiting, M.F., Whiting, A.S., Hastriter, M., and Dittmar, K. (2008). A molecular phylogeny of fleas (Insecta: Siphonaptera): origins and host associations. *Cladistics*, **24**, 677–707.

Witalinski, W. (2000). *Aclerogamasus stenocornis* sp. n., a fossil mite from the Baltic amber (Acari: Gamasida: Parasitidae). *Genus*, **11**, 619–26.

Zink, R.M. (2002). Methods in comparative phylogeography, and their application to studying evolution in the North American aridlands. *Integrative and Comparative Biology*, **42**, 953–59.

Zink, A.R., Spigelman, M., Schraut, B., Greenblatt, C.L., Nerlich, A.G., and Donoghue, H.D. (2006). Leishmaniasis in ancient Egypt and Upper Nubia. *Emerging Infectious Diseases*, **12**, 1616–17.

CHAPTER 3

Phylogeography and historical biogeography of obligate specific mutualisms

Nadir Alvarez, Doyle McKey, Finn Kjellberg, and Martine Hossaert-McKey

3.1 Historical perspective

Evolutionary processes usually take place in a dynamic geographical context (Schaal and Olsen 2000; Whittaker *et al.* 2005; Diniz Filho *et al.* 2008). As a consequence, patterns of genetic variation are typically strongly structured in space and time (Hewitt 2001). Thanks to the recent development of multiple types of molecular markers, genetic information has been analysed in numerous organisms in order to disentangle processes occurring at the intra-specific level. Since the origin of phylogeography as a new field of research in evolutionary biology, lineage genealogies have been widely analysed to deduce the evolutionary history of populations and species (Avise 2001). Patterns addressed in such studies give information about the spatial structure of gene pools and on the values of diversity estimators among the different sampling sites (Emerson and Hewitt 2005). By combining these, it is possible to infer recent and ancient population processes, including survival in restricted refugia, colonization routes, and recent bottlenecks (Provan and Benett 2008). Nonetheless, with recent methodological improvements and conceptual advances, the classical framework of phylogeography has given birth to the field of comparative phylogeography (Bermingham and Moritz 1998; Knowles 2004), in which within-species patterns of genetic variation are compared among species sharing some aspects of their ecological niche, of their spatial distribution or of their environmental history. For instance, in

the last decade, phylogeographic histories of diverse biota (e.g. desert vertebrates in Baja California [Riddle *et al.* 2000]), Alpine plants in the European Alps (Alvarez *et al.* 2009)) have been extensively studied in order to elucidate major effects of landscape features, such as topographical variation or bedrock types, on the genetic structure of species (Schönswetter *et al.* 2005; Pepper *et al.* 2008). However, it is only very recently that researchers have begun to investigate the comparative phylogeography of mutualistic systems. Diffuse or generalist systems (i.e. multi-partner mutualisms, as for instance in the pollinating interactions between honeybees and angiosperms (Proctor *et al.* 1996) or even in Mullerian mimicry rings (Elias *et al.* 2008)) seem much too complex to be apprehended concretely by this approach. In contrast, one-to-one mutualisms allow tests of simple hypotheses, such as addressing whether or not the partners' phylogeographic histories are congruent or not. In parasitic interactions, one species (the host) is not dependent on the other (the parasite). It is thus allowed to migrate freely, and may therefore demonstrate a spatial genetic structure different from that of the dependent species. How likely this is may depend on the relative dispersal capacities of the two species, or on whether the parasite evolves adaptations that ensure vertical transmission and thus codispersal with its host (Yamamura 1993; Rózsa 2000). In contrast, in specific mutualistic interactions, each partner benefits from association with the other, and the relationship may even be obligate: one species cannot survive without the other. However, one must keep in mind that interactions lie on a continuum from mutualism to parasitism (Ewald 1987) and that many 'mutualisms' could be parasitic under certain ecological conditions (Hernandez and Barradas 2003). Nonetheless, the tight relationship required by reciprocal dependence is more likely to lead to the parallel migration of gene pools of both partners. In contrast to host–parasite systems (where selection favours hosts that escape their parasites in space and time), in mutualisms selection favours traits that facilitate partner encounter. The hypothesis that populations of mutualists are costructured has been tested, and related hypotheses examined, over the last few years. In this chapter, we will examine several recent case studies, in order to identify major patterns in comparative phylogeographies of specific mutualistic partners.

3.2   Studying interactions from the mutualistic point of view

Biogeographic histories of specific parasites have long been used as sources of information about phylogeny of hosts (Brooks and McClennan 1991). More recently, biogeographic patterns have also been used to infer genealogy of lineages within host species (Nieberding and Olivieri 2007; see Chapter 5 of this volume). For instance, gene genealogies of the prokaryote digestive parasite *Helicobacter pylori* have revealed insights into human phylogeography (Linz *et al.* 2007). This approach has also been applied to the study of mutualisms at several scales of time and space (e.g. Downie and Gullan 2005; Jousselin *et al.* 2009). The most striking example certainly lies in the reconstruction of genealogies of vertically-inherited cytoplasmic symbionts in eukaryotes (i.e. mitochondrion and/or chloroplast) to infer the evolutionary history of their hosts.

Much work in the evolutionary biology of interspecies interactions has been inspired by the theory of the geographic mosaic of coevolution, developed by Thompson (1999). However, most researchers have approached this question from a rather static point of view, emphasizing variation in space but without taking into account variation in time and especially changes driven by climatic oscillations,

and how the two may interact. To our knowledge, only one study has explored the intra-specific variation of morphotypes and neutral genetic variation in a framework combining Thompson's theory and phylogeography (Toju and Sota 2008). This study, dealing with an antagonistic interaction between the far-eastern Palaearctic Ericales *Camellia japonica* (Theaceae) and its specific seed predator, the weevil *Curculio camelliae* (Curculionidae), shows that local selection has had a higher impact than historical factors (i.e. phyletic relationship as revealed by phylogeography) in driving the distribution of morphological variation. However, this study examined only one of the myriad ways in which population history, as estimated by phylogeography, and coevolutionary processes could interact. For example, greater dispersal capacity may be selected for along colonization fronts (Bialozyt *et al.* 2006). Could this affect benefit/cost ratios associated with a mutualism? Do mutualists become less lavish (and do parasites become more virulent) at colonization fronts? In the context of specific mutualisms, there would be much to gain by combining the two perspectives, phylogeography and geographic mosaics in coevolution. Future research could profitably focus on integrating Thompson's theory into the temporal framework, for instance, by measuring traits such as the specificity of the interaction (e.g. inequalities in dispersal capacities might weaken associations) or the level of 'mutualism' (along the mutualism–parasitism continuum) throughout the geographic ranges of interacting species.

3.3   Consequences of different life-history traits on phylogeographies of mutualistic partners

From the principles of population genetics, one could imagine that life-history differences should affect the respective phylogeographic patterns of associated mutualists, by influencing genetic signals of population histories, as well as contemporary population functioning (Althoff *et al.* 2007). Here, we discuss the potential role of a character in which associated partners often show an extremely marked contrast: the size of organisms, which is highly correlated with differences in other life-history traits, such as generation time, population size, and dispersal capacity. As a striking example, one can imagine how huge is this discrepancy in the mutualistic associations of termites and bacteria or protozoa in termite guts, associations that allow these animals to profit from the digestion of cellulose and other plant fibres. Similarly, in plant–insect mutualisms, protecting or pollinating insects are often much smaller in size, have much larger and denser populations, and show much shorter life-cycles than their host plants. Such an important discrepancy in size and other life-history traits means that mutualistic interactions cannot be treated within the classical framework of theoretical models of coevolution. Indeed, such models often consider the interacting species as obeying the same adaptive rules at similar time scales. Classical models of coevolution—mutualistic or parasitic—are generally envisioned within the Red Queen hypothesis. However, because of large population size and short generation time the 'smaller' species should undergo adaptation much faster than the long life-cycle species and hence the smaller species should win the race. A sounder perspective is probably to consider that the larger species (the host species) constitutes the habitat of the smaller species. The structure of this habitat will determine which traits will be selected and how intra-specific competition will operate, and hence how microevolutionary processes in the smaller species will play out. The coevolutionary selection processes will operate more slowly (i.e. on a longer time scale), as host structures will only be selected on how their modification locally affects evolutionarily stable strategies within the smaller species. As a consequence, phylogeography of the host will simply enable retracing the history of the habitat of the smaller species. In contrast, the phylogeography of the smaller partner will retrace how that species has tracked habitat (host) expansion and retraction (or how its own expansion and retraction were affected, in response to events that affected it, but not the host). Recent range expansion of the host, reflected by its genetic structuring, will probably often be associated with signatures of recent range expansion of the smaller associate. But this holds true as long as the term 'recent' applies to the genetic and demographic processes within both species. When the smaller species has higher dispersal capacity and shorter generation time than its host, we expect

that the range expansion signature in the smaller species will be lost much more rapidly than in its host. This will lead to strong disconnection of phylogeographic signal between species pairs in some mutualisms and strong similitude in other mutualisms. Another factor that could impact the system is rapid lineage replacement within the smaller species. If this occurs, traces of recent expansion will be found in the more dispersive species. In order to evaluate the first hypothesis and detect the occurrence of the second scenario, we will review case studies from the literature and from our ongoing research in fig-wasp interactions, as well as in ant-plant protective mutualisms.

The ant-plant *Leonardoxa africana africana* and its ant associates in the Atlantic coastal forests of Cameroon illustrate several patterns that may be general in the comparative phylogeography of mutualist pairs. The swollen, hollow twigs (ant domatia) of this small rainforest understory tree are the sole nesting site of the ant *Petalomyrmex phylax*. A single *Petalomyrmex* colony occupies all the domatia of a host tree. The ants also obtain their food from the host, directly via foliar nectar and probably indirectly as well, from symbiotically associated fungi they tend within domatia (Defossez *et al.* 2009). *Petalomyrmex* workers constantly patrol the plant's young leaves, protecting them from attack by insect herbivores (Gaume *et al.* 1997). The mutualism between *Leonardoxa* and *Petalomyrmex* is parasitized by another host-specific ant, *Cataulacus mckeyi*, which occupies a small but variable proportion of host trees. All three members of this symbiotic system present a striking signal of recent southward progression, probably reflecting recolonization after a late-Holocene dry period that led to retraction of dense forests, followed by their re-expansion (Léotard *et al.* 2009). The pattern is marked for microsatellite markers by a southward decrease of two classical measures of genetic diversity (allelic richness and expected heterozygosity), but also for a measure of the progressive local recovery of genetic diversity after bottleneck effects (allelic variance in the number of motif repeats is progressively recovered by addition or loss of one motif at a time; hence variance in allele size is recovered more slowly than allele richness). In *Petalomyrmex*, values of classical measures of gene diversity have equalized along the transect. However, a signature of a bottleneck is still observed (Fig. 3.1). Hence along this recent colonization front, the phylogeographic signature of the colonisation process is still very strong in the host plant but only partly visible in the associated ants (Léotard *et al.* 2009). In both ant species, populations at the colonization front present more dispersive traits, a feature which is predicted to come at a cost for the mutualism. We may surmise that at the colonization front, the ants' biology and hence the functioning of the mutualism cannot be fully understood if population expansion associated with global change is not taken into account.

Phylogeographic congruence has also been revealed in the case of *Roridula* carnivorous plants and their hemipteran mutualists *Pameridea* at the scale of South Africa (Anderson *et al.* 2004). Whereas *Roridula* is able to capture insect prey on its sticky leaves, it does not secrete any digestive enzymes. Conversely, *Pameridea* bugs patrol the

plant without being harmed by its trapping-system and consume all prey captured by the plant. Once these are eaten, digested, and their remains excreted by the bugs, the leaves are able to absorb the remaining organic matter. In this spectacular mutualism, spatial analyses of the genetic variation in both plant and insect populations revealed three congruent clusters (Fig. 3.2).

Another example where the phylogeographies of two species show a trend to congruence is that of lycaenid butterflies and *Iridomyrmex* ants in Australia (Eastwood *et al.* 2006), in which ants protect caterpillars, while the latter provide food rewards to the ants. In this system, the two partners are relatively similar in body size, as well as in associated life-history traits (see above), although there is some asymmetry in the respective pressures for habitat structuring (e.g. female butterflies look for ant colonies in order to lay eggs close to their protecting partners, but the ants do not look for butterflies) as well as in the dispersal abilities of the

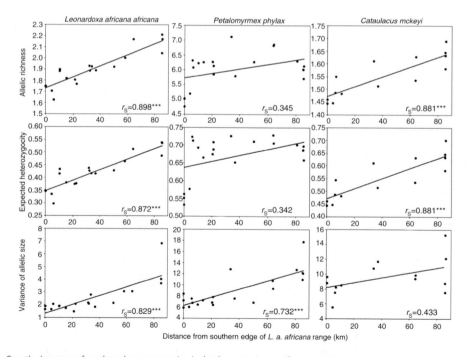

Figure 3.1   Genetic signatures of southward range expansion in the three-species specific association, between the plant *Leonardoxa a. africana*, its mutualistic ant *Petalomyrmex phylax*, and the parasitic ant *Cataulacus mckeyi*. In the three species we observe a decrease of genetic diversity towards the colonization front. However, this effect is only visible for variance of allele size in *P. phylax*, suggesting that the signal will be lost much more rapidly in that species than in the two other. Modified after Léotard *et al.* 2009.

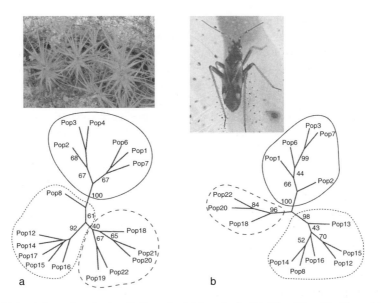

Figure 3.2 Neighbour-joining phylogram based on Cavalli–Sforza chord distance for *Roridula* (A) and *Pameridea* (B) populations from South Africa. Populations from a given region are circled with the same kind of line (both in plants and in insects). Numbers at nodes indicate the percentage of bootstrap samples with such nodes; only values of more than 40 per cent are shown. Modified after Anderson *et al.* (2004). Photo credits: Wikipedia (GNU FDL, copyright).

two partners (e.g. butterflies are likely to disperse over larger distances than do ants). As a consequence, small-scale genetic structure is not congruent between the two partners—and the ants' host plant might also play a role in structuring habitats—whereas barriers to gene flow facing the two organisms are very similar at the continental scale. Similarly, in one of the rare studies on marine mutualists, Thompson *et al.* (2005) examined phylogeographic congruence in the predator–protection mutualism between gobies and their associated shrimps in coral reef lagoons of Indo-Pacific islands. In this obligate interaction, shrimps build burrows in which gobies hide from predators and shrimps get the benefit of being warned by the gobies about the presence of potential predators. The two partners reveal a mostly congruent phylogenetic structure, at least regarding the distribution of the most common clade.

Another example of large-scale congruence in phylogeographic patterns is that of *Ficus* species. *Ficus* produce urn-shaped inflorescences called figs, the inside of which is lined by tiny unisexual flowers. When the female flowers are receptive, specific mutualistic pollinating wasps are attracted to the figs. Loaded with pollen, they enter the fig, pollinate and oviposit in some of the ovules. Some weeks later the wasps emerge into the fig cavity, at a time when the fig's male flowers have matured and released pollen. Winged female wasps become loaded with pollen (or actively load pollen in specialized pollen pockets, depending on the species), mate with wingless males, and leave the fig in search of a new receptive fig of the same species in which to oviposit. In half of *Ficus* species all individuals produce figs that host wasps and produce seeds. These species are monoecious. In other *Ficus* species some individuals produce pollen and pollen vectors (pollinating wasps), and no seeds while other individuals produce seeds but no wasps and no pollen. Such species are functionally dioecious. In one monoecious species, *Ficus racemosa*, we have analysed spatial genetic structure using microsatellite markers (A. Bain *et al.*; unpublished data). We observe genetic differentiation between India, the rest of continental Asia, Borneo, and Australia. However, within India and within continental Asia, genetic differentiation is very limited. For the pollinating wasps, morphological observations suggest that Indian wasps are somewhat different from all the others. Wasps from the rest of continental Asia form one genetic entity, Bornean ones another, and Australian ones yet another (N. Kobmoo et al.; unpublished data). Hence in this case, the broad scale phylogeographic structure of plant populations and those of the mutualistic pollinator are similar or perhaps even congruent. This could be more or less expected because pollinators carry pollen and probably regularly travel rather far, driving homogenization of the host's regional gene pools. Within this general pattern, preliminary data (N. Kobmoo and J. Cook; unpublished data) suggest that the wasps from Australia, but not the host fig, have gone through a very strong recent bottleneck. No genetic data are available for dioecious *Ficus*, for which we may, however, predict a very different pattern. Many dioecious *Ficus* are small understory species structured in patches. In such species male trees produce figs quite frequently, allowing local survival of the pollinating wasps. Endemism is much higher than in monoecious species. This is very striking if one does a simple survey, using Flora Malesiana (Berg and Corner 2005), of the number of monoecious and dioecious species endemic to large islands such as Borneo. Of 43 monoecious *Ficus* spp. found on Borneo, only 3 spp. (7.5 per cent) are endemic to this island. In contrast, of 99 dioecious *Ficus* spp. on Borneo, 48 (48.5 per cent) are endemic to Borneo. These observations suggest limited gene flow in dioecious *Ficus* and hence stronger genetic structuring than in monoecious species. Because male trees allow local survival of wasp populations, simple reasoning would suggest that both dioecious *Ficus* species and their pollinating wasp populations would present similar structuring. A further reason why monoecious *Ficus* species would be expected to show weaker genetic structuring is suggested by the fact that the pollinators of these species, most of which are forest-canopy hemi-epiphytes, drift in the wind over large distances above the canopy, while pollinators of the predominantly understory-restricted dioecious figs seem to remain below the canopy, where wind is limited (Harrison and Rasplus 2006). However, during a strong El Niño event on Borneo, all pollinating wasps became regionally extinct in different sites from the northern edge of the island (Harrison 2000). The wasps

took several months to Recolonize these places. Their host trees, of course, persisted. The great difference in life-history traits of fig and wasp thus introduces a striking asymmetry into the interaction: figs can wait for pollinators, but the converse is not true. As a consequence, during exceptional climatic events, genetic structuring of the wasps, but not that of the host plant, may be completely reshuffled on a regional scale. This may lead to discrepancies between regional-level genetic structuring of dioecious figs and their pollinating wasps. In monoecious figs, we only expect to observe very limited genetic structuring on a regional scale, and hence such climatic events may have limited incidence on patterns of costructuring.

Phylogeographic patterns might be expected to be highly incongruent in other cases where the difference between partners in size (and in correlated traits such as dispersal capacity or generation time) is large (as in both *Ficus* and *Leonardoxa*), and in the absence of a recent colonization-front syndrome. This is for instance the case in the leaf-cutting ant *Trachymyrmex septentrionalis* and its two microbial symbionts, a fungus on which the ant is nutritionally dependent, and *Pseudonocardia* bacteria living on the cuticle of the ants that produce antibiotics which protect the mutualist fungus from a specific fungal pathogen. Despite predominantly vertical cotransmission of the symbionts, there was no phylogeographic concordance between the ant host and either of its obligate microbial symbionts (Mikheyev et al. 2008). However, genetic structure among populations of *Pseudonocardia* bacteria was significantly correlated with that among populations of the fungal mutualist. Two hypotheses could account for this pattern, assuming that symbionts may have been transmitted horizontally to some extent. First, microbial symbionts may be better dispersers and could have effected long-distance dispersal events between different Pleistocene refugia. Second, shorter generation time in the microbial symbionts (leading to more rapid equilibration of their population genetic structure following northern recolonization) may have erased any signals of more ancient expansion events. As a result, the phylogeography of the ants is nowadays incongruent with that of the microorganisms, although their respective patterns of spatial genetic structure might have shown some overlap in the past. In this example, genetic structure of the ant host appears to have been shaped by historical forces, while that of the two microbial symbionts may reflect to a greater degree the impact of ecological interactions, in which these two species, which interact indirectly since they share the same host-habitat, might have followed similar population processes.

3.4 Main driving factors

These different case studies show how life-history traits and biogeographic patterns can affect differentially the respective phylogeographies of mutualistic species. Here, we propose that two main issues—first the level of dimorphism in organism size and other life-history traits, and second the recentness of colonization-front syndromes—are likely to predict whether or not we might expect congruence (see Table 3.1). When the colonization-front syndrome reflects a process still occurring today, or that occurred sufficiently recently, the phylogeographic histories of partners are expected to be congruent. In contrast, when current distributions have been stable across a large number of generations, we expect two different outcomes: first, incongruence, when organism sizes are very different between mutualists (meaning that dispersal abilities and generation times are also very different), and second, varying levels of congruence when organism sizes are similar. Congruence should be greater, for example, if dispersal capacities of both partners are low, and lower if dispersal capacities of both partners are high. Congruence should also be high if there is codispersal and vertical transmission, but the *Trachymyrmex* case suggests that even in cases where vertical transmission appears

to be the principal dispersal mode, exceptional events could have a huge impact at a large scale. Indeed, some patterns at very large scales may result from exceptional long-distance dispersal events, and these events sometimes cannot be predicted by the 'usual' dispersal capacity of the considered organism (Higgins *et al.* 2003).

These considerations should also be adapted to the scale at which the phylogeographies are considered: at a small scale, congruence is less likely to be detectable than at a large scale, due to incomplete lineage sorting and multidirectional migrations. However, many other factors are also likely to affect the level of congruence between mutualists. For instance:

- kind of transmission (vertically vs. horizontally transmitted);
- number of species involved (strictly pairwise mutualism vs. diffuse interactions in multipartner mutualisms);
- tightness of the interaction (facultative vs. obligate);
- biological organization of mutualisms (symbiotic vs. non-symbiotic);
- tightness of the interaction across the partners' distribution ranges (constant interaction vs. host-switching over the geographic range);
- latitudinal distribution, and hence, the extent to which glacial cycles produce extremely marked climatic fluctuations that impede the development of coadaptive processes (temperate vs. tropical regions);
- connectivity between populations (organized in metapopulations or in a classical populational framework); and
- distribution of individuals/populations (continuous vs. patchy distributions).

Further modelling and empirical research is needed to predict how these different variables might influence the comparative phylogeographies of mutualist partners. The future of comparative phylogeography is also to incorporate niche-based modelling, using both contemporary and paleoecological data, to confirm that the inferred past co-occurrences in space and time, and inferred parallel dispersal pathways, were possible in the inferred time frame.

Table 3.1 Prediction of congruence or incongruence between phylogeographies of mutualistic partners, as a function of the dimorphism in the organisms' sizes and of the recentness of the colonization-front syndrome.

		Dimorphism in organisms sizes	
		small	large
Colonization edge syndrome	recent	congruence	congruence
	ancient	congruence/incongruence	incongruence

References

Althoff, D.M., Svensson, G.P., and Pellmyr, O. (2007). The influence of interaction type and feeding location on the phylogeographic structure of the yucca moth community associated with *Hesperoyucca whipplei*. *Molecular Phylogenetics and Evolution*, **43**, 398–406.

Alvarez, N., Thiel-Egenter, C., and Tribsch, A. *et al.* (2009). History or ecology? Substrate type as a major driver of spatial genetic structure in Alpine plants. *Ecology Letters*, **12**, 632–40.

Anderson, B., Olivieri, I., Lourmas, M., and Stewart, B.A. (2004). Comparative genetic structures of *Roridula* and *Pameridea*: cospeciation through vicariance. *Evolution*, **58**, 1730–47.

Avise, J.C. (2001). *Phylogeography: The History and Formation of Species*. Harvard University Press, Cambridge, Massachusetts.

Berg, C.C. and Corner, E.J.H. (2005). *Moraceae (Ficus). Flora Malesiana, Series I (Seed plants,) Volume 17/Part 2*. National Herbarium of the Netherlands, Leiden.

Bermingham, E. and Moritz, C. (2008). Comparative phylogeography: concepts and applications. *Molecular Ecology*, **7**, 367–69.

Bialozyt, R., Ziegenhagen, B., and Petit, R.J. (2006). Contrasting effects of long distance seed dispersal on genetic diversity during range expansion. *Journal of Evolutionary Biology*, **19**, 12–20.

Brooks, D. and McLennan, D. (1991). *Phylogeny, Ecology, and Behaviour. A Research Program in Comparative Biology*. University of Chicago Press, Chicago.

Defossez E., Selosse, M.-A., Dubois, M.P. *et al.* (2009). Ant-plants and fungi: a new threeway symbiosis. *New Phytologist*, **182**, 942–49.

Diniz Filho, J.A.F., Telles, M.P.C., Bonatto, S.L. *et al.* (2008). Mapping the evolutionary twilight zone: molecular markers, populations and geography. *Journal of Biogeography*, **35**, 753–63.

Downie, D.A. and Gullan, P.J. (2005). Phylogenetic congruence of mealybugs and their primary endosymbionts. *Journal of Evolutionary Biology*, **18**, 315–24.

Eastwood, R., Pierce, N.E., Kitching, R.L., and Hughes, J.M. (2006). Do ants enhance diversification in lycaenid butterflies? Phylogeographic evidence from a model myrmecophile *Jalmenus evagoras*. *Evolution*, **60**, 315–27.

Elias, M., Gompert, Z., Jiggins, C., and Willmott, K. (2008). Mutualistic interactions drive ecological niche convergence in a diverse butterfly community. *PLoS Biology*, **6**, e300.

Emerson, B. and Hewitt, G. (2005). Phylogeography. *Current Biology*, **15**, R367–71.

Ewald, P.W. (1987). Transmission modes and evolution of the parasitism-mutualism continuum. *Annals of the New York Academy of Sciences*, **503**, 295–306.

Gaume, L., McKey, D., and Anstett, M.C. (1997). Benefits conferred by "timid" ants: active anti-herbivore protection of the rainforest tree *Leonardoxa africana* by the minute ant *Petalomyrmex phylax*. *Oecologia*, **112**, 209–16.

Harrison, R.D. (2000). Repercussions of El Niño: drought causes extinction and the breakdown of mutualism in Borneo. *Proceedings of the Royal Society of London B*, **267**, 911–15.

Harrison, R.D. and Rasplus, J.Y. (2006). Dispersal of fig pollinators in Asian tropical rain forests. *Journal of Tropical Ecology*, **22**, 631–39.

Hernandez, M.J. and Barradas, I. (2003). Variation in the outcome of population interactions: bifurcations and catastrophes. *Journal of Mathematical Biology*, **46**, 571–94.

Hewitt, G. (2001). Speciation, hybrid zones and phylogeography - or seeing genes in space and time. *Molecular Ecology*, **10**, 537–49.

Higgins, S.I., Nathan, R. and Cain, M.L. (2003). Are long-distance dispersal events in plants usually caused by nonstandard means of dispersal? *Ecology*, **84**, 1945–56.

Jousselin, E., Desdevises, Y., and d'Acier, A.C. (2009). Fine-scale cospeciation between *Brachycaudus* and *Buchnera aphidicola*: bacterial genome helps define species and evolutionary relationships in aphids. *Proceedings of the Royal Society of London B*, **276**, 187–96.

Knowles, L.L. (2004). The burgeoning field of statistical phylogeography. *Journal of Evolutionary Biology*, **17**, 1–10.

Léotard, G., Debout, G., Dalecky, A. *et al.* (2009). Range expansion drives dispersal evolution in an equatorial three-species symbiosis. *PLoS One*, **4**, e5377.

Linz, B., Balloux, F., Moodley, Y. *et al.* (2007). An African origin for the intimate association between humans and *Helicobacter pylori*. *Nature*, **445**, 915–18.

Mikheyev, S.A., Vo, T. and Mueller, G.U. (2008). Phylogeography of post-Pleistocene population expansion in a fungus-gardening ant and its microbial mutualists. *Molecular Ecology*, **17**, 4480–88.

Nieberding, C. and Olivieri, I. (2007) Parasites: proxies for host history and ecology? *Trends in Ecology & Evolution*, **22**, 156–65.

Pepper, M., Doughty, P., Arculus, R., and Scott Keogh, J. (2008). Landforms predict phylogenetic structure on one of the world's most ancient surfaces. *BMC Evolutionary Biology*, **8**, 152.

Proctor, M., Yeo, P., and Lack, A. (1996). *The Natural History of Pollination*. Timber Press, Portland, Oregon.

Provan, J. and Bennett, K.D. (2008). Phylogeographic insights into cryptic glacial refugia. *Trends in Ecology and Evolution*, **23**, 564–71.

Riddle, R.R., Hanna, D.J., Alexander, L.F., and Jaeger, J.R. (2000). Cryptic vicariance in the historical assembly of a Baja California peninsular desert biota. *Proceedings of the National Academy of Sciences of the USA*, **97**, 14438–44.

Rózsa, L. (2000). Spite, xenophobia, and collaboration between hosts and parasites. *Oikos*, **91**, 396–400.

Schaal, B.A. and Olsen, K.M. (2000). Gene genealogies and population variation in plants. *Proceedings of the National Academy of Sciences of the USA*, **97**, 7024–29.

Schimper, A.F.W. (1883). Über die Entwicklung der Chlorophyllkörner und Farbkörper. *Botanische Zeitung*, **41**, 105–14, 121–31, 137–46, 153–62.

Schönswetter, P., Stehlik, I., Holderegger, R. and Tribsch, A. (2005). Molecular evidence for glacial refugia of mountain plants in the European Alps. *Molecular Ecology*, **14**, 3547–55.

Thompson, A.R., Thacker, C.E., and Shaw, E.Y. (2005). Phylogeography of marine mutualists: parallel patterns of genetic structure between obligate goby and shrimp partners. *Molecular Ecology*, **14**, 3557–72.

Thompson, J.N. (1999). Specific hypotheses on the geographic mosaic of coevolution. *American Naturalist*, **153**, S1–14.

Toju, H. and Sota, T. (2008). Phylogeography and the geographic cline in the armament of a seed-predatory weevil: effects of historical events vs. natural selection from the host plant. *Molecular Ecology*, **15**, 4161–73.

Wallin, I.E. (1923). The mitochondria problem. *American Naturalist*, **57**, 255–61.

Whittaker, R.J., Araújo, M.B., Jepson, P., Ladle, R.J., Watson, J.E.M., and Willis, K.J. (2005). Conservation biogeography: assessment and prospect. *Diversity and Distributions*, **11**, 3–23.

Yamamura, N. (1993). Vertical transmission and evolution of mutualism from parasitism. *Theoretical Population Biology*, **44**, 95–109.

CHAPTER 4

Biogeography, humans, and their parasites

Pascale Perrin, Vincent Herbreteau, Jean-Pierre Hugot, and Serge Morand

4.1 Introduction

Homo sapiens is obviously the most investigated species regarding its infectious diseases. Humans are parasitized by a large number of macro- or microparasite species and are certainly the most infected mammals on earth. In fact, more than 1,400 parasite species are listed as human pathogens (Cleaveland *et al.* 2001; Woolhouse and Gowtage-Sequeria 2005). Among them, at least 60 per cent are zonootic, that is, they can also infest wild or domestic animals (Taylor *et al.* 2001). This highlights how much knowledge concerning parasitological medicine (including veterinary medicine) and infectious diseases has been accumulated through the centuries. Documenting and understanding ecological, historical, and biogeographical associations between humans and parasites has been the subject of several studies (from May 1958 to Wolfe *et al.* 2007) although more emphasis has been placed on emerging and re-emerging infectious diseases (Jones *et al.* 2008).

As in any other animal species, pathogens have imposed strong selection on genetics, behavioural ecology, and potentially cultural and social structures of humans. In this chapter, we aim to summarize some of the recent advances in the evolutionary and ecological relationships between humans and their pathogens. We also reinvestigate the geography of human parasites and the geography of human defences.

4.2 A long coevolutionary history

As in any other host, humans have gained their parasites either through descent, that is, from a common ancestor, or by acquiring from other host species residing in the same location. Several studies have explored how cospeciation processes may explain the presence of specific parasites in humans (Hugot *et al.* 2003). Evidence comes from cophylogenetic investigations. For example, nematodes such as pinworms (Hugot 1999), fungi such as *Pneumocystis* spp. (Hugot *et al.* 2003), or lice (Reed *et al.* 2007) that specifically infect humans were inherited by descent from the common ancestors of *Homo sapiens* and its close relatives. However, as highlighted by Cleaveland *et al.* (2001) and Woolhouse and Gowtage-Sequeria (2005), most parasite species infecting modern humans were not inherited through descent, that is, by these cospeciation events, but came from domestic and wild animals and have evolved following several types of human–animal relationships, from hunting to domestication.

Cleaveland *et al.* (2001) used a database of 1,922 pathogen species causing infectious diseases in domestic animals and in humans and showed that 32.9 per cent were bacteria, 26 per cent helminths, 17.1 per cent fungi, 7.5 per cent protozoa, and 16.4 per cent viruses and prions. In this database, 1,415 pathogens cause diseases in humans, 616 in livestock, and 374 in domestic carnivores. Looking at host range, the database showed that pathogens capable of infecting two or more hosts predominate among human pathogens (> 60 per cent), but even more so among pathogens of domestic mammals (77 per cent in livestock and 90 per cent in carnivores). This pattern suggests that humans share numerous pathogens with domestic animals, but also that domestic animals may share numerous pathogens with wild animals. This also questions the respective influences of evolutionary history and ecology in the patterns of pathogen diversity in humans and animals.

Nevertheless, it appears that very few studies have investigated the respective contributions of history and geography in shaping the pathogen community of a given host species. Similarly to what was shown in humans, Pedersen *et al.* (2005) estimated that over 60 per cent of the pathogens of wild primates can infect multiple host species. Recently, Davies and Pedersen (2008) have explored patterns of pathogen sharing between primate hosts, taking into account the relative importance of phylogeny and ecology for explaining the community structures of pathogens. Occurrences of pathogen species were obtained from the Global Mammal Parasite Database (Nunn and Altizer 2005; http://www.mammalparasites.org), with 415 pathogen species identified from 117 wild primate species. The results of the study of Davies and Pedersen (2008) showed that infectious diseases are more often shared between those pairs of primate species that are phylogenetically related but also that live in the same geographical region. These findings suggest that frequent pathogen host shifts between close relatives and inheritance of pathogens from a common ancestor were the likely explanations of the observed pattern. However, if this pattern applies for most of the pathogens (helminths, bacteria, protozoans), it seems to be very different for viruses. Geographical overlap among neighbouring primate hosts is likely more important in determining virus sharing rather than phylogenetic proximity. This may be a consequence of rapid evolutionary dynamics within viral lineages, which may allow host jumps across between less related primate hosts.

The findings of Davies and Pedersen (2008) are in agreement with those of Nunn *et al.* (2003), who showed the existence of a latitudinal gradient in virus diversity (but not in most other pathogens) in primates, potentially explained by an increase in virus sharing among primates at low latitudes. The reasons why related primates may share pathogens can be attributed to similar life-history traits, immunological responses, and immunogenetic backgrounds, which may facilitate pathogens to exploit phylogenetically

related hosts. The reasons for host shifts, even among phylogenetically distant hosts, lie in geographical proximity, which influences contact rates and opportunities for pathogen spillover and subsequently pathogen transfer.

Quite importantly, Davies and Pedersen (2008) showed that the phylogenetic pattern of pathogen sharing with humans is the same as that among wild primates, even if wild primates represent a small (but not insignificant) sample of potential pathogen reservoir species. The study of Davies and Pedersen emphasized the importance of geography, leading to the exploration of the geography of pathogen richness and distribution.

4.3 The geography and ecology of human parasites

Very few studies have investigated the pattern of geographic distribution of human parasites, with the exception of Guernier *et al.* (2004), in which a latitudinal gradient of pathogen diversity was depicted with low latitudes characterized by high species diversity of human pathogens. Here, we reinvestigate this pattern using the checklist of Ashford and Crew (1998) of 402 parasite species (helminths, arthropods, and protozoans) in humans. Ashford and Crew (1998) also summarized and categorized the main information on taxonomy, pathogenicity, biological cycle, and geographic distribution of these parasites. Using this well-documented checklist, we aimed at analyzing the distribution of these parasites in relation to parasite taxonomy and to mammal hosts with which humans share parasites as well as distribution of parasites according to major biogeographical areas.

Parasite species included in the checklist belong to the following groups: cestodes, trematodes, nematodes, acanthocephalans, insects, other arthropods, and protozoans. For definition of mammalian taxonomic groups, we followed Waddel *et al.* (1999). We categorized human parasites into the following categories: (a) parasite, the regular host of which is a non-human mammal, but in particular circumstances it may switch to humans (human becoming a substituted definitive or secondary host); (b) human-specific parasite with the majority of closely related species found in another mammalian group;

(c) parasite transmitted to humans via intermediate (secondary) mammalian host; and (d) generalist parasite, using both humans and other mammalian species as hosts. We recorded as 'human parasite' the parasites for which humans are the single and specific definitive host, and/or the main source of transmission to other humans. We distinguished these parasites from those defined as 'primate parasites', which are encountered in primates and only exceptionally in humans.

We quantified the distribution of parasite species in the major biogeographic realms: Ethiopian, Oriental, Australian, Palaearctic, Nearctic, and Neotropical. Two types of counts were done. The first includes parasite species recorded in several different areas or in all listed areas (cosmopolitan parasites), with these parasites counted for each area. The second is limited to parasite species present in a single area (endemic of this area).

The most successful parasite group, in term of species richness, in humans is trematodes with 118 species (representing 29 per cent of the total number of species), followed by nematodes with 108 species (27 per cent), protozoans with 73 species (18 per cent), and cestodes with 53 species (13 per cent). Insects were represented by 31 species (8 per cent) and other arthropods by 14 species (4 per cent) with 5 species belonging to pentastomids and 9 species to Acari. Very few acanthocephalans were described (5 species or 1 per cent).

Comparison with the general number of parasite species recorded in each parasite group shows a general correlation between species richness and the number of parasites species recorded in humans. The numbers of parasite species recorded in animals are between 6,000 and 8,000 for trematodes and nematodes, 2,500 for cestodes, 100 for pentastomid, and 1,140 for acanthocephalans (Hugot *et al.* 2001).

4.3.1 Distribution of human parasites among main mammal groups

Mammal groups that share most parasite species with humans are carnivores with 124 parasite species (31 per cent), followed by ruminants and pigs with 83 species (21 per cent) and rodents with 66 species (17 per cent) (Fig. 4.1). Strict specific human

parasites are represented by only 58 species (15 per cent). Other primate parasites comprise 46 species (12 per cent). Nine parasite species (2 per cent) are shared with equid hosts and four parasite species (1 per cent) may be considered as shared either with bats or marsupials. Within carnivores, most parasite species are shared with dogs or cats (71 species or 57 per cent, and 54 species or 44 per cent, respectively), with only 14 species (11 per cent) shared with seals and 6 species (5 per cent) with ursids. Within ruminants and pigs, 33 species (40 per cent) and 34 species (41 per cent) are shared with bovid and suid hosts respectively, and 2 species (2.4 per cent) with camelids. Finally, 26 species (39 per cent) are shared with commensal rats or mice.

Different host groups may share the same or closely related parasite species for different reasons. First, closely related hosts might inherit closely related parasites from their common ancestor (transmission by descent). Second, parasites may switch from one host to another host if the hosts are sympatric. The nature of the relationships between host species is important in determining their ability to share and to exchange parasites. As mentioned above, two main factors are first, intensity and duration of potential contacts (direct or indirect) and second, similarity in their life traits.

Relationships between carnivores and humans, especially between humans and dogs, might be influenced by these factors. Humans and canids were chasing the same prey, using similar hunting strategies, and cohabiting long before dogs became domesticated. Moreover, dogs were domesticated before other mammals (Horwitz and Smith 2000). Cats were domesticated later than dogs, but we should notice that several parasite species are shared by dogs and cats (43 species). Ruminants and pigs are considered among the oldest domesticated groups (after dogs) and prior to domestication they were usually hunted for food. The rank of the rodents in this classification is not astonishing: since they were, and are still, used for meat and also due to their importance as vectors and/or reservoirs for human diseases. They have lived as human commensals from time immemorial, feeding on food storage, waste, and rubbish. These close relationships with humans may have reached an apogee when humans started to employ agricultural

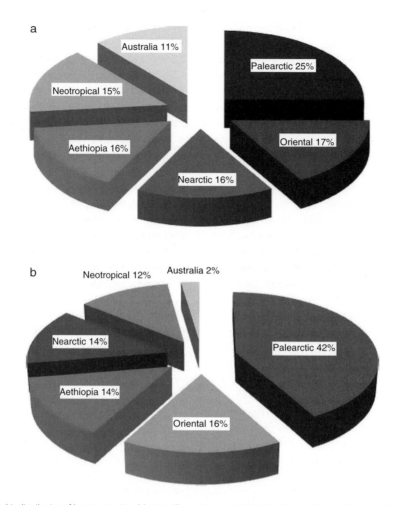

Figure 4.1   Geographic distribution of human parasites (a) using all parasite species including those with a worldwide distribution, and (b) using only species described as endemic of each region. Data from Ashford and Crewe (1998).

methods more regularly and as a result formed more permanent settlements. The weak influence of horses is more surprising. Horses were domesticated between 6,000 years and 3,000 years later than dogs and cattle (Horwitz and Smith 2000). However, before their domestication, they were frequently hunted for food. The same observation is true for camels. Nevertheless, this may be due to the small geographic areas in which camels and horses were originally found, and therefore hunted or domesticated by humans, compared with cattle or dogs, which were associated with humans over a much wider geographic zone.

4.3.2   Distribution of human parasite species among biogeographic realms

Whichever way we count the distribution of human parasites between the geographic realms, the Palaearctic exhibits the highest parasite species diversity: 244 (25 per cent) or 91 (42 per cent) (Fig. 4.1a). Oriental, Ethiopian, and Nearctic realms have roughly similar richness from 17 per cent to 15 per cent. The Neotropics and Australia are the parasite poorest. Counting endemic species only does not change the pattern but only increases the contrasts (Fig. 4.1b).

The predominance of the Palaearctic is surprising. Generally, tropical areas are considered as favouring the appearance, development, and prevalence of parasitic diseases (Guernier *et al.* 2004; Prugnolle *et al.* 2006). Different factors may influence parasite diversity and richness in a particular area, such as age of colonization by humans and population densities. Recent human colonization may explain low diversity in parasite species in Australia and the Neotropics because both areas were colonized relatively recently and still have relatively low human population densities. In contrast, the Palaearctic and Oriental realms were colonized much earlier, reached high population densities, and were characterized by much contact with other areas. The Nearctic realm has a different ranking position depending on the way parasites were counted. When cosmopolitan (and introduced) parasites are considered, it ranks close to the Oriental realm (Fig. 4.1a). But, when only endemic parasites are counted, it ranks below the Oriental realm. This suggests that a prevailing part of human parasite species in the Nearctic comes from parasite species introduced by human migration. Finally, the relatively modest rank of Ethiopia, considered to be the origin of our species, is more intriguing.

4.3.3   Pathogen impact of animal domestication

Interpreting geographic distribution and the pattern of shared human-animal parasites needs to take into account the domestication process. Archaeological studies carried out in tropical and subtropical areas of Asia, northern and central Africa and Central America, suggest rapid and large-scale domestication of plants and animals between 10,000–7,000 cal years BP (Gupta 2004). During this interval, and in all these regions, an intense humid phase and equable climates were detected in paleosurveys (Gupta 2004). However, the potential sources of plants and animals suitable for domestication seem not to be randomly distributed on continents. It appears that the majority of domestic animals originated from the Middle East, central, south-west, and southern Asia (Diamond 1997; Gupta 2004; Driscoll *et al.* 2007; Larson *et al.* 2005; Naderi *et al.* 2008) (Fig. 4.2a). Thus, the origins of animal domestication may explain the relative

poor parasite diversity recorded in the Ethiopian realm (see above).

Several studies highlight the impact of domestication on human health. For example, Horwitz and Smith (2000) illustrated the association between animal and plant domestication and deterioration in health of both humans and their herds. A significant deterioration in the health status of early Neolithic populations was observed in comparison to the hunter and gatherer populations that preceded them in the Southern Levant. The evidence for changes in diet is from skeletal remains and dentition as described in Abu-Hureyra in Syria (Molleson *et al.* 1993, 1994).

The contribution of human–animal contacts to the spread of pathogens can be illustrated by the correlation between the time since domestication and the number of diseases shared with humans (Fig. 4.2b). Recently domesticated species, such as poultry and camels, share fewer diseases with humans compared to the earliest domesticated animal, such as dogs, which share the most with humans (see also above).

4.4   The phylogeography of parasites

More information is required on the exact geographic distribution of parasites and pathogens, but the origins and dispersion of human pathogens are even less well known. The use of molecular markers and the development of new phylogenetic methods may allow us to explore them.

Phylogeography is a recent science that aims to explain the actual geographic distribution of species and populations. More precisely, phylogeography investigates the processes governing the geographical distributions of lineages within and among closely related species following the definition of Avise (2000). Whereas numerous studies aimed at comparing host and parasite phylogenies have been performed (Page 2003), phylogeographic studies of pathogens and parasites are few (Holmes 2004). Phylogeographic studies of parasites may explain the phylogeography of the hosts, as they often mirror each other (Nieberding *et al.* 2004; Wirth *et al.* 2005; Nieberding and Olivieri 2007).

A variety of human pathogens such as human polyomavirus JCV (Agostini *et al.* 1997), the human

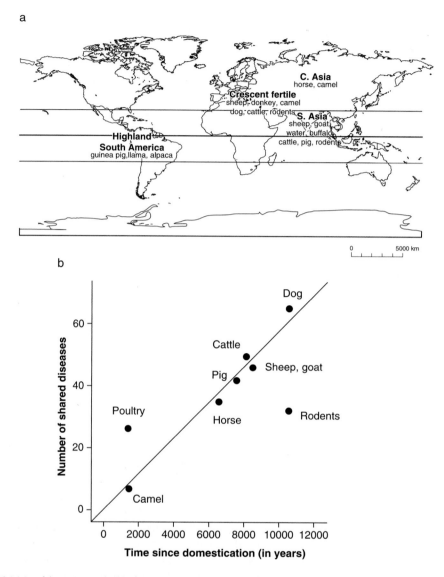

Figure 4.2   (a) Origins of domestic animals (data from Gupta 2004; Driscoll *et al.* 2007; Larson *et al.* 2005; Naderi *et al.* 2008). Fertile Crescent, South and Central Asia appear to be the centres of earliest domestication of animals. (b) Relationship between time since animal domestication and the number of pathogens shared with humans (data from McNeil 1976; Horwitz and Smith 2000) (*p*=0.021).

T-cell lymphotropic virus I (HTLV-I) (Miura *et al.* 1994), *Mycobacterium tuberculosis* (Kremer *et al.* 1999), *Haemophilus influenzae* (Musser *et al.*, 1990), the human pathogenic fungus *Histoplasma capsulatum* (Kasuga *et al.* 1999), and *Helicobacter pylori* (Ghose *et al.* 2002; Falush *et al.* 2003; Wirth *et al.* 2004) show geographic subdivisions. These parasites may have accompanied humans during their ancient

and recent migrations and investigation of their population structures may help to understand human evolutionary history (Wirth *et al.* 2005) and the evolutionary relationships between humans and their pathogens.

Helicobacter pylori is a bacterium that colonizes the stomachs of most humans and can cause chronic gastric pathogeny. Wirth *et al.* (2004) showed that

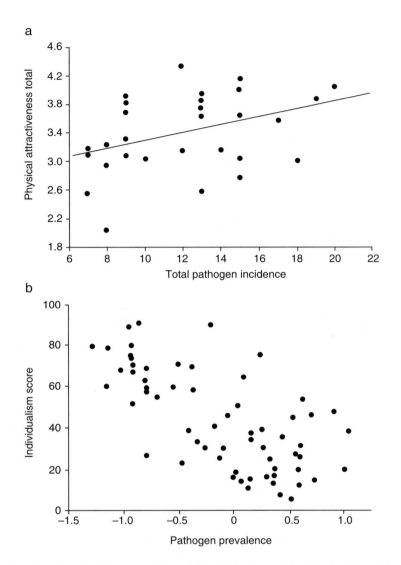

Figure 4.3   Relationship between (a) mate preference and pathogen incidence in human populations (data from Gangestad and Buss 1993), $p=0.04$; and (b) pathogen prevalence and individualism (redrawn after Fincher *et al*. 2008), $p<0.001$.

DNA sequences from *Helicobacter pylori* can distinguish between closely related human populations. For example, *H. pylori* from Buddhists and Muslims in Ladakh (India) differ in their population-genetic structure. This bacterium is also divided into several populations with distinct geographical distributions at a more global scale (Falush *et al*. 2003). Molecular studies and phylogenetic reconstructions showed that these populations derived from ancestral populations that have arisen in Africa, Central Asia,

and East Asia. Subsequent worldwide spread can be attributed to human migrations such as the prehistoric colonization of Polynesia and the Americas, the Neolithic introduction of farming to Europe, the Bantu expansion within Africa, and the slave trade. This kind of study may help to explain the evolutionary and recent history of humans, but pathogens can, by their selective pressures, impose strong selection on the behaviour, social organization, and genetics of humans.

4.5   The geography of selection by parasites

4.5.1   Parasites, pathogens, and human behaviour

Human mate preference and pathogens

Heritable differences in pathogen resistance may explain the evolution of selection for mate preferences, if mate preferences are based on qualities that help to discriminate individuals in relation to their pathogen resistance (Hamilton and Zuk 1982; Andersson 1994). Gangestad and Buss (1993) investigated this hypothesis in humans. Attractiveness is a criterion of mate choice in humans and seems to be linked to pathogen resistance. Gangestad and Buss (1993) tested how the rating scores of the importance of attractiveness, in several societies, were related to pathogen prevalence. For this, they used a database of individuals from 37 different societies located on 6 continents and 5 islands, where each subject rated the importance of 18 attributes (from 0=irrelevant or unimportant to 3=indispensable) as criteria for mate selection. Pathogen stress was indexed following the procedures of Low (1990) with seven pathogens (leishmania, trypanosomes, malaria, schistosomes, filariae, spirochetes, and leprosy) were taken into consideration using medical and public health sources. Each of them was rated by Low (1990) on a 3-point scale for frequency; the individual scores were summed and gave a total pathogen stress (or pressure) score. Gangestad and Buss (1993) showed a positive relationship between pathogen prevalence and the value people place on physical attractiveness (Fig. 4.3a). This positive relationship between pathogen stress and mate preference is consistent with the hypothesis that pathogen resistance may have a significant contribution to individual differences regarding mate quality among humans. The results of Gangestad and Buss (1993) also support a sexual selection hypothesis. In pathogen-rich environments, humans may have evolved by assessing pathogen resistance in potential mates on the basis of physical attractiveness.

Several conditions should be met to explain this relationship, according to Gangestad and Buss (1993) and following the 'good gene' hypothesis (Hamilton and Zuk 1982). First, selection pressures among prehuman primates should have promoted mechanisms enabling individuals to track pathogen stress and, in relation to this, to adjust their mate choice criteria. Second, selection pressures should favour individuals with genotypes predisposing mate choice on the basis of attractiveness, particularly in pathogen-rich areas. Third, the association between pathogen and mate choice criteria should be caused by the greater variance in parasite impact in pathogen-rich areas. A greater variance in parasite impact may favour attractiveness as a criterion of mate choice.

Monogamy, polygyny, and diseases

Sexual selection has a considerable influence on the mating strategies and the economy of sex (Andersson 1994). Human societies are very diverse and also exhibit considerable variation in monogamy versus polygamy. As pathogens may affect sexual selection, one may question if pathogen pressure or stress may be linked with this aspect of social organization.

In the Standard Cross-Cultural Sample (SCCS, see Murdock and White 1969), 83.8 per cent of societies are socially polygynous, and 28 per cent of societies have more than 40 per cent of marriages being polygynous. As emphasized by Ember *et al.* (2007), polygyny does not automatically benefit men and may seldom benefit women. Low (1990) and Ember *et al.* (2007) questioned the reasons why polygyny was so common in 'traditional' societies and why it has become less common in the 'modern' world. Using the SCCS, Low (1990) tested the hypothesis that polygyny should be linked to high exposure to pathogens in accordance with Hamilton's criteria (Hamilton 1980; Hamilton and Zuk 1982). Indeed, Low (1990) confirmed the existence of such relationship.

Ember *et al.* (2007), following Low (1990), suggested that polygyny is likely to become prevalent if there are more females than males. This male mortality/sex ratio interpretation requires conditions where biased sex ratios are associated with appreciable polygyny. Pathogen stress (and war) is one factor that may bias the sex ratio. The study of Ember *et al.* (2007) also using the SCCS indicated the existence of two main independent cross-cultural predictors of appreciable nonsororal polygyny (co-wives are not sisters): high male mortality in war and high pathogen stress. Whereas high male mortality at war may result in high mortality in males, the pathogen stress was not related to a sex-biased mortality in the study. However, it is suggested that pathogen stress may favour nonsororal polygyny, in that this system may maximize genetic variation and disease resistance in progeny (Ember *et al.* 2007). Finally, this study also showed that high male mortality in war is a better predictor of polygyny in nonstate societies, whereas high pathogen stress is a better predictor in more densely populated state societies.

The studies of Low (1990) and Ember *et al.* (2007) emphasized the importance of an extrinsic factor (pathogen pressure) and social complexity (density and state organization) in the differences observed among societies concerning their mating strategies.

Parental effort and parasitism
Animal life history may reflect two major decisions: when to reproduce and what amount of care to invest in each offspring (Trivers 1974). These decisions depend on the costs and benefits of mating and parenting effort, which may be influenced by environmental risk (Stearns 1992) and pathogen pressures. Humans should not be an exception. Quinlan (2007) hypothesized that humans should show reduced parental effort in environments where parenting cannot improve offspring survival. He examined the relationships between environmental risk factors and parenting effort among societies.

Pathogen stress, as well as armed conflict, are extrinsic causes of mortality, which are predicted to be inversely related to parental effort. When the risks of extrinsic mortalities increase, energy allocations should shift from parental effort to mating effort; to the production of additional offspring or somatic effort (self-preservation) in order to enhance fitness.

For testing this prediction, Quinlan (2007) also used the data from the SCCS and showed that maternal care was inversely associated with famine and warfare. Interestingly, he showed a quadratic association with pathogen stress. The parental effort, but also age at weaning, grows up as patho-

gen stress increases to moderate levels, but decreases at higher levels. This curvilinear association between parental effort and pathogen stress may reflect the saturation point of parental care as a function of environmental risks.

The study of Quinlan (2007), as compared to those of Low (1990) and Ember *et al.* (2007) suggested that pathogens may favour sexual selection through female mate choice for heritable immunity, leading to higher levels of polygyny associated with reduced paternal investment. Also, pathogen stress imposes strong selection on immune genes and may increase the benefits of genetic diversity in offspring.

Collectivism, individualism, and parasitism
The most astonishing result comes from the recent study of Fincher *et al.* (2008), who suggested that collectivism, contrary to individualism, may serve as an antiparasitic defence. The hypothesis is that collectivism is more likely to emerge and persist within populations that have been submitted to a greater prevalence of pathogens. Fincher *et al.* (2008) showed a strong correlation between pathogen stress and an indicator of collectivism versus individualism (Fig. 4.3b). Although the socio-biological interpretation of their results can be largely criticized, their study highlighted the importance of the social complexity in determining the pathogen risks and pressures (Bordes *et al.* 2007).

4.5.2 Parasites, pathogens, and human genetics

Humans evolved in a hostile microbial environment and we have explained above how pathogen stress may play a key role in human mating strategies and on the social organization of human societies. We will now discuss how pathogens have imposed a strong selection on the genetics of humans and particularly in genes implicated in defence, as exemplified in the study of Prugnolle *et al.* (2006). In this study, a positive relationship between HLA class I genetic diversity and pathogen richness in human populations was demonstrated.

The neolithization process has probably favoured the emergence of virulent pathogens. Farming and sedentary lifestyle, coupled with domestication, have provided new potential sources of diseases (see above). Increased human population densities have allowed much more frequent contacts, and therefore zoonotic disease transmissions, between animals and humans. There is evidence that both measles and pertussis originated from domesticated animals, although it is still difficult to explain the emergence of other pathogens, (such as tuberculosis, falciparal malaria, smallpox, and taenid tapeworms) as a direct result of neolithization (Pearce-Duvet 2006). Indeed, emergences of some human diseases such as tuberculosis, falciparal malaria, and dysentery may have pre-dated the dawn of agriculture. Moreover, recent data suggest that humans may have transmitted tuberculosis to bovids (Gibbons 2008). The acquisition of pathogens by humans may have occurred through multiple routes and the rise of agriculture is simply one piece of this puzzle, with other factors being related to climatic and/or ecological changes (Diamond 1997).

Several studies on infectious diseases have demonstrated the importance of host genetics in determining susceptibility/resistance to infectious pathogens in humans (Hill 2006). The explosion of genomics and intensive sequencing gives us unprecedented opportunity for understanding the genetic basis of susceptibility and resistance to infectious diseases. Here, we will focus on two infectious diseases, malaria and tuberculosis.

Genes and malaria

The best described example of association between infectious disease and host genetics concerns malaria and haemoglobinopathies, and particularly the sickle cell disease (Weatherall and Clegg 2001). Haldane (1949) was the first to correlate the frequencies of some common mutant alleles concerning globin synthesis with the world distribution of malaria endemicity. Disorders of globin synthesis are found commonly in regions of high malaria occurrence and there is a clear correlation between the distribution of malaria and the distribution of mutation carriers for HbS (Haemoglobin S) and thalassemias. Heterozygosity for HbS was shown to be strongly associated with protection against *Plasmodium falciparum* parasite, the more severe malarial pathogen. Similar protection has recently been described for HbC and HbE, G6PD-A deficiency and some types of α-thalassemias. The selec-

tion hypothesis of common genetic disorders such as HbS, beta-thalassemia, G6PD due to infectious agents is now well supported (Tishkoff *et al.* 2001).

One other association is linked to the absence of the Duffy antigen (FY-null) on the surface of red blood cells, which prevents the entry of *Plasmodium vivax* into erythrocytes. The FY null variant, also called the DARC-negative variant disrupts the Duffy antigen chemokine receptor (DARC) promoter and alters a GATA-1 binding site, which inhibits DARC expression on erythrocytes and therefore prevents DARC-mediated entry of *P. vivax* into red blood cells. As this blood group is very common in West Africa, infection by *P. vivax* in this zone does not regularly cause severe illness, even if some cases of mortality have been described (Rogerson and Carter 2008).

High levels of tumour necrosis factor α (TNF-α) cytokine, secreted from white blood cells, were found in the serum in some forms of severe malaria (TNFα-308) (McGuire *et al.* 1997). CD36 has a role in the endothelial adhesion of infected erythrocytes and their clearance by the immune system. CD36 deficiency has been associated with both susceptibility and resistance to forms of severe malaria. The capacity of the parasite to evolve and to generate high variations in its genome could explain this result. The same type of result was observed for the ICAM[Kilifi] variant of the ICAM1 (InterCellular Adhesion Molecule 1 [CD54]).

Genes and tuberculosis

If leprosy and tuberculosis differ greatly in their clinical phenotypes some similarities appear in the roles of some candidate genes. NRAMP1 (SLC11A1) has been found associated with tuberculosis in different populations (Bellamy *et al.* 1998). Vitamin D receptor (present in macrophages and in B and T lymphocytes) deficiencies were found to be associated with increased risk of tuberculosis (Wilkinson *et al.* 2000).

Mapping disease occurrences and allele frequencies

We decided to analyse the world distribution of prevalence of malaria and tuberculosis (Fig. 4.4) and to correlate them with the world distribution of allelic frequencies of some associated-disease alleles and polymorphic alleles (HLA alleles) (Fig. 4.5, Table

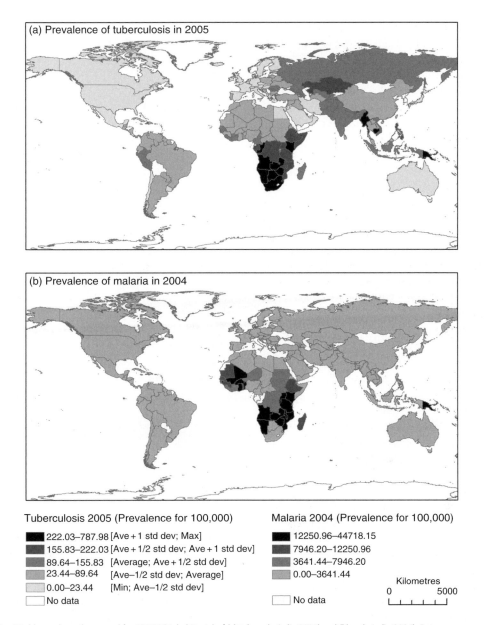

Figure 4.4   World prevalence (expressed for 100,000 inhabitants) of (a) tuberculosis (in 2005) and (b) malaria (in 2004). Data on case numbers from the Gideon database (http://www.GIDEONonline.com/); and population estimates from http://www.census.gov/ipc/www/idb/tables.html.

4.1). To avoid spurious correlations, Cavalli-Sforza *et al.* (1994) pooled human populations according to linguistic or ethnographic grouping. We followed this by attributing same allele frequency for countries where similar linguistic or ethnic groups occurred (references and data can be sent on request to the senior author). We focused on malaria for the year 2004 and on tuberculosis for the year 2005. Malarial prevalence was very variable between countries, with some countries recording over 30 per cent malaria prevalence in 2004 such as Uganda, Zambia, Papua New Guinea, and Mozambique. Values of tuberculosis prevalence showed less variation among countries. Significant correlation

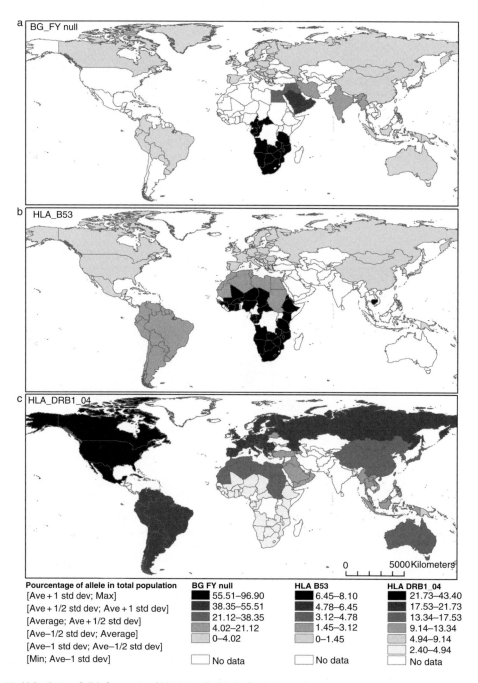

Figure 4.5   World distribution of allele frequencies of (a) BG-Fy null allele (Duffy Blood Group) as percentages (data from Cavalli-Sforza *et al.* 1994), (b) HLA-B53, (c) HLA-DRB1*04 HLA-B53 and HLA-DRB1*04 are given in allele frequency percentages. Data for (b) and (c) from the dbMHC database (www.ncbi.nlm.nih.gov/gv/mhc/).

Table 4.1 Statistically significant ($p<0.05$) correlation coefficients found between the prevalence of tuberculosis (2005) (GIDEON database) or malaria (2004) (GIDEON database) and the frequency of polymorphic or variant alleles (references and data can be sent on request to the senior author).

Tuberculosis 2005			Malaria 2004		
Polymorphic or variant alleles	Coefficient correlation	Significance (pairs nb)	Polymorphic or variant alleles	Coefficient correlation	Significance (pairs nb)
			CFΔ508	−0.3140	0.043 (42)
CCR5Δ32	−0.4411	0.0000 (75)	CCR5Δ32	−0.3535	0.003 (70)
G6PD A−	0.4437	0.0000 (63)	G6PD A−	0.4359	0.001 (59)
BG–FY null	0.6427	0.0000 (76)	BG–FY null	0.5461	0.000 (71)
HLA–A24	−0.2539	0.007 (87)			
HLA–A29 (dbMHC)	0.4036	0.000 (101)	HLA–A29 (dbMHC)	0.3874	0.000 (95)
HLA–A29	0.2849	0.007 (87)	HLA–A29	0.2511	0.024 (81)
HLA–B27 (db MHC)	−0.2934	0.003 (102)	HLA–B27 (db MHC)	−0.2815	0.005 (96)
HLA–B27	−0.2830	0.008 (86)			
HLA–B35 (dbMHC)	−0.3296	0.000 (89)	HLA–B35 (dbMHC)	−0.3230	0.001 (108)
HLA–B*35	−0.3596	0.001 (86)	HLA–B35	− 0.3880	0.000 (80)
HLA–B*53 (dbMHC)	0.5992	0.000 (89)	HLA–B53 (dbMHC)	0.5483	0.000 (84)
HLA–DRB1*04 (dbMHC)	−0.5031	0.000 (113)	HLA–DRB1*04 (dbMHC)	−0.5184	0.000 (108)
HLA–DRB1*07 (dbMHC)	−0.2158	0.022 (113)	HLA–DRB1*07 (dbMHC)	−0.2067	0.032 (108)
			HLA–DRB1*10 (dbMHC)	0.2251	0.035 88)
			HLA–C14 (dbMHC)	−0.2585	0.014 (90)

CFΔ508= Cystic Fibrosis, deletion 508; CCR5Δ32= Human (C-C motif chemokine receptor 5, deletion 32; G6PDA-= Glucose-6-Phosphate-Deshydrogenase, A- variant; BG-FY null = Duffy Blood Group, null; HLA-A24, -A29 = polymorphic allele of the locus A of HLA class I genes; HLA-B27, -B35, -B53= polymorphic alleles of the locus B of HLA class I B genes; HLA-C14= polymorphic allele of the locus C of HLA class I genes; HLA-DRB1*04, -DRB1*07 and -DRB1*10= polymorphic alleles of the locus D (subregion DR) of HLA class II genes. Data from Cavalli-Sforza *et al.* 1994; dbMHC, data from the dbMHC database; data for CFΔ508 and data for CFΔ508 and CCR5Δ32 from Bobadilla *et al.* 2002 and Dean *et al.* 2002.

coefficients were found between the prevalence of these two diseases and the frequencies of some polymorphic or variant alleles (Table 4.1).

The prevalence of malaria is significantly correlated with the frequency of several alleles such as G6PD-A-, FY null allele, HLA-B*35, and HLA-B*53. For each, previous cohort analysis suggested a protective effect (Ruwende and Hill 1998; Zimmerman *et al.* 1999; Ghosh 2008). Our results were concordant with these findings except for HLA-B*35 allele, for which the associated correlation coefficient was negative.

For tuberculosis, none of the three alleles that have been suspected to confer some resistance or susceptibility were found to be associated with the geographic occurrences of these diseases (Table 4.1).

However, some results reveal a striking feature. For nine alleles, significant correlation coefficients were found for both diseases. Indeed, prevalence values of tuberculosis and malaria were significantly correlated (0.41, $p<0.001$). Some significant

correlations may be the result of malaria selection (e.g. G6PD-A- allele).

The genetic markers that best predict the prevalence of malaria in 2004 are HLA-B*53 and HLA-DRB1*10. One other genetic marker that have strongly significant correlation is BG-FY null allele (Table 4.1).

The subset of genetic markers that best predict the prevalence of tuberculosis in 2005 were BG-FY null and G6PD-A. BG-FY null was the main predictor of tuberculosis prevalence. G6PD-A had also a very high but negative correlation (Table 4.1). Two other genetic markers presented strongly significant negative correlations: HLA-B35 and CCR5Δ32 (Dean *et al.* 2002).

Intriguing associations are found here, but further works are required to ascertain their importance. The classic tool is linkage analysis in which highly polymorphic markers are used to identify chromosomal regions that segregate with disease susceptibility within families. Some other approaches are needed to enlighten these associations by comparing allele frequencies in diseased individuals with those of controls recruited from the same ethnic group, or by studying intra-familial association in which the distribution of genotypes among the index cases are compared with those predicted from parental genotypes (Kwiatkowski 2000). The key difficulty is that precise allelic association may change according to populations and/or may show a lack of statistical power. Adequate controls and sample size are needed to clarify some apparent inconsistencies. Without any doubt, it seems that natural selection by infectious diseases has contributed to the maintenance of the remarkable allelic diversity of HLA class I and class II loci.

Paleomicrobiology could help us to analyse the temporal and geographical spread of infectious diseases and to trace the genetic evolution of the pathogens themselves. For example, more than ten studies have been performed on *Mycobacterium tuberculosis*. Some presumptions of the molecular presence of tuberculosis in the earliest urban societies were given (Drancourt and Raoult 2005), and recently Hershkovitz *et al.* (2008) described the molecular characterization of 9000-year-old *Mycobacterium tuberculosis* from the Neolithic settlement of Atlit-Yam in the Eastern Mediterranean. However, few studies have been completed on human and animal remains, so the opportunity to test the hypothesis of a coevolution rather than a direct transmission from bovines to humans as proposed by Brosch *et al.* (2002) is not yet possible. Finally, detection of *Plasmodium falciparum* DNA was described in a Roman baby (Abbott 2001; Salares and Gomzi 2001).

4.6 Conclusion

Humans are exceptional in having colonized more diverse habitats and continents than any other vertebrate species. During their migrations they came to share new habitats with a great diversity of other living species. Progressively, they establish very close relationships with some of them. The great diversity of their parasites is related to this unique story.

If all the different factors suggested as the explanation for variations within human parasite

diversity—richness, geographic distribution, and closeness with other mammal parasites—are considered, one is dominant: time. The longer humans have inhabited an area and associated with different species, the higher the diversity of their parasites. Other factors having an influence are human population density and mobility.

In 1969, the Surgeon General of the World Health Organization claimed that it was 'time to close the book on infectious diseases, declare the war on pestilence won, and shift national resources to such chronic problems as heart disease'. However, the following decades were the time of emerging infectious diseases, mostly caused by a tremendous global change. The human–environment relationship is consistently disturbed by major changes in land use, migration, and population pressure, all of which act on the coadaptation between humans and their infectious diseases and manifest in the appearance or diffusion of new infectious diseases. Infectious diseases specific to humans appear to be uniformly distributed, whereas zoonotic infectious diseases seem more localized in their geographical distribution (Smith *et al.* 2007; Jones *et al.* 2008). What will be the future distribution of infectious diseases? We are far from giving accurate scenarios, but according to Smith *et al.* (2007), it is likely that due to globalization infectious diseases will also become uniformly distributed.

A multidisciplinary approach is needed, together with a multiscale study for a better understanding of the emergence and re-emergence of infectious diseases. Amalgamation of modern, paleomicrobiological and paleoenvironmental data will help us to achieve this goal.

References

Abbott, A. (2001). Earliest malaria DNA found in Roman baby graveyard. *Nature*, **412**, 847.

Agostini, H.T., Yanagihara, R., Davis, V., Ryschkewitsch, C.F., and Stoner, G.L. (1997). Asian genotypes of JC virus in Native Americans and in a Pacific Island population: markers of viral evolution and human migration. *Proceedings of the National Academy of Sciences of the USA*, **94**, 14542–46.

Andersson, M. (1994). *Sexual Selection*. Princeton University Press, Princeton.

Ashford, R.W. and Crewe, W. (1998). *The Parasites of Homo Sapiens: An Annotated Checklist of the Protozoa, Helminths and Arthropods for Which we are Home*. Liverpool School of Tropical Medicine, Liverpool.

Avise, J.C. (2000). *Phylogeography: The History and Formation of Species*. Harvard University Press, London.

Bellamy, R., Ruwende, C., Corrah, T., McAdam, K.P., Whittle, H.C., and Hill, A.V. (1998). Variations in the NRAMP1 gene and susceptibility to tuberculosis in West Africans. *New England Journal of Medicine*, **338**, 640–4.

Bobadilla J.L, Macek M. JR, Fine J.P., and Farrell P.M. (2002). Cystic fibrosis: a worldwide analysis of CFTR mutations—correlation with incidence data and application to screening. *Human Mutation* **19**, 575–606.

Bordes, F., Blumstein D.T., and Morand, S. (2007). Rodent sociality and parasite diversity. *Biology Letters*, **3**, 692–94.

Brosch, R., Gordon S.V., Marmiesse, M. *et al.* (2002). A new evolutionary scenario for the Mycobacterium tuberculosis complex. *Proceedings of the National Academy of Sciences of the USA*, **99**, 3684–89.

Cavalli-Sforza, L.L., Menozzi, P., and Piazza, A. (eds.) (1994). *The History and Geography of Human Genes*. Princeton University Press, Princeton.

Cleaveland, S., Laurenson, M.K., and Taylor, L.H. (2001). Diseases of humans and their domestic mammals: pathogen characteristics, host range and the risk of emergence. *Philosophical Transaction of the Royal Society of London B*, **356**, 991–99.

Davies, T.J. and Pedersen, A.B. (2008). Phylogeny and geography predict pathogen community similarity in wild primates and humans. *Proceedings of the Royal Society of London B*, **275**, 1695–701.

Dean, M., Carrington, M., and O'Brien, S.J. (2002). Balanced polymorphism selected by genetic versus infectious human disease. *Annual Review of Genomics and Human Genetics* **3**, 263–92.

Diamond, J. (1997). *Guns, Germs and Steel: The Fates of Human Societies*. Norton, New York.

Drancourt, M. and Raoult, D. (2005). Paleomicrobiology: current issues and perspectives. *Nature Reviews Mircobiology*, **3**, 23–35.

Driscoll, C.A., Menotti-Raymond, M., Roca, A.L. *et al.* (2007). The Near-Eastern origin of cat domestication. *Science*, **317**, 519–23.

Ember, M., Ember, C.R., and Low, B.S. (2007). Comparing explanations of polygyny. *Cross-Cultural Research*, **41**, 428–40.

Falush, D., Wirth, T., Linz, B. *et al.* (2003). Traces of human migrations in *Helicobacter pylori* populations. *Science*, **299**, 1582–85.

Fincher, C.L., Thornhill, R., Murray, D.R., and Schaller, M. (2008). Pathogen prevalence predicts human cross-cultural variability in individualism/collectivism. *Proceedings of the Royal Society of London B*, **275**, 1279–85.

Gangestad, S.W. and Buss, D.M. (1993). Pathogen prevalence and human mate preferences. *Ethology and Sociobiology*, **14**, 89–96.

Ghose, C., Perez-Perez, G.I., Dominguez-Bello, M.G., Pride, D.T., Bravi, C.M., and Blaser, M.J. (2002). East Asian genotypes of *Helicobacter pylori* strains in Amerindians provide evidence for its ancient human carriage. *Proceedings of the National Academy of Sciences of the USA*, **99**, 15107–11.

Ghosh, K. (2008). Evolution and selection of human leukocyte antigen alleles by *Plasmodium falciparum* infection. *Human Immunology*, **69**, 856–60.

Gibbons, A. (2008). Tuberculosis jumped from humans to cows, not vice versa. *Science*, **320**, 608.

Guernier, V., Hochberg, M.E., and Guégan, J.-F. (2004). Ecology drives the worldwide distribution of human diseases. *PloS Biology*, **2**, 740–45.

Gupta, A.K. (2004). Origin of agriculture and domestication of plants and animals linked to early Holocene climate amelioration. *Current Science*, **87**, 54–59.

Haldane, J.B.S. (1949). The rate of mutations of human genes. *Hereditas Supplement*, **35**, 267–73.

Hamilton, W.D. (1980). Sex versus non-sex versus parasite. *Oikos*, **35**, 282–90.

Hamilton, W.D. and Zuk, M. (1982). Heritable true fitness and bright birds: a role for parasites? *Science*, **218**, 384–87.

Hershkovitz I., Donoghue H.D., Minnikin D.E. *et al.* (2008). Detection and molecular characterization of 9000-year-old *Mycobacterium tuberculosis* from a Neolithic settlement in the Eastern Mediterranean. *PloS ONE*, **3**, e3426.

Hill, A.V.S. (2006). Aspects of genetic susceptibility to human infectious diseases. *Annual Review of Genetics*, **40**, 469–86.

Holmes, E.C. (2004). The phylogeography of human viruses. *Molecular Ecology*, **13**, 745–56.

Horwitz, L.K. and Smith, P. (2000). The contribution of animal domestication to the spread of zoonoses: a case study from the Southern Levant. *Anthropozoologica*, **31**, 77–84.

Hugot, J.P. (1999). Primates and their pinworm parasites: the Cameron hypothesis revisited. *Systematic Biology*, **48**, 523–46.

Hugot, J.P., Baujard, P., and Morand, S. (2001). Biodiversity in helminths and nematodes as a field of study: an overview. *Nematology*, **3**, 1–10.

Hugot, J.P., Demanche, C., Barriel, V., Dei-Cas, E., and Guillot, J. (2003). Phylogenetic systematics and evolution of the *Pneumocystis* parasite on primates. *Systematic Biology*, **52**, 735–44.

Jones, K.E., Patel, N.G., Levy, M.A., Storeygard, A., Balk, D., Gittleman, J.L., and Daszak P. (2008). Global trends in emerging infectious diseases. *Nature*, **451**, 990–93.

Kasuga, T., Taylor, J.W., and White, T.J. (1999). Phylogenetic relationships of varieties and geographical groups of the human pathogenic fungus *Histoplasma capsulatum* Darling. *Journal of Clinical Microbiology*, **37**, 653–63.

Kremer, K., van Soolingen, D., Frothingham, R. *et al.* (1999). Comparison of methods based on different molecular epidemiological markers for typing of *Mycobacterium tuberculosis* complex strains: interlaboratory study of discriminatory power and reproducibility. *Journal Clinical Microbiology*, **37**, 2607–18.

Kwiatkowski, D. (2000). Science, medicine and the future. Susceptibility to infection. *British Medical Journal*, **321**, 1061–10655.

Larson, G., Dobney, K., Albarella, U. *et al.* (2005). Worldwide phylogeography of wild boar reveals multiple centers of pig domestication. *Science*, **307**, 1618–21.

Low, B. (1990). Marriage systems and pathogen stress in human societies. *American Zoologist*, **30**, 325–39.

May, J.M. (1958). *The Ecology of Human Disease*. M.D. Publications, New York.

McGuire, W., Knight, J.C., Hill, A.V., Allsopp, C.E., Greenwood, B.M., and Kwiatkowski, D. (1997). Severe malarial anemia and cerebral malaria are associated with different tumour necrosis factor promoter alleles. *Human Molecular Genetics*, **6**, 1357–60.

McNeil, W.H. (1976). *Plagues and People*. Anchor Press, New York.

Miura, T., Fukunaga, T., Igarashi, T. *et al.* (1994). Phylogenetic subtypes of human T-lymphotropic virus type I and their relations to the anthropological background. *Proceedings of the National Academy of Sciences of the USA*, **91**, 1124–27.

Molleson, T.I., Jons, K., and Jons, S. (1993). Dietary changes and the effects of food preparation on microwear patterns in the Late Neolithic of Abu Hureyra, Northern Syria. *Journal of Human Evolution*, **24**, 455–68.

Molleson, T. (1994). The eloquent bones of Abu Hureyra. *Scientific American*, **271**, 70–75.

Murdock, G.P. and White, D.R. (1969). Standard crosscultural sample. *Ethnology*, **8**, 329–69.

Musser, J.M., Kroll, J.S., Granoff, D.M. *et al.* (1990). Global genetic structure and molecular epidemiology of encapsulated *Haemophilus influenzae*. *Reviews of Infectious Diseases*, **12**, 75–111.

Naderi, S., Rezaei, H.R., Pompanon, F. *et al.* (2008). The goat domestication process inferred from large-scale mitochondrial DNA analysis of wild and domestic individuals. *Proceedings of the National Academic Science of the USA*, **105**, 17659–64.

Nieberding, C., Morand, S., Libois, R., and Michaux, J.R. (2004). A parasite reveals cryptic phylogeographic history of its host. *Proceedings of the Royal Society of London B*, **271**, 2559–68.

Nieberding, C. and Olivieri, I. (2007). Parasites: proxies for host genealogy and ecology? *Trends in Ecology and Evolution*, **22**, 156–65.

Nunn, C.L. and Altizer, S.M. (2005). The global mammal parasite database: an online resource for infectious disease records in wild primates. *Evolutionary Anthropology*, **14**, 1–2.

Nunn, C.L., Altizer, S., Jones, K.E., and Sechrest, W. (2003). Comparative tests of parasites species richness in primates. *American Naturalist*, **162**, 597–614.

Page, R.D.le (ed.) (2003) *Tangled Trees: Phylogeny, Cospeciation, and Coevolution*. University of Chicago Press, Chicago.

Pearce-Duvet, J.M.C. (2006). The origin of human pathogens: evaluating the role of agriculture and domestic animals in the evolution of human disease. *Biological Review of Cambridge Philosophical Society*, **81**, 369–82.

Pedersen, A.B., Altizer, S., Poss, M., Cunningham, A.A., and Nunn, C.L. (2005). Patterns of host specificity and transmission among parasites of wild primates. *International Journal for Parasitology*, **35**, 647–57.

Prugnolle, F., Manica, A., Charpentier, M., Guégan, J.-F., Guernier, V., and Balloux, F. (2006). Pathogen-driven selection and worldwide HLA class I diversity. *Current Biology*, **15**, 1022–27.

Quinlan, R.J. (2007). Human parental effort and environmental risk. *Proceedings of the Royal Society of London B*, **274**, 121–25.

Reed, D.L., Light, J.E., Allen, J.M., and Kirchman, J.J. (2007). Pair of lice lost or parasites regained: the evolutionary history of anthropoid primate lice. *BMC Biology*, **5**, 7.

Rogerson, S.J. and Carter, R. (2008). Severe *vivax* malaria: newly recognised or rediscovered? *PloS Medicine*, **5**, e136.

Ruwende, C. and Hill, A. (1998). Glucose-6-phosphate deshydrogenase deficiency and malaria. *Journal of Molecular Medicine*, **76**, 581–88.

Salares, R. and Gomzi, S. (2001). Biomolecular archaeology of malaria. *Ancient Biomolecules*, **3**, 195–213.

Smith, K.F., Sax, D.F., Gaines S.D., Guernier, V., and Guégan, J.F. (2007). Globalization of human infectious disease. *Ecology*, **88**, 1903–10.

Stearns, S.C. (1992). *The Evolution of Life Histories*. Oxford University Press, Oxford.

Taylor, L. H., Latham, S. M., and Woolhouse, M.E.J. (2001). Risk factors for human disease emergences. *Philosophical Transaction of the Royal Society of London B*, **356**, 983–89.

Tishkoff, S.A., Varkonyi, R., Cahinhinan, N. *et al.* (2001). Haplotype diversity and linkage disequilibrium at human G6PD: recent origin of alleles that confer malarial resistance. *Science*, **293**, 455–62.

Trivers, R.L. (1974). Parent-offspring conflict. *American Zoologist*, **14**, 249–64.

Waddel, P.J., Okada, N., and Hasegawa, M. (1999). Towards resolving the interordinal relationships of placental mammals. *Systematic Biology*, **48**, 1–5.

Weatherall, D.J. and Clegg J.B. (2001). *The Thalassaemia Syndromes*, 4th Edn. Blackwell, Oxford.

Wilkinson, R.J., Llewelyn, M., Toossi, Z. *et al.* (2000). Influence of vitamin D deficiency and vitamin D receptor polymorphisms on tuberculosis among Gujarati Asians in west London: a case-control study. *Lancet*, **355**, 618–21.

Wirth, T., Meyer, A., and Achtman, M. (2005). Deciphering host migrations and origins by means of their microbes. *Molecular Ecology*, **14**, 3289–306.

Wirth, T., Wang, X., Linz, B. *et al.* (2004). Distinguishing human ethnic groups by means of sequences from *Helicobacter pylori*: lessons from Ladakh. *Proceedings of the National Academy of Sciences of the USA*, **101**, 4746–51.

Wolfe, N.D., Panosian Dunavan, C., and Diamond, J. (2007). Origins of major human infectious diseases. *Nature*, **447**, 279–83.

Woolhouse, M.E.J. and Gowtage-Sequeria, S. (2005). Host range and emerging and reemerging pathogens. *Emerging Infectious Diseases*, **11**, 1842–47.

Zimmerman, P.A., Woolley, I., Masinde, G.L. *et al.* (1999). Emergence of FY*A(null) in a *Plasmodium vivax*-endemic region of Papua New Guinea. *Proceedings of the National Academy of Science of the USA*, **96**, 13973–77.

CHAPTER 5

The use of co-phylogeographic patterns to predict the nature of host–parasite interactions, and vice versa

Caroline Nieberding, Emmanuelle Jousselin, and Yves Desdevises

5.1 Introduction

Interactions among organisms are a fundamental driving force of species diversification (Brooks and McLennan 2003; Thompson 2005). Historically, interest has mostly focused on identifying a common evolutionary history between associated organisms which has led to co-phylogenetic analyses (Page 1994a; Peek *et al.* 1998; Desdevises *et al.* 2002; Clayton and Johnson 2003). Several model systems have driven most attention such as pocket gophers and their associated lice (Hafner *et al.* 1994, 2003; Rich *et al.* 1998), phytophagous insects and their host plants (Farrell and Mitter 1990; Yokoyama 1995; Becerra and Venable 1999; Percy *et al.* 2004), some mutualistic interactions (Moran *et al.* 1995; Clark *et al.* 2000; Degnan *et al.* 2004; Hosokawa *et al.* 2006), as well as specialized interactions between plants and pollinators (Weiblen and Bush 2002; Kawakita *et al.* 2004; Althoff *et al.* 2007; Silvieus *et al.* 2008; Jousselin *et al.* 2008). Specifically, these studies aimed at detecting a pattern of cospeciation between the associated organisms. Parallelisms between host and parasite phylogenies were seen as illustrations of what was coined as 'Fahrenholz's rule' (Fahrenholz 1913).

Yet, in most cases, host–parasite interactions do not lead to congruent demographic/phylogeographic/phylogenetic patterns: 'Fahrenholz's rule' is actually rarely observed. The seminal example of generalized cospeciation between pocket gophers and their lice remains an exception: these lice are extremely specialized on their host and cannot easily disperse to another host (Hafner and Page 1995). The search for topological congruence between host and parasite phylogenies has nonetheless led to the development of complex methodologies for comparing trees. Most methods aim at maximizing the chance to observe cospeciation, or specify as null hypothesis that cospeciation is the main force driving the diversification of symbionts. Because cospeciation is clearly not the rule in the evolutionary history of interspecific interactions, these methods appear limited and often fail when trying to estimate optimal coevolutionary scenarios or when dealing with complex associations that involve several parasitic lineages (Jackson 2004, 2005).

Based on biogeographical patterns of interacting organisms (Brooks 1981; Weckstein 2004; Brooks and Ferrao 2005) and on our own experience using host–parasite associations (involving fish, rodents, aphids, figs, and their respective symbionts) (Desdevises *et al.* 2000, 2002; Nieberding *et al.* 2004, 2007, 2008; Desdevises 2007; Jousselin *et al.* 2008, 2009), it appears that some species-specific ecological traits and their geographical variation greatly influence the outcome of co-differentiation scenarios (Brown *et al.* 1997; Thompson 1999, 2005; Gomulkiewicz *et al.* 2000; Thompson and Cunningham 2002; Balakrishnan and Sorenson 2007; Craig *et al.* 2007; Whiteman *et al.* 2007) (see Section 5.5). Traits such as species distribution range, relative abundance of symbionts, population size, dispersal ability, and the presence of species closely related to the interacting organisms, all influence the probability to observe cospeciation events (Nieberding *et al.* 2008). The effects of ecological traits on co-phylogenetic scenarios have been discussed qualitatively (see references above), and some experimental work has tested the effect of some ecological traits in ensuring host specificity (e.g. dispersal ability in bird mites (Clayton *et al.* 2003)), but studies that quantitatively test the role of these traits in generating co-differentiation scenarios are, to our best knowledge, missing.

5.2 Objectives

None of the existing co-phylogenetic programs reconstructing the history of associations between interacting organisms such as hosts and parasites (H–P) incorporates ecological traits (ETs) of these organisms, nor their geographic variation over the species distribution ranges. Here we aim at producing a co-phylogenetic program that integrates ETs and their geographical variation in order to:

(1) identify which ETs determine the observed co-phylogenetic pattern of diverse interacting organisms, and vice versa; and
(2) predict the evolutionary outcome of H–P interactions based on the identification of the important ETs.

Therefore, we aim at including ETs in tests of co-differentiation and test their contribution in producing congruence or incongruence in the histories of hosts and parasites. Depending on the interacting organisms, relevant ETs are listed and ways to quantify them are described (see Section 5.5).

In order to include the geographical dimension, and more specifically the regional variation, of ETs across species distribution ranges in our analysis, we design the co-phylogenetic analysis of host–parasite interactions at the phylogeographical level. One species of parasite is confronted to one species of host, so that the datasets of hosts and parasites consist of specimens (i.e. sequences) sampled in the same geographical locations over the two species distribution ranges. If the parasite species parasitizes more than one host species, successive analyses will be computed for each H–P species pair, and the results obtained for each H–P pair can then be compared (this is discussed in Section 5.6).

5.3 Relevance of ParaFit as the supporting source program

Our objective here is to estimate the amount of common history of interacting taxa sharing the same geographical area. We deal with two phylogenetic trees, between which the leaves are obligatorily connected via their common areas of occurrence. If the topologies of the trees are perfectly congruent, the evidence of their common history is straightforward. However, they are generally at least slightly different, requiring the use of co-phylogenetic methods (such as Brooks Parsimony Analysis (Brooks 1981), TreeMap (Page 1994), TreeFitter (Ronquist 1995), and ParaFit (Legendre *et al.* 2002)) to measure the importance of co-differentiation in the evolving association. Among these programs ParaFit seems to be the most relevant for analysing co-phylogeographic patterns of H–P interactions for the reasons below.

(1) Classical co-phylogenetic methods consider that an 'associate' is tracking a 'host' (Page and Charleston 1998), while in phylogeography one is interested in comparing phylogenetic trees for several clades living in the same areas, to assess if they display the same phylogeography. Testing whether one tree is 'fitting' the other (like parasites on hosts), is a different process than testing whether the two trees are equivalent entities (like different species in the same area). In the first case, the host tree is considered as fixed and the method tries to fit the associate tree on it (which is then in a large part 'dependent' on the host tree), by adequately mixing different types of evolutionary events (i.e. cospeciation, sorting, duplication, and switching). These events concern only the associate tree (except cospeciation for obvious reasons), and this philosophy is at the root of methods like reconciled trees (Page 1993, 1994a) or parsimony-based tests (Brooks 1981; Ronquist 1995). A global fit method, such as ParaFit (Legendre *et al.* 2002), only aims at assessing the global congruence between two trees without considering events, but also without considering these trees differently. This has important consequences on the reconstructed scenarios and on the way null hypotheses are generated via randomization.
(2) Conversely to other co-phylogenetic methods, ParaFit is not designed to maximize cospeciation and

is rather conservative in observing cospeciation (Type I error is adequate, Type II error is rather low; Legendre *et al.* 2002); further, it does not aim at reconstructing co-diversification scenarios (i.e. identifying host switch, lineage sorting, and duplication), which is not the question we want to address here.

(3) ParaFit assesses the congruence level between hosts and parasites based on a matrix of genetic distances between DNA sequences (or other types of datasets) and as such is not limited to fixed phylogenetic topologies. This is very interesting at the phylogeographic level at which most terminal branches, representing individuals from the same location, display no strong phylogenetic support and are thus better described using genetic distances rather than their (unstable, swapping) position on the phylogenetic tree.

5.4 Methodology

The methodology proposed below is based on the existing ParaFit program described in Legendre *et al.* (2002). ParaFit is based on the combination of three matrices, one representing the phylogenetic relationships between parasites (matrix B), one between hosts (matrix C), while the third matrix describes the pattern of presence/absence of parasites on their hosts (matrix A). These three matrices are used to produce a matrix D and compute its 'Trace' value from which the statistical estimation of the level of H–P congruence is assessed.

The main modification we propose in ParaFit is to change the association matrix A so that it incorporates *a priori* the ETs associated to the studied organisms and their geographic variation. This allows a statistical estimation of the H–P level of congruence based on explicit values for relevant ETs. The ETs are used to differentially weight the value of the links (from 0 to 1) of the association matrix over the distribution range of the interacting species. We expect that the lower the value of the link, the lower the probability of long-term common differentiation between the two organisms in the part of the tree where these species have been sampled. This would allow the identification of the part (i.e. geographical areas) of the trees responsible for the potential congruence in the association, and/or the ET mainly responsible for the observed

(in)congruence (see Section 5.6 for specific hypotheses and tests using the program).

Practically, the modifications of the matrices A, B, and C consist of:

(1) Matrix B (and matrix C) contains the phylogeographic (instead of phylogenetic) relationships between individuals (instead of species) of parasites (or hosts, respectively). Thus each matrix contains information from one species (of parasite, or host). The matrix summarizes the genetic distances between sequences representing individuals from different geographic locations. The columns in the matrix B and rows in the matrix C list the geographic locations of origin in the same order so that the genetic distances between pairs of hosts and pairs of parasites match each other when converging in matrix A (Fig. 5.1).

(2) Matrix A contains information about the ETs linking hosts and parasites (see the list in Section 5.5) instead of the presence/absence matrix between host and parasite. This is possible because the H–P associations are given by the position of the taxa in the matrices B and C, then the taxa from the same geographic origin are at corresponding positions in the host and parasite matrices: location L1 is at line 1 in matrices B and C, location L2 at line 2, etc (Fig. 5.1).

Matrix A is formed by a combination of a series of matrices $A_1, A_2, \ldots A_n$, each matrix representing one ET. Each ET is represented by a value varying between 0 (ET minimizing the probability to observe long-term H–P common differentiation or congruence) to 1 (ET maximizing H–P congruence) (for example, if ET_1 is host specificity, then in matrix A_1, the value of ET is close to 1 if the parasite is a specialist, and close to 0 if the parasite is a generalist). The value of each ET in matrix A_n can vary geographically, in which case the numbers in matrix A vary accordingly from 0 to 1, or are stable across the whole species geographic ranges, in which case the number in matrix A_n is constant (Fig. 5.2a).

The final A_{fin}, which is used to estimate the statistical values 'ParaFitGlobal', 'Trace', and 'ParaFitlink1 and 2', is obtained from the following procedure. The value of each box in matrix A_{fin} is the sum of the values in the homologous box for each matrix A_1 to A_n, divided by the number of matrices A that have been associated (Fig. 5.2b). As such, the matrix A_{fin} summarizes the

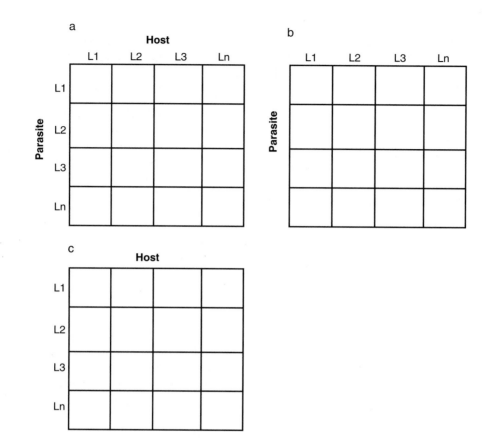

Figure 5.1 Representation of the three matrices and the changes we propose compared to Figure 2 in Legendre *et al.* (2002). (a) Matrix A is matrix encoding values (between 0 and 1) of various ecological traits (ET$_1$ to ET$_x$ in matrices A$_1$ toA$_x$). (b) Matrix B represents principal coordinates describing the phylogeographical tree of parasites. (c) 'Matrix C' represents principal coordinates describing the phylogeographical tree of hosts. L1 to Ln are location 1, location 2, . . . , location *n*, according to the geographical origin of samples 1 to n. The association of the host and parasite taxa (sequences) is given by their position on the matrices B and C.

values of the ET$_s$ that are geographically locally relevant to reinforce or weaken the H–P association over evolutionary time, that is, reinforce or weaken the hypothesis of H–P phylogenetic congruence. This is fundamentally different from the original ParaFit program, and is the main novelty of our approach.

5.5 List of relevant ecological traits and their implementation in matrices A$_1$ to A$_n$

The ETs listed below are selected based on a literature search. We classify ETs into two categories:

(1) traits that do not vary geographically; and
(2) traits that vary along the distribution range of the organisms involved in the interaction.

Practically, for each ET (ET$_n$) presented below, one needs to divide all values of the matrix A$_n$ by the highest value in the matrix A$_n$ so that each value of the matrix is set between 0 and 1.

5.5.1 Traits varying geographically

The co-occurrence of host and symbionts in the same area (Huyse and Volckaert 2005; Criscione and Blouin 2007; Whiteman *et al.* 2007) and high abundance of symbionts, obviously reduces the chances of lineage extinction (Clayton *et al.* 2004). The abundance of the 'preferred host' in comparison with alternative hosts has been mentioned as an important ET in several cospeciation studies; host

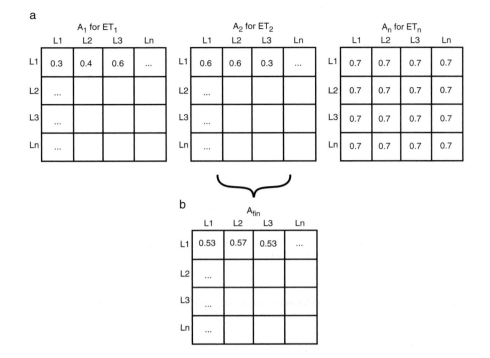

Figure 5.2   Relationships between matrices A_1 to A_n and ecological traits ET_1 to ET_x. (a) Matrices A_1 and A_2 encode ET_1 and ET_2 that vary geographically across host and parasite distribution range. In contrast, matrix A_n is constant across the host and parasite distribution range. (b) The final matrix A_{fin} summarizes the values for the different ET_1 to ET_n considered. For example, the value in box {L1, L1} of matrix A_n equals the sum of the values in box of matrix A1 {L1,L1} and matrix A2 {L1,L1} and matrix An {L1,L1}, divided by the number n of matrices A. This can be represented by: {An;L1, L1} = ({A1;L1, L1} +{A2;L1, L1} + {An;L1, L1})/n.

shifts are just more likely to occur on geographically available taxa (Roy 2001; Clayton *et al.* 2004), 'failure to switch host' can simply be a consequence of the absence of alternative hosts rather than an inability to establish on novel hosts (Brooks and Ferrao 2005). Thus, integrating matrices of presence/absence of alternative hosts and symbionts over their geographic range could be a good first step towards incorporating biogeographic data into co-phylogenetic studies (Brooks and Ferrao 2005). These matrices may incorporate abundance data. The presence/abundance of alternative hosts could also consider their phylogenetic proximity with the 'preferred' hosts to give a clearer estimate of the chances of successful colonization of a new host, as host switches are supposedly more likely between related hosts (Reed and Hafner 1997). All these considerations have been summarized as quantitative data in the following ET_1 to ET_9.

ET_1. '1 / number of recorded host species'
This is an alternative or complementary estimator of host specificity; the value of this ET increases accordingly to the expected congruence within the HP association. As alternative host species do not have the same distribution range, this ET can vary geographically.

ET_2. 'presence of main host (H_1) × abundance'
These two variables are linked as they both increase the probability for a parasite to encounter its host at a local scale, which determines the risk of lineage sorting. This ET varies geographically as the abundance of any species does (e.g. Nieberding *et al.* 2008), and a high value of this ET favours H–P congruence (the index is close to 1). If there is no main host and alternative hosts can be equally relevant, the analysis should be performed independently for each H_n–P pair.

ET_3. '1/ (presence of alternative host (H_n) × abundance × phylogenetic proximity between H_n and H_1)'

The same couple of variables can be important if parasite specialization on H_1 is not perfect and alternative species (H_2 to H_n) can host the parasite, especially when H_n is closely related to H_1. In that case, an increasing availability of H_n and increased closeness of H_n to H_1 both decrease the chance that H_1 and the parasite display congruent phylogenies, which is why the term 'presence of alternative host (H_n) × abundance' is displayed at the denominator.

ET_4. 'presence of fossil evidence of main host (H_1) × abundance'

The logic is similar to ET_2, but concerns fossil data for host taxa (e.g. Nieberding *et al.* 2008).

ET_5. '1/(presence of fossil evidence of alternative host (H_n) × abundance × phylogenetic closeness between H_n and H_1)'.

The logic is similar to ET_3, but concerns fossil alternative host taxa.

ET_6. specialization index: 'host genetic diversity/ parasite genetic diversity'

(See Poulin and Mouillot 2005; Rohde and Rohde 2008). Host specificity, which is necessary (but not sufficient) for long-term H–P co-differentiation (Reed and Hafner 1997; Clayton *et al.* 2004) is taken into account in most cospeciation methods via a matrix of association (Stevens 2004). It can be represented by this specialization index. It is thought that a generalist parasite displays a higher level of genetic diversity than a specialist (Kaci-Chaouch *et al.* 2008), because the presence of genetic diversity allows the parasite to adapt more easily to new hosts (host shift). A high value of the specialization index indicates a specialist (low value of parasite diversity), which favours congruence (the index is close to 1).

ET_7. 'geographic distance among localities/parasite dispersal ability'

Although dispersal ability is not believed to vary significantly over the distribution range of a species (but see ET_1), the importance of dispersal ability varies according to the distance between the sampled localities. Distances among localities obviously vary geographically, so that the values taken by this ET vary geographically as well. Comparative studies of wing and body lice associated with doves suggest that differences in dispersal capacity may account for differences in host/lice cospeciation scenarios (Clayton *et al.* 2001; Johnson and Clayton 2003; Clayton and Johnson 2003). Moreover, rare events, such as long dispersal events, that are difficult to measure on ecological times, can have a high impact on an evolutionary time scale (Brooks and McLennan 1991; Brooks and Ferrao 2005). We expect that if the ratio of 'geographic distance among localities/parasite dispersal ability' increases, it increases the chance that congruence will be observed between the local H–P phylogenies, and the reverse is true.

ET_8. 'population size of symbionts/population size of hosts'

Large population size in symbionts compared to their hosts could limit the opportunity for genetic structuring of the symbiont population (Whiteman *et al.* 2007), which, as any other traits that limit structuring, is predicted to limit host/symbiont cospeciation (Clayton *et al.* 2004; Whiteman *et al.* 2007). In plant–pathogen interactions where the pathogen disperses via spores, average dispersal distances can vary according to climatic variables (wind, rainfall) (Roy 2001), in aquatic symbiotic interactions where the symbionts disperse via currents, again their dispersal ability will vary regionally (Jones *et al.* 2006). When the symbiont needs a vector (such as an insect), dispersal and probability of colonization of a new host vary with vector abundance, which can be governed by regional factors (Roy 2001). Dispersal can be simply measured as dispersal distances and frequencies. Clayton and Johnson (2003) suggested that the level of population genetic structure could be indicative of the dispersal ability of the lice they studied, and studies of marine systems also infer dispersal ability from population structure (e.g. Jones *et al.* 2006). We could also use indirect measures such as variables that influence dispersal ability: mode of transmission (with or without a vector), or ability to overcome geographic barriers (see Clayton *et al.* 2004 for a discussion on these variables).

ET_9. '*number of alternative symbionts co-occurring on the same host and availability of the symbiont ecological niche*'

Both are important factors when investigating processes behind codiversification patterns. If several symbionts use the same niche, competitive exclusion could limit host switches (Lenormand and Raymond 2000; Clayton *et al.* 2004). On the other hand, it could also lead to lineage extinction and disrupt phylogenetic congruence. Diversification of ecological niches within a host has also been mentioned as a factor favouring duplication, hence decreasing congruence between hosts and symbionts phylogenies (Weiblen and Bush 2002; Lopez-Vaamonde *et al.* 2003; Jousselin *et al.* 2008). The search for an index of the ability of the symbiont to establish on a novel host, based on the width of its ecological niche (a trait that should not vary geographically), the occurrence of other symbionts able to compete for that niche (a trait that does vary geographically), or a combination of those traits, is probably a worthy task.

Note that the ET 'parasite abundance' is not considered here because it can lead to either increased congruence if high abundance prevents lineage sorting, or to decreased congruence if high abundance favours host switch. We consider the 'fossil parasite abundance' in the same way.

5.5.2 Traits not varying geographically

These ETs are constant for each H–P pair over the whole distribution range of both species.

ET_{10}. '*sex ratio for* H_1'

One assumes that males in most species are more dispersive than females and as such increase the rate of parasite transmission between populations (Poulin 2007).

ET_{11}. '*generation time P/H*'

Predictions here are not straightforward, but a low value for this trait indicates that P has a shorter generation time than H and therefore a potentially faster adaptation of parasites to their hosts or to a new host. In plant–phytophagous insect associations, fast evolution in insects is seen as facilitating the colonization of new host plants (Agosta 2006),

fast evolution of the parasites could also favour intra-host speciation (duplication) (Stefka and Hypsa 2008). Thus we assume here that a high value of this ET could increase the probability of congruence between the species pair.

ET_{12}. '*life-cycle/free dispersing phase of the parasite*'

Life-cycle can be coded as 0.1 to 1 if there are 10 to 0 intermediate hosts, as a direct cycle increases the chance of congruence (numerator), while the existence and the length of a free dispersing phase decreases the probability to observe congruence (denominator) (Roy 2001). Vertical (i.e. transmission from parents to offspring) versus horizontal (i.e. transmission between unrelated individuals) parasite transmission, one of the main factors determining the probability of congruence (Clayton and Johnson 2003; Clayton *et al.* 2004; Balakrishnan and Sorenson 2007), are included in this ET. Vertical transmission indeed indicates that the life-cycle is direct and that no free dispersive phase exists; horizontal transmission is usually due to the existence of intermediate hosts and/or free dispersal stage.

5.6 Application: questions of interest and statistical procedure

Based on the matrices A_{fin}, B, and C, matrix D can be computed as originally designed in ParaFit, and the corresponding statistics ParaFitGlobal, ParaFitLink1, and ParaFitLink2 can be computed as well (eq. 2, 4 and 6, respectively in Legendre *et al.* 2002).
We want to address three specific questions using this modified program.

(1) Do we observe a global signal of congruence between the H–P phylogeographic trees? This is similar to the question asked originally by the designers of ParaFit, and should be answered using the same procedure.
(2) Which ET(s) has (have) a major effect on the observed signal of congruence (or incongruence)? To answer this question, one can modify and use the statistical procedure described in 'ParaFit statistics for tests of individual host-parasite association links' (p. 219). Instead of computing the Trace (k) based on the suppression on one single value in matrix A, one can measure Trace (k) after removing

the matrix A_x corresponding to the ET_x being tested and recalculating matrix A_{fin} without the contribution of matrix A_x. The contribution of matrix A_x, representing ET_x, will be given following the described procedure in Legendre *et al.* (2002) without further change.

(3) Does the parasite display a significantly higher level of congruence with one or another host species, in different geographical regions? Here we aim at comparing the level of regional congruence between different pairs of H–P associations. For this, we need to perform the statistical analysis of congruence between the parasite and additional host species that we suspect are alternative hosts for the parasite at least in part of its distribution range (see Nieberding *et al.* 2008 for an example). We also need to develop a statistic similar to Trace (k) that will be in this case limited to part of matrix A_{fin}, so that it represents the H–P congruence level for a particular geographical region: Trace (k1) represents region 1, Trace (k_2) represents region 2, ... Trace (k_n) represents region n (Fig. 5.3). The comparison of the value of Trace (k_n) obtained in the analyses of congruence for the species pairs H_1-P and H_2-P will then reveal whether H_1 displays a phylogeographical pattern significantly more congruent with the parasite than H_2, or not.

5.7   Statistical issues

One main problem concerns the null hypothesis testing. The original ParaFit computation randomizes the matrix A containing the links: this is equivalent to randomly assigning any parasite to any host, then destroying the co-phylogenetic structure, if any. Here, because the values of the links can be different for each individual association, they cannot be meaningfully permuted: each given link value corresponds to a given individual association. We should instead randomize the trees, via their patristic distance matrices represented in the matrices B and C containing principal coordinates in the original paper. Both trees should be randomized in order not to give a priority to any tree but a choice may be given to the user. However, trees have to be randomized in a proper way depending on their type so that tree structures are conserved. In this respect, a specific permutation procedure must be used whether the tree is additive (triple-permutation: Lapointe and Legendre 1992) or ultrametric (double-permutation: Lapointe and Legendre 1990, 1991). In cases where raw genetic distance matrices are used, a simple permutation procedure such as in the Mantel test (1967) can be used (see also Legendre *et al.* 1994). These modifications would

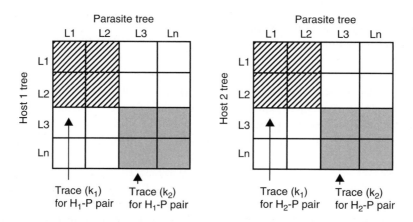

Figure 5.3   Representation of matrix D for two H-P pairs, H_1-P on the left and H_2-P on the right. The matrix D encodes the fourth-corner parameters derived from the crosses between the matrices B and C. In both H-P matrices, Trace (k_1) is calculated from the dashed area (representing the geographical region 1), by taking the square of the sum of values in these boxes. Trace (k_2) is similarly calculated from the grey region (representing the geographical region 2). The comparison of trace (k_1) between H_1-P and H_2-P pairs can reveal whether the parasite displays a significantly higher level of phylogeographical congruence with one or the other host species.

generate a meaningful distribution of the null hypothesis of no congruence between trees. The fourth-corner statistic has recently been expanded to integrate a permutation of rows in the matrix B and of columns in the matrix, C allowing for triple permutations (Dray and Legendre 2008).

5.8   Perspectives

The implementation of this modified ParaFit program will permit testing whether our basic assumption holds in the program: that a higher value in matrix A_{fin} correlates with a higher probability of long-term common evolution between the associated organisms. This can be done by comparing the outcome of the program when given simulated datasets producing perfect co-phylogenies and respectively high or low values of ETs. Because large datasets on various ecological traits are needed to make use of our approach, we will similarly use simulation studies to test the range of sensitivity of the program to:

(1) detect significant differences in levels of (in) congruence in different parts of the species geographic ranges (i.e. in different parts of the matrix D, Fig. 5.3); and

(2) detect the effect of a single ET on the observed differences in levels of (in)congruence over the species geographic ranges.

We will test the ability of the program to distinguish among alternative host species for being responsible for the pattern of congruence observed in part of the parasite distribution range. Altogether, this should help us select among competing ecological traits to better understand how diverse species interactions lead to different evolutionary outcomes, and identify among competing hosts the one(s) responsible for the pattern of differentiation observed in the parasite.

References

Agosta, S.J. (2006). On ecological fitting, plant-insect associations, herbivore host shifts, and host plant selection. *Oikos*, **114**, 556–65.

Althoff, D.M., Svensson, G.P., and Pellmyr, O. (2007). The influence of interaction type and feeding location on the phylogeographic structure of the yucca moth community associated with *Hesperoyucca whipplei*. *Molecular Phylogenetics and Evolution*, **43**, 398–406.

Balakrishnan, C.N. and Sorenson, M.D. (2007). Dispersal ecology versus host specialization as determinants of ectoparasite distribution in brood parasitic indigobirds and their estrildid finch hosts. *Molecular Ecology*, **16**, 217–29.

Becerra, J.X. and Venable, D.L. (1999). Macroevolution of insect-plant associations: the relevance of host biogeography to host affiliation. *Proceedings of the National Academy of Sciences of the USA*, **96**, 12626–31.

Brooks, D.R. (1981). Hennig's parasitological method: a proposed solution. *Systematic Zoology*, **30**, 229–49.

Brooks, D.R. and Ferrao, A.L. (2005). The historical biogeography of coevolution: emerging infectious diseases are evolutionary accidents waiting to happen. *Journal of Biogeography*, **32**: 1291–99.

Brooks, D.R. and McLennan, D.A. (1991). *Phylogeny, Ecology, and Behavior. A Research Program in Comparative Biology*. University of Chicago Press, Chicago.

Brooks, D.R. and McLennan, D.A. (2003). Extending phylogenetic studies of coevolution: secondary Brooks parsimony analysis, parasites, and the Great Apes. *Cladistics*, **19**, 104–19.

Brown, J.M., Leebens-Mack, J., Thompson, J.N., and Pellmyr, O. (1997). Phylogeography and host association in a pollinating seed parasite *Greya politella* (Lepidoptera: Prodoxidae). *Molecular Ecology*, **6**, 215–24.

Clark, M.A., Moran, N.A., Bauman, P., and Wernegreen, J.J. (2000). Cospeciation between bacterial endosymbionts (*Buchnera*) and a recent radiation of aphids (*Uroleucon*) and pitfalls of testing for phylogenetic congruence. *Evolution*, **54**, 517–25.

Clayton, D.H. and Johnson, K.P. (2003). Linking coevolutionary history to ecological process: doves and lice. *Evolution*, **57**, 2335–41.

Clayton, D.H., Al-Tamimi, S., and Johnson, K.P. (2001). The ecological basis of coevolutionary history. In RDM Page, ed. *Tangled Trees: Phylogeny, Cospeciation and Coevolution*, pp. 310–42. University of Chicago Press, Chicago.

Clayton, D.H., Bush, S.E., Goates, B.M., and Johnson, K.P. (2003). Host defense reinforces host-parasite cospeciation. *Proceedings of the National Academy of Sciences of the USA*, **100**, 15694–99.

Clayton, D.H., Bush, S.E., and Johnson, K.P. (2004). Ecology of congruence: past meets present. *Systematic Biology*, **53**, 165–73.

Craig, T.P., Itami, J.K., and Horner, J.D. (2007). Geographic variation in the evolution and coevolution of a tritrophic interaction. *Evolution*, **61**, 1137–52.

Criscione, C.D. and Blouin, M.S. (2007). Parasite phylogeographical congruence with salmon host evolutionarily significant units: implications for salmon conservation. *Molecular Ecology*, **16**, 993–1005.

Degnan, P.H., Lazarus, A.B., Brock, C.D., and Wernegreen, J.J. (2004). Host-symbiont stability and fast evolutionary rates in an ant-bacterium association: cospeciation of *Camponotus* species and their endosymbionts, *Candidatus blochmannia*. *Systematic Biology*, **53**, 95–110.

Desdevises, Y. (2007). Cophylogeny: insights from fish-parasite systems. *Parasitologia*, **49**, 125–8.

Desdevises, Y., Jovelin, R., Jousson, O., and Morand, S. (2000). Comparison of ribosomal DNA sequences of *Lamellodiscus* spp. (Monogenea, Diplectanidae) parasitizing *Pagellus* (Sparidae, Teleostei) in the North Mediterranean sea: species divergence and coevolutionary interactions. *International Journal for Parasitology*, **30**, 741–6.

Desdevises, Y., Morand, S., Jousson, O., and Legendre, P. (2002). Coevolution between *Lamellodiscus* (Monogenea) and Sparidae (Teleostei): the study of a complex host-parasite system. *Evolution*, **56**, 2459–71.

Dray, S. and Legendre, P. (2008). Testing the species traits-environment relationships: the fourth-corner problem revisited. *Ecology*, **89**, 3400–12.

Fahrenholz, H. (1913). Ectoparasiten und Abstammungslehre. *Zoologischer Anzeiger*, **41**, 371–74.

Farrell, B. and Mitter, C. (1990). Phylogenesis of insect/plant interactions: have *Phyllobrotica* leaf beetles (Chrysomelidae) and the lamiales diversified in parallel. *Evolution*, **44**, 1389–403.

Gomulkiewicz, R., Thompson, J.N., Holt, R.D., Nuismer, S.L., and Hochberg, M.E. (2000). Hot spots, cold spots, and the geographic mosaic theory of coevolution. *American Naturalist*, **162**, s80–s93.

Hafner, M.S. and Page, R.D.M. (1995). Molecular phylogenies and host-parasite cospeciation—gophers and lice as a model system. *Philosophical Transactions of the Royal Society of London B*, **349**, 77–83.

Hafner, M.S., Sudman, P.D., Villablanca, F.X., Spradling, T.A., Demastes, J.W., and Nadler, S. (1994). Disparate rates of molecular evolution in cospeciating hosts and parasites. *Science*, **265**, 1087–90.

Hafner, M.S., Demastes, J.W., Spradling, T.A., and Reed, D.L. (2003). Cophylogeny between pocket gophers and chewing lice. In RDM Page, ed. *Tangled trees: phylogeny, cospeciation and coevolution*, pp. 195–220. University of Chicago Press, Chicago.

Hosokawa, T., Kikuchi, Y., Nikoh, N., Shimada, M., and Fukatsu, T. (2006). Strict host-symbiont cospeciation and reductive genome evolution in insect gut bacteria. *PLoS Biology*, **4**, 1841–50.

Huyse, T. and Volckaert, F.A.M. (2005). Comparing host and parasite phylogenies: *Gyrodactylus* flatworms jumping from Goby to Goby. *Systematic Biology*, **54**, 710–18.

Jackson, A.P. (2004). A reconciliation analysis of host switching in plant-fungal symbioses. *Evolution*, **58**, 1909–23.

Jackson, A.P. (2005). The effect of paralogous lineages on the application of reconciliation analysis by cophylogeny mapping. *Systematic Biology*, **54**, 127–45.

Johnson, K.P. and Clayton, D.H. (2003). Coevolutionary history of ecological replicates: comparing phylogenies of wing and body lice to columbiform hosts. In RDM Page, ed. *Tangled Trees: Phylogeny, Cospeciation and Coevolution*, pp. 262–86. University of Chicago Press, Chicago.

Jones, B.W., Lopez, J.E., Huttenburg, J., and Nishiguchi, M.K. (2006). Population structure between environmentally transmitted vibrios and bobtail squids using nested clade analysis. *Molecular Ecology*, **15**, 4317–29.

Jousselin, E., Desdevises, Y., and Coeur d'acier A. (2009). Fine-scale cospeciation between *Brachycaudus* and *Buchnera aphidicola*: bacterial genome helps define species and evolutionary relationships in aphids. *Proceedings of the Royal Society of London B*, **276**, 187–96.

Jousselin, E., van Noort, S., Berry, V., Rasplus, J.-Y., Ronsted, N., Erasmus, J.C., and Greeff, J.M. (2008). One fig to bind them all: host conservatism in a fig wasp community unravelled by cospeciation analyses among pollinating and nonpollinating fig wasps. *Evolution*, **62**, 1777–97.

Kaci-Chaouch, T., Verneau, O., and Desdevises, Y. (2008). Host specificity is linked to intraspecific variability in the genus *Lamellodiscus* (Monogenea). *Parasitology*, **135**, 607–16.

Kawakita, A., Takimura, A., Terachi, T., Sota, T., and Kato, M. (2004). Cospeciation analysis of an obligate pollination mutualism: have Gochlidion trees (Euphorbiaceae) and pollinating *Epicephala* moths (Gracillariidae) diversified in parallel? *Evolution*, **58**, 2201–14.

Lapointe, F.-J. and Legendre, P. (1990). A statistical framework to test the consensus of two nested classifications. *Systematic Zoology*, **39**, 1–13.

Lapointe F.-J. and Legendre, P. (1991). The generation of random ultrametric matrices representing dendrograms. *Journal of Classification*, **178**, 177–200.

Lapointe, F.-J. and Legendre, P. (1992). A statistical framework to test the consensus among additive trees (cladograms). *Systematic Biology*, **41**, 158–71.

Legendre, P., Lapointe, F-J and Casgrain, P. (1994). Modeling brain evolution from behavior: a permutational regression approach. *Evolution*, **48**, 1487–99.

Legendre, P., Desdevises, Y. and Bazin, E. (2002). A statistical test for host-parasite coevolution. *Systematic Biology*, **51**, 217–34.

Lenormand, T. and Raymond, M. (2000). Analysis of clines with variable selection and variable migration. *American Naturalist*, **155**, 70–82.

Lopez-Vaamonde, C., Godfray, H.C.J. and Cook, J.M. (2003). Evolutionary dynamics of host-plant use in a genus of leaf-mining moths. *Evolution*, **57**, 1804–21.

Mantel, N. (1967). The detection of disease clustering and a generalized regression approach. *Cancer Research*, **27**, 209–20.

Moran, N.A., vonDohlen, C.D., and Baumann, P. (1995). Faster evolutionary rates in endosymbiotic bacteria than in cospeciating insect hosts. *Journal of Molecular Evolution*, **41**, 727–31.

Nieberding, C. and Olivieri, I. (2007). Parasites: proxies for host genealogy and ecology? *Trends in Ecology and Evolution*, **22**, 156–65.

Nieberding, C., Morand, S., Libois, R., and Michaux, J.R. (2004). A parasite reveals cryptic phylogeographic history of its host. *Proceedings of the Royal Society of London B*, **271**, 2559–68.

Nieberding C., Durette-Desset M.-C., Libois R. *et al.* (2008). Geography and host biogeography matter for understanding the phylogeography of a parasite. *Molecular Phylogenetics and Evolution*, **47**, 538–54.

Page, R.D.M. (1993). Parasites, phylogeny and cospeciation. *International Journal for Parasitology*, **23**, 499–506.

Page, R.D.M. (1994a). Maps between trees and cladistic analysis of historical associations among genes, organisms, and areas. *Systematic Biology*, **43**, 58–77.

Page, R.D.M. (1994b). Parallel phylogenies: reconstructing the history of host-parasite assemblages. *Cladistics*, **10**, 155–73.

Page, R.D.M and Charleston, M.A. (1998). Trees within trees: phylogeny and historical associations. *Trends in Ecology and Evolution*, **13**, 356–59.

Peek, A.S., Feldman, R.A., Lutz, R.A., and Vrijenhoek, R.C. (1998). Cospeciation of chemoautotrophic bacteria and deep sea clams. *Proceedings of the National Academy of Sciences of the USA*, **95**, 9962–66.

Percy, D.M., Page, R.D.M, and Cronk, Q.C. (2004). Plant-insect interactions: double-dating associated insect and plant lineages reveals asynchronous radiations. *Systematic Biology*, **53**, 120–27.

Poulin R. (2007). *Evolutionary Ecology of Parasites*, 2nd edn. Princeton University Press, Princeton.

Poulin, R. and Mouillot, D. (2005). Combining phylogenetic and ecological information into a new index of host specificity. *Journal of Parasitology*, **91**, 511–14.

Reed, D.L. and Hafner, D.J. (1997). Host specificity of chewing lice on pocket gophers: a potential mechanism for cospeciation. *Journal of Mammalogy*, **78**, 999–1007.

Rich, S.M., Light, M.C., Hudson, R.R., and Ayala, F.J. (1998). Malaria's eve: evidence of a recent population bottleneck throughout the world populations of *Plasmodium falciparum*. *Proceedings of the National Academy of Sciences of the USA*, **95**, 4425–30.

Rohde, K. and Rohde, P. (2008). How to measure ecological host specificity. *Vie et Milieu*, **58**, 121–24.

Ronquist, F. (1995). Reconstructing the history of host-parasite associations using generalised parsimony. *Cladistics*, **11**, 73–89.

Roy, B.A. (2001). Patterns of association between crucifers and their flower-mimic pathogens: host jumps are more common than coevolution or cospeciation. *Evolution*, **55**, 41–53.

Silvieus, A.I., Clement, W.L., and Weiblen, G.D. (2008). Cophylogeny of figs, pollinators, gallers and parasitoids. In K. Tilmon, ed. *The Evolutionary Biology of Herbivorous Insects: Specialization, Speciation and Radiation*, pp. 225–39. University of California Press, Berkeley.

Stefka, J. and Hypsa, V. (2008). Host specificity and genealogy of the louse *Polyplax serrata* on field mice, *Apodemus* species: a case of parasite duplication or colonisation? *International Journal for Parasitology*, **38**, 731–41.

Stevens, J. (2004). Computational aspects of host-parasite phylogenies. *Briefings in Bioinformatics*, **5**, 339–49.

Thompson, J.N. (1999). Specific hypotheses on the geographic mosaic of coevolution. *American Naturalist*, **153**, 1–14.

Thompson, J.N. (2005). *The Geographic Mosaic of Coevolution*. University of Chicago Press, Chicago and London.

Thompson, J.N and Cunningham, B.M. (2002). Geographic structure and dynamics of coevolutionary selection. *Nature*, **417**, 735–38.

Weckstein, J.D. (2004). Biogeography explains cophylogenetic patterns in toucan chewing lice. *Systematic Biology*, **53**, 154–64.

Weiblen, G.D. and Bush, G.L. (2002). Speciation in fig pollinators and parasites. *Molecular Ecology*, **11**, 1573–78.

Whiteman, N.K., Kimbal, R.T., and Parker, P.G. (2007). Co-phylogeography and comparative population genetics of the threatened Galapagos hawk and three ectoparasite species: ecology shapes population histories within parasite communities. *Molecular Ecology*, **16**, 4759–73.

Yokoyama, J. (1995). Insect-plant coevolution and cospeciation. In R. Arai, M. Kato and Y. Doi, eds. *Biodiversity and Evolution*, pp. 115–30. National Science Museum Foundation, Tokyo.

PART II

Ecological Biogeography and Macroecology

CHAPTER 6

Marine parasite diversity and environmental gradients

Klaus Rohde

6.1 Introduction

A brief review of our knowledge of marine biodiversity, including parasite diversity, shows that we know little, due in particular to the scarcity of studies of deep-sea and meiofaunal organisms. This chapter, after a brief discussion of older studies which established the existence of zoogeographical regions and some patterns in the geographical distribution of marine parasites, describes latitudinal gradients in parasite diversity, reproductive strategies and host ranges, longitudinal gradients in diversity (centres of diversity and oceanic barriers between them), and depth gradients. The marine environment is less heterogeneous than terrestrial and freshwater habitats, and therefore more suitable for evaluating the causes of gradients. An attempt at such an evaluation is made. Important gaps in our knowledge are discussed and suggestions are made for future studies.

6.2 Parasite diversity in the oceans

6.2.1 How many parasite species do we know?

Rohde (2002) reviewed species diversity of various marine animals and their parasites and methods used in studying them. Largely because of our poor knowledge of deep-sea diversity and diversity of the meiofauna, even at the level of free-living species, parasite diversity remains poorly understood. Even parasites of marine vertebrates and invertebrates in surface waters are far from exhaustively known. Although fish belong to the better known groups, many fish species have never been thoroughly examined for parasites or not examined at all. Every year many new species of trematodes, monogeneans, and protistans, to mention only some groups, are described. For example, according to Cribb (2002), trematodes have been recorded from only 62 of the 159 species of the large family of marine fishes, the Epinephelidae: most species were sampled from single localities, and not a single tropical species has been thoroughly examined. Exhaustive surveys of more or less 'all' parasite groups of marine fishes have been conducted in some northern seas, whereas surveys of parasites in most seas deal with particular groups, depending on the expertise available. Thus, monogeneans and trematodes of fishes along the eastern coast of Australia are much more thoroughly (though far from completely) known than, for example, Myxozoa and Microsporidia, which are known from less than perhaps 1 per cent of the species. Parasite surveys of surface invertebrates, if they have been done at all, are even more sketchy. The recent survey of host and parasite diversity by Dobson *et al.* (2008) is restricted to helminths of vertebrates, because 'we have no credible way of estimating how many parasitic protozoa, fungi, bacteria, and viruses exist'. Proceeding from earlier estimates of Rohde (1982) and Poulin and Morand (2000, 2004), the authors estimate that 75,000–300,000 helminth species parasitize the approximately 45,000 vertebrate species on earth. Poulin and Morand had estimated that there are at least 50 per cent more helminth than vertebrate species. Few molecular studies to reveal cryptic parasitic species have been made, such as the one by Glennon *et al.* (2008) on monogeneans of elasmobranchs, which could lead to an underestimate of parasitic species. On the other hand, a limited knowledge of host specificity would lower estimates, if a greater range of host species was found to be infected than known at present.

According to Hoberg (2005), there are more than 300 species of seabirds, but only 700 species of digeneans, eucestodes, nematodes, and acanthocephalans are known from 165 bird species (Hoberg 1996; Storer 2000, 2002), in other words, the parasite fauna of about half of all the host species is completely unknown.

Estimates of total marine species diversity vary widely (see discussion in Rohde 2002); at the higher end of estimates are those by Lambshead (1993), according to whom only about 160,000 marine species have been described, but the total macrofauna could reach 10 million species (1×10^7), with the total meiofauna 'an order of magnitude higher'. The coastal meiofauna of only one island, Sylt in the North Sea, has been examined thoroughly, where 652 meiofaunal species have been recorded to date. But even there, many species particularly of flagellates, foraminiferans, ciliates, and nematodes, but also of platyhelminths, remain to be described, and the estimated total of undescribed species is about 200 (Armonies and Reise 2000). Global richness of meiofauna must be very large indeed, considering that widespread species (at least of one of the major group, the Platyhelminthes) seem to be rare. Of 259 meiobenthic turbellarians collected by Faubel from sandy exposed beaches along the Australian East Coast, only 2 species were found both in tropical northern Queensland and south of temperate Sydney (Faubel, personal communication, cited in Rohde 2002). I am not aware of any studies which have made a serious attempt to discover parasites in meiofaunal invertebrates.

The fauna of the deep-sea, which covers about 65 per cent of the earth's surface, is also largely unknown. Practically nothing is known about diversity of deep-sea protistans and, according to Lambshead et al. (1994), only 10 studies of species diversity of deep-sea nematodes, one of the largest groups of invertebrates in that environment, had been conducted at the time, in other words nematode diversity was known from less than a square metre of sea bed. No significant increase in our

knowledge has been achieved since. Some recent studies, some of which also discuss possible causes of deep-sea diversity, are those of Levin et al. (2001), Stuart et al. (2003), Rex et al. (2005), de Mesel et al. (2006), and Copley et al. (2007).

With regard to parasites, best but still poorly known are parasites of some deep-sea fishes, and next to nothing is known about parasites of deep-sea invertebrates. According to Bray (2005): 'A huge proportion of the area covered by the deep-sea has not been sampled at all, let alone for parasites. No data is available from trenches. The data from hydrothermal vents and cold-seeps is rudimentary. The life-cycles, host relationships, distribution and zoogeography of deep-sea parasites have not been studied'. Our incomplete knowledge of diversity in the oceans explains why biogeographical patterns including gradients have been documented only for few groups, for example, for some parasites of teleost fishes.

6.2.2 Oceanic characters that may affect parasite diversity

It has been known for a long time that warmer seas are generally richer in species of hosts and parasites than colder seas, and that the Indo-Pacific is richer than the Atlantic Ocean. This is discussed in detail in separate sections below. According to a recent study by Pascual et al. (2007), parasite communities of cephalopods and exploited fish were poorest in instable water masses, that is, in convergent waters and in upwelling systems.

6.2.3 Host and parasite traits that may affect parasite density and diversity

A considerable number of studies have attempted to correlate host parameters such as body size, distributional range, feeding habits, habitat, and schooling habits, with parasite diversity. Poulin (1997) and Poulin and Morand (2000, 2004) reviewed some findings on metazoans parasitizing other metazoans and concluded that both host and parasite traits may affect patterns in parasite diversity. Concerning host traits, the most prevalent trend found in these studies is that parasite diversity is greater in hosts with a large body size, population density, or geographical range, all of which increase the likelihood that an infective parasite stage encounters a host. George-Nascimento *et al.* (2004) using assemblages of trophically transmitted endoparasitic helminths of 131 vertebrate hosts, found that the total number of parasite individuals per host and the total volume of parasites per host increased with body size of hosts. In a recent study, Luque and Poulin (2008) examined metazoan parasite assemblages of 651 Neotropical fish species. Parasite diversity was tested using species richness and the taxonomic distinctness of parasite assemblages. The latter was shown to be more important than the former, with higher values in benthic-demersal fish and in fish at higher trophic levels, with higher values for all freshwater parasites and marine ectoparasites, and lower values for marine endoparasites at higher temperatures. Higher values were also found in Pacific than in Atlantic fish. Munoz *et al.* (2007), after correcting for host phylogeny, found a positive correlation of mean richness, abundance, and biovolume of ectoparasites but not of endoparasites per host species with body size. They explain this by a possible increase in parasite colonization and recruitment by the increased surface area of larger fish. But because endoparasites usually have indirect transmission, many other variables may be involved.

Concerning the effect of parasite traits on diversity patterns, Poulin and Morand (2000, 2004) stated that 'such effects are hard to isolate'. A factor involved may be body size of the parasite, because habitat heterogeneity is scale-dependent and is probably greater for smaller parasites, perhaps favouring diversification of smaller over larger species. Also, complex life-cycles may constrain diversification, although the number of genera per family and the number of species per genus in nematodes of vertebrates did not differ between species with complex and simple (one host) life-cycles (Morand 1996). Poulin and Morand (2004) conclude that it is too early to assess the importance of life-cycles in affecting parasite speciation and diversification.

Further, Poulin and Morand (2004) examined the relation between epidemiology and parasite diversity, considering the basic reproductive rate R_0 (the average number of new cases of infection (or new parasites) that arise from one infectious host (or one individual parasite), if introduced into a population of uninfected hosts). This rate allows predictions about diversity of directly transmitted macroparasites, assuming that certain conditions are met. One such prediction is that high population density and/or a long life span of a host should enhance parasite diversity in or on it. The prediction has been verified in some studies on mammals and freshwater fishes, but so far not for marine hosts. Virulence of a parasite, on the other hand, probably has no effect on parasite diversity, because it varies between populations.

6.3 Some early studies of the geographical distribution of marine parasites: zoogeographical regions, differences between the Atlantic and Indo-Pacific, and parasite endemicity at remote oceanic islands

Rohde (1993 and references therein) has reviewed earlier studies. Based on an evaluation of 216 genera of Monogenea and 420 genera of Digenea, Lebedev (1969) distinguished 10 zoogeographical regions, although not all seas were included because of the lack of data. A number of studies have shown a greater diversity of hosts and their parasites in the Indo-Pacific than the Atlantic Ocean (Parukhin 1975 for nematodes; Lebedev 1969 for trematodes and monogeneans) (Fig. 6.1). There is also convincing evidence that the species and genus composition of various parasite groups differs between the two ocean systems (Lebedev 1969), although there may be a high degree of 'sharing' between certain seas. Thus, although the majority of trematode species on the Caribbean side of the Panamanian land-bridge are different from those on the Pacific side, a considerable number are shared by both regions (Sogandares-Bernal 1959). Even greater is the number of helminth species found both in the northern Atlantic and the northern Pacific (Strelkov 1960). Briggs (1966) has demonstrated that endemicity of marine animals at remote oceanic islands (at least 300 miles from the nearest continent) is very low in the north and middle Atlantic, greater in the south Atlantic and Pacific, and very high in sub-Antarctic waters. He suggested that this is due to differences in Pleistocene temperature reductions: endemic animals that were exposed to great temperature drops were eliminated. Manter (1967) could show the same phenomenon, although for fewer localities, for trematodes of marine fishes.

Concerning the differences between the Indo-Pacific and Atlantic, several explanations are possible: the greater diversity of the Indo-Pacific than the Atlantic Ocean could be caused by the greater area or the greater age of the former, or a combination of both. Rohde (1993), comparing the species richness of Monogenea and their taxonomy between the northern Atlantic and northern Pacific, suggested that not area but age is important. The reason is that most species of Monogenea infecting the gills of teleost fishes in northern waters, in both oceans, belong to the Gyrodactylidae, which are extremely rare in warm seas and are therefore unlikely to have emigrated from lower latitudes (Fig. 6.2). The relative species diversity of Monogenea species in the North Pacific is more than double that in the North Atlantic. The North Pacific is not larger than the North Atlantic, nor is structural complexity larger (as indicated by the length of the coastline and the number of islands): the length of the major rivers draining into the North Pacific and their discharge rate and annual discharge volume are even smaller than those of the North Atlantic. Therefore, neither differences in

area nor in heterogeneity can explain the differences in diversity. The only explanation left is the older age of the North Pacific Ocean.

6.4 Marine gradients in parasite diversity, host ranges, and reproductive strategies

The marine environment, although far from homogeneous because of underground mountain ranges, currents, and surface disturbances caused by storms, etc, nevertheless is less heterogeneous than terrestrial and freshwater habitats. For this reason, distribution patterns are less affected by local conditions, in other words, gradients of various kinds should be clearer, and a careful examination of marine gradients may yield causal explanations which are more difficult to evaluate in terrestrial/ freshwater environments.

6.4.1 Latitudinal gradients in diversity

By far the clearest gradients in the oceans are latitudinal gradients in diversity. Such gradients are well documented for surface waters, and there is some evidence that they also occur in the deep-sea (e.g. Lambshead *et al.* 2002 for deep-sea nematodes). These gradients have been in existence for long geologic periods, although their steepness underwent changes (see

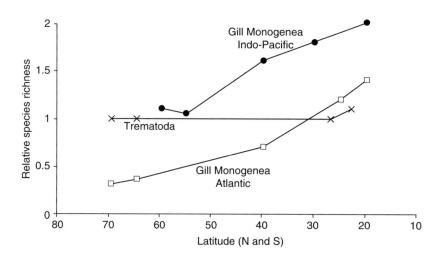

Figure 6.1 Relative species diversity (number of parasite species per host species) of digeneans and monogeneans of marine teleosts at various latitudes. After Rohde (2005b). Reproduced with permission of CSIRO Publishing.

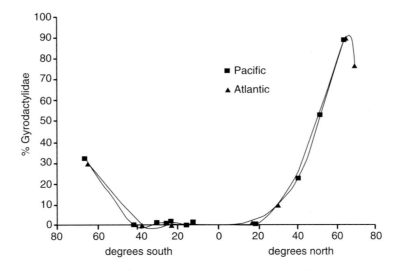

Figure 6.2 Percentages of marine Gyrodactylidae (relative to all gill Monogenea) on the gills of marine fishes at different latitudes. After Rohde (2005b). Reproduced with permission of CSIRO Publishing.

Powell 2007 for brachiopods in the late Paleozoic ice age and Brayard *et al.* 2006 for Triassic ammonoids).

Marine latitudinal diversity gradients of parasites were last reviewed by Rohde (2002, 2005b). They are best documented for metazoan ecto- and endoparasites of marine fishes: diversity peaks in the tropics, but, whereas species richness of mono-geneans increases at a greater rate towards the equator than that of host species, the number of parasite species per host species of trematodes does not change significantly with latitude (Fig. 6.1). Hence, the great number of endoparasite species in tropical seas is entirely or almost entirely due to the much greater diversity of hosts. Poulin and Rohde (1997), using the method of phylogenetically inde-pendent contrasts, have shown that this difference remains after correction for phylogeny. Hence, infra-community richness of ectoparasites but not of endoparasites is greater in the tropics (Rohde and Heap 1998).

6.4.2   Causes of latitudinal diversity gradients

There are numerous papers dealing with possible causes of the gradients, and there is a wide range of opinions. Recent reviews are by Rohde (1992, 1999), Gaston (2000), Willig *et al.* (2003), and Mittelbach *et al.* (2007). Much of the evidence in many of the

papers including reviews comes from non-marine systems.

One explanation often given (and sometimes believed the major and perhaps only one), is the larger area of the tropics, which must result in greater diversity. Rosenzweig (1995), following Terborgh (1973), emphasized that tropical areas are generally larger than those of high latitude and that this is the explanation for latitudinal gradients. Gorelick (2008), using analytic geometry, demonstrated the obvious, namely that all other parameters being equal, larger areas provide more niches and consequently can hold more species. Applied to the latitudinal gradi-ent, it can be explained by the convergence of lines of meridian at the poles. 'There simply is less area at high latitudes, which means fewer niches', leading to reduced species richness. Using analytic geometry of a cone, he similarly showed that species numbers should decrease linearly with altitude. A re-analysis of a published latitudinal dataset on marine benthos between 62 °S and 63 °N supported his conclusion. However, Rohde (1997, 1998) has pointed out that the only continent with a larger tropical area is Africa, and diversity is greater even in tropical regions with much smaller area, such as southern and south-eastern Asia. Area can therefore not give a general explana-tion for the gradients. The same applies to some seas. Nevertheless, area may be more important in the oceans, because in the Indo-Pacific (but not in the Atlantic) tropical seas are larger than colder seas.

Gaston (2000) gave an excellent outline of the major distributional patterns of biodiversity and their explanations. He writes that: 'no single mechanism adequately explains all examples of a given pattern...that the patterns....may vary with spatial scale, that processes operating at regional scales influence patterns at local ones, and that the relative balance of causal mechanisms means that there will invariably be variations in and exceptions to any given pattern'.

Among the major models attempting to explain latitudinal gradients, discussed by Gaston (2000), is the 'mid-domain' model. It implies that species closer to the geographic midpoint of distribution can extend further away from the midpoint than species having their midpoint closer to some boundary, be it climatic or geographic. This will result in denser species-packing around the midpoint. Another explanation is area: according to the well documented species/area relationship, larger areas can accommodate more species. However, as pointed out by Gaston (2000), it is insufficient as the sole model; additional assumptions have to be met to give an explanation. These assumptions are: low availability of productivity/energy at high latitudes would reduce diversity even if the area was large; zonal bleeding of species from large tropical into smaller non-tropical regions would smooth out the gradient; and high climatic variability at high latitudes would increase the area actually occupied by species, because they need to have greater environmental tolerances. But as pointed out by Gaston (2000), not only explanations based on area, but others as well, fail to give explanations on their own, and for most mechanisms critical tests have not been conducted. Gaston (2000) also doubts the often made assumption that the same mechanism must explain latitudinal gradients in different taxa and in different environments, such as the oceans and forests.

Gaston (2000) devotes much space in his discussion to species–energy relationships. In plants, water and heat correlate best with diversity, in animals, it is measures of heat. The explanation he emphasizes is that more energy leads to greater biomass in an area, which facilitates coexistence of more individuals and therefore of more species at densities necessary to maintain viable populations. He points out various weaknesses of this model: it does not necessarily apply to plants; many taxa use only a small proportion of the available energy; temporal variations in energy supply are often not considered; and at regional scales extant diversity cannot have been produced by extant conditions. Other explanations of the diversity–energy relationship he refers to include, for example, the assumption that the relationship reflects physiological constraints on species distribution. However, Gaston (2000) emphasizes that there is no strong support for any of these explanations, and that for this reason there may often be no causal relationship between species richness and energy, it may often only be a covariate of some other factor.

Concerning explanations of gradients not only along latitude, but also along depth, elevation etc., Gaston (2000) concludes that 'although multiple factors doubtless contribute', it is likely that, if a factor was found contributing along one axis, it will also be contributing along others.

It is important to note that Gaston (2000) does not mention at all the well documented direct temperature effects on mutation rates and on generation times, in other words, he does not mention the effects temperature must have on evolutionary speed (see below).

Mittelbach *et al.* (2007) reviewed (what they believe are) the two major hypotheses that attempt to explain the gradients, that is the 'time and area hypothesis', and the 'diversification rate hypothesis'. According to the former, tropical climates are older and larger, thereby permitting a higher degree of diversification. According to the latter, there is faster diversification in the tropics because of faster speciation, which may result from a variety of causes such as increased opportunities for reproductive isolation, or faster molecular evolution, or the increased importance of biotic interactions. Among the many mechanisms proposed to explain latitudinal variation on diversification rates are genetic drift, climate change, speciation types, area, physiological tolerances and dispersal limitation, evolutionary speed, and biotic interactions: there is some evidence for all or most of them, but evidence is not conclusive. The authors examine evidence for latitudinal variation in diversification rates from a number of paleontological studies and find that: '...data for many taxonomic groups are consistent with the hypothesis that the latitudinal diversity gradient is a result of higher rates of diversification in the tropics compared with temperate and polar regions, and the available evidence suggests that this difference is due to higher origination rates and lower extinction rates in the tropics.'

Phylogenetic studies discussed by the authors are more controversial. Whereas some studies found that bird clades from lower latitudes diversified faster than those from higher latitudes, another study (on hylid frogs) did not find any differences or even the opposite effect. The authors further discuss methods to study the evolution of reproductive isolation, which, however, have not yet been applied to latitudinal gradients. Finally, Mittelbach *et al.* (2007) point out that, if speciation rates increase towards the equator, so should the degree of genetic divergence among populations, which can be considered as an indicator of incipient speciation. Martin and McKay (2004) have indeed found evidence for this in vertebrates, and Phillimore *et al.* (2007) reported that high subspecies richness in birds positively correlated with large breeding range size, island dwelling, occurrence in montane regions, habitat heterogeneity, and low latitude. Martin and Tewksbury (2008), using all species with 5 or more subspecies from 12 of the most diverse families of birds in the world, demonstrated that there are consistently more subspecies at lower latitudes across all families, in both hemispheres, and in all continents examined. Although there was an effect of area on the number of subspecies within species, area did not explain the greater number of subspecies at lower latitude. The authors conclude that 'global patterns of subspecies support the idea that phenotypic differentiation of populations is greater at lower latitudes within species', and that 'if subspecies density provides an index of rates of incipient speciation, then our results support evolutionary hypotheses for the latitudinal diversity gradient that invoke higher tropical speciation rates.' Although similar studies on marine parasites have not been made, the bird data suggest that incipient speciation among parasites, as

indicated by cryptic species, may be more common at low than at high latitudes.

Many of the factors discussed in the two reviews support the hypothesis of effective evolutionary time (Rohde 1992), according to which diversity should be greatest in areas with high temperature and with a long undisturbed history. Temperature is important, because higher temperatures lead to greater mutation rates, faster selection and shorter generation times, all of which accelerate speciation; a long undisturbed history permits accumulation of species over time. Much evidence is available for the hypothesis. In particular, the metabolic theory of ecology (Brown *et al.* 2004) is in agreement with it (Figs. 6.3 and 6.4).

6.4.3   Latitudinal gradients in host ranges

It is useful to distinguish host range and host specificity. Different authors have used different definitions (discussion and references in Rohde and Rohde 2005). We define host range as the number of host species infected by a parasite species, irrespective of how frequently and intensely the various species are infected. Host specificity takes intensity and/or prevalence of infection into account as well. Several indices have been proposed to measure specificity, some ('ecological' specificity indices) using only information on frequencies (prevalences) and intensities of infection, others using phylogenetic information on the relatedness of host species as well.

Data that permit the evaluation of latitudinal differences in host ranges/specificities are scarce. Best known are such differences for monogeneans and digeneans of marine teleosts. Host ranges of Monogenea are narrow at all latitudes, that is, few host species are utilized at any latitude. In contrast, host ranges of Digenea are much greater in the tropics (Fig. 6.5). The difference between the two groups disappears when ecological host specificity is considered: both the endoparasitic digeneans and the ectoparasitic monogeneans have a strong host preference for very few host species at all latitudes. Caution, however, is necessary; only one large survey of digeneans in cold northern waters permits calculation both of host ranges and host specificity.

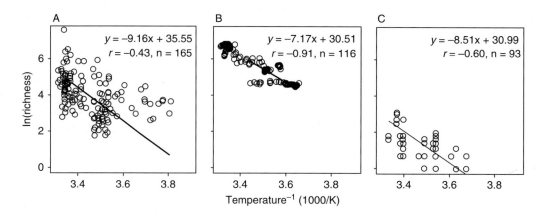

Figure 6.3 The effect of mean annual water temperature on species richness of (a) fish, (b) numbers of marine prosobranch snails per latitudinal degree band along the continental shelves of North and South America, and (c) ectoparasite species per host of marine teleost fish from the Antarctic to the Tropics. After Allen *et al.* (2002). Reproduced with permission of the authors and the American Association for the Advancement of Science.

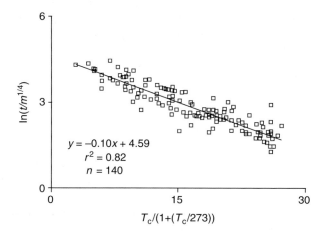

Figure 6.4 The effect of incubation temperature on mass-corrected embryonic development time for marine fishes in the field. After Gillooly *et al.* (2002). Reproduced with permission of the authors and MacMillan Publishers.

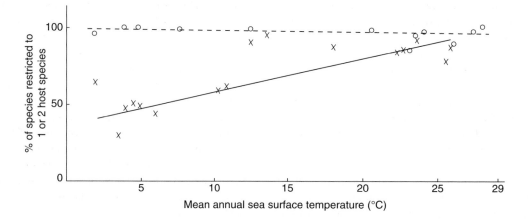

Figure 6.5 Host ranges of marine Monogenea (x) and Digenea (o) at various sea surface temperatures. Ordinate: host ranges as a percentage of species found in one or two host species. After Rohde (1978). Reproduced with permission of Springer Verlag.

6.4.4   Latitudinal gradients in reproductive strategies

Latitudinal gradients for reproductive strategies have been described for marine invertebrates, and are referred to as 'Thorson's rule' (Mileikovsky 1971), according to which invertebrates in cold, high latitude, seas tend to produce small numbers of offspring, often by viviparity or ovoviviparity; eggs are large containing much yolk and larvae likewise are large with little ability of active dispersal; and larvae develop in egg capsules or by brooding. In contrast, invertebrates in tropical waters tend to produce large numbers of small eggs which hatch early, giving rise to large numbers of pelagic, planktotrophic widely dispersing larvae. A similar phenomenon has been described for gill monogeneans (Rohde 1985): Gyrodactylidae, small viviparous worms, predominate at high latitudes, but are very rare in tropical waters (Fig. 6.2). Juveniles infect other host fish by contact transfer. Larvae of other monogeneans, predominant in warm waters, are in contrast small and typically disperse and find other hosts by active swimming.

6.4.5   Causes of latitudinal gradients in reproductive strategies

Several explanations for Thorson's rule have been given (see discussion in Rohde 1985); according to the most widely accepted hypothesis, phytoplankton blooms at high latitudes do not last long enough to permit development of small planktotrophic larvae. However, this explanation does not apply to the Monogenea, because monogenean larvae do not feed on phytoplankton. Rohde (1985, 1999) proposed an explanation based on direct temperature effects on reproductive rates and host finding: at low temperatures the proportion of larvae reaching and infecting hosts is reduced due to slower physiological processes including sensory processes, which makes it more difficult to locate and infect hosts; and at low temperatures reproductive rates are reduced. In the Gyrodactylidae, a strategy has evolved that guarantees a suitable habitat: offspring remains on the same host individual, and infection of other hosts is by contact transfer.

6.4.6   Longitudinal gradients in diversity

Longitudinal gradients in diversity have been documented much less than latitudinal gradients, although there is much circumstantial evidence. They have existed for long periods. For example, changes in longitudinal gradients have been documented for Triassic ammonoids. They were present in the Tethys, with increasing diversity eastwards, but disappeared during the Spathian, at the end of the Early Triassic (Brayard *et al.* 2006).

It has become clear that there are some centres of diversity with much greater species numbers than elsewhere. A primary centre is South-east (SE) Asia including the Malaysian, Indonesian, and Philippine islands. A secondary centre is the Caribbean, where species numbers are enhanced but less so than in SE Asia. Also, it has become clear that there are barriers to dispersal in the oceans, which inhibit a 'levelling' out of the gradients. The most effective such barrier appears to be the East Pacific Barrier which reduces the dispersal of many species from the western to the eastern Pacific; a second barrier, the New World Barrier represented by the South and North American continent, has been much less effective. All these phenomena have been particularly well documented for some scombrid fishes and their monogenean and copepod ectoparasites (Rohde and Hayward 2000). Very clearly, species diversity of the largely coastal scombrid genera *Scomberomorus* and *Grammatorcynus* and their parasites is greatest in the tropical/subtropical western Pacific with the tropical/subtropical western Atlantic (the Caribbean) acting as a secondary centre of diversity (Fig. 6.6). An evaluation of the geographical distribution strongly supports the hypothesis that coastal scombrid fishes and their parasites spread between the Indo-West Pacific and Atlantic Oceans through the Tethys Sea, and between the western Atlantic and the eastern Pacific, before establishment of the Central American landbridge. The East Pacific barrier prevented dispersal of species across the western and eastern Pacific. For species of *Scomber*, which are coastal/open ocean species, the East Pacific Barrier has been less effective.

Similar studies of other fish groups and their parasites are scarce. However, a comparison of

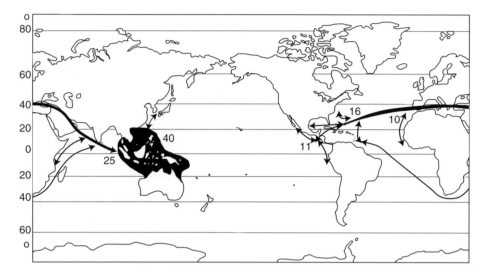

Figure 6.6 Extant geographical interrelationships of scombrid fishes of the genera *Scomberomorus* and *Grammatorcynus* and their copepod and monogenean parasites. Note: arrows do not indicate routes of dispersal, but interrelationships of genera. Numbers indicate species numbers of hosts plus parasites. Primary centre of diversity in the West Pacific, secondary centre in the Caribbean. There has been no dispersal across the East Pacific Barrier. Modified after Rohde and Hayward (2000). Reproduced with permission of Elsevier.

parasites of rabbit fish, *Siganus* spp. from coastal eastern Australia and East Africa has shown a much greater species richness of ectoparasites, but not of endoparasites, in Australian waters: metazoan ectoparasites 19 vs. 7 or 8 species, endoparasite 7 vs. 5 species (Kleeman 2001; Martens and Moens 1995; Geets *et al.* 1997).

6.4.7 Causes of longitudinal gradients

Belasky (1996) explains extant longitudinal gradients (e.g. in corals) by a deepening of the mean sea surface temperature (SST) and thermocline towards the east, and the lack of islands which would facilitate larval dispersal across the Pacific. With regard to scombrid fishes and their parasites, the largely coastal genera *Scomberomorus* and *Grammatorcynus* cannot spread across the vast ocean stretches of the eastern Pacific, but the less coastal species of *Scomber* apparently can.

6.4.8 Depth gradients in diversity

Bray (2005) has reviewed our knowledge of deep-sea parasites, that is, of parasites in the ocean

below the level of the continental shelf margin or shelf-slope break, which is usually at a depth of around 200 m. This brief account is based on Bray's (2005) review. For example, de Buron and Morand (2002) reported that 57 per cent of metazoan parasites reported in waters deeper than 1,000 m were digeneans, 25 per cent crustaceans (80 per cent copepods), 10 per cent cestodes, 4 per cent acanthocephalans, 2 per cent nematodes, and 2 per cent monogeneans. Protistans (including myxozoans) amount to less than 17 per cent. According to Rohde (1993), who used extensive surveys of surface and deep-sea fishes from one area, off the coast of south-eastern Australia, relative species diversity (number of parasite species per host species) of Monogenea is about five times greater in surface than in deepwater (for details and references see Rohde, 1993).

According to Bray (2005), the typical shelf fauna may reach down to more than 1000 m at low latitudes, and deep-sea parasites may occur at a few hundred metres depth in polar regions. Free-living deepsea animals typically live over a much greater depth range (eurybathy) than shallow water species (stenobathy).

6.4.9 Causes of depth gradients

Possible causes of the diversity gradients with depth are underlying gradients in physical parameters, such as temperature, light penetration, pressure, and food supply. At the thermocline about 800–1,300 m deep, temperature is around 4°C and decreases little below that depth. Pressure increases by one atmosphere for each 10 m of depth, water density by about 0.5 per cent per 1,000 m, sufficient to affect reaction rates of, for example, enzymes (Bray 2005). The deep-sea is without light (except for that produced by deep-sea organisms by bioluminescence) and photosynthesis below 200 m is poor, consequently, the main food source is detritus sinking down from the surface, which may be seasonal. According to Bray (2005), many organisms may be attracted by large food-falls, which act as venues for mating and transmission. Other such venues may be hydrothermal vents and cold-seeps.

A comprehensive hypothesis which does not only attempt to explain latitudinal gradients in diversity, but depth gradients as well, is Rohde's (1992) hypothesis of effective evolutionary time which has two components: evolutionary speed explained by direct effects of temperature on mutation rates, speed of selection, and generation times; and the more or less unchanged conditions under which ecosystems have existed over evolutionary time. The first component, evolutionary speed, has been considered by many authors considering latitudinal diversity gradients, the second component is rarely mentioned (see, for example, the review by Mittelbach et al. (2007) discussed above). It cannot be ignored, however, if we consider the high diversity in the deep sea. Although temperatures are low, a long, relatively unchanged history has permitted the accumulation of species. In addition, the large area of the deep sea may be important.

6.5 Some novel approaches

In the following I discuss some novel approaches which may contribute to understanding the causes of diversity patterns.

6.5.1 Simulations using an agent-based model

A single method, statistical or otherwise, even if based on comprehensive datasets, cannot be expected to solve the problems of evolutionary ecology, including

that of diversity gradients, on its own. An explanation will become more convincing if it is strongly supported by several quite different approaches. For elucidating causal factors responsible for latitudinal gradients in species diversity, latitudinal ranges, and habitat width, Rohde and Stauffer (2005), Stauffer and Rohde (2006), and Stauffer et al. (2007) have applied the Chowdhury ecosystem model (Chowdhury and Stauffer 2005; Stauffer et al. 2005, 2007; Stauffer 2006). It is one of the most complex agent-(individual) based (Billari et al. 2006) ecological models (Pekalski 2004; Grimm and Railsback 2005). In agent-based models, instead of using differential equations describing the fate of whole populations, each individual is treated separately, that is, each individual has its own random birth and death. In the Chowdhury model, simulations over millions of years—evolutionary time scales at the level of species and above—are made, as are simulations over short ecological time scales at the level of populations.

Rohde and Stauffer (2005) found that simulations using the Chowdhury model led to faster speciation in the tropics and to latitudinal gradients in diversity; the complexity of food-webs increased with time and at a faster rate in the tropics. Best results were obtained when many niches were kept empty. Stauffer and Rohde (2006) showed that latitudinal ranges of species are greater and not smaller in the tropics, contradicting Rapoport's rule. Stauffer et al. (2007), examining the niche dimension habitat width, found no support for the latitude–niche breadth hypothesis, according to which tropical species have narrower niches (see references in Vázquez and Stevens 2004). Rather, habitats, measured by comparing species numbers in small and large areas at a particular locality, are generally wider and not narrower in the tropics. This implies that there must be much overlap between habitats of the numerous tropical species. These results are in good agreement with empirical data analysed with a variety of methods (Rohde 1992, 1999; Rohde et al. 1993; Vázquez and Stevens 2004 and references therein).

6.5.2 Historical (phylogenetic) zoogeography of hosts and parasites

Since the classic book by Hennig (1966) was translated into English, numerous studies have used phylogenetic systematics combined with studies of the geographical distribution of organisms, to arrive at conclusions concerning the geographical origin of certain taxa, and their dispersal from the point of origin. Many studies, using the same approach, also focused on coevolution and historical biogeography of hosts, parasites, and biotas. General accounts have presented and discussed methods such as Brooks Parsimony Analysis (BPA), and Secondary BPA (Brooks 1990; Brooks and McLennan 2002), PACT (Wojcicki and Brooks 2005) or Maximum Cospeciation (e.g. Paterson and Banks 2001; Page 2003) that are used to explore the intricacies of complex biological associations and geographic distributions. Hoberg and Klassen (2002) give a detailed list (68 references) of phylogenetic studies of marine host–parasite assemblages, as well as of papers describing their theoretical foundations. A study dealing with a specific group is that by Hoberg (1986), who examined the evolution and historical biogeography of a species of the cestode *Alcataenia* and their hosts, the Alcidae and Laridae (Charadriiformes). Hoberg and Adams (1992, 2000) studied the phylogeny, historical biogeography, and ecology of cestodes of the genus *Anophryocephalus* and their hosts, holarctic pinnipeds. They found that phylogenies of the parasite and hosts were highly incongruent, supporting the hypothesis that colonization of hosts by parasites, rather than coevolution, was the most important factor determining parasite diversification. They postulated *Phoca* (*Pusa*) spp. in the Atlantic as the most basal host. Hosts with their parasites expanded into the North Pacific via the Arctic, where parasites radiated among *Phoca* spp. and later colonized otariid seals. Klassen (1994a, b) analysed the distribution of species of the monogenean *Haliotrema* on boxfishes (Ostraciidae) and concluded that hosts and parasites have repeatedly entered the Caribbean from the Pacific. Hoberg (1996) discussed the faunal diversity of parasite assemblages of marine birds and their historical, ecological, and biogeographical interactions.

We have to address the question of how much these studies have contributed to providing evidence for diversity gradients, and whether they have contributed and can contribute to evaluating causes of the gradients. Hoberg and Klassen

(2002) write, referring to databases for the distribution of digeneans and monogeneans in marine teleosts, that 'a phylogenetic context for such data is critical for understanding the history of faunal assemblage, the interaction of dispersal and vicariance, and the evolutionary relationships of taxa within and among identifiable faunal provinces'. But they also write: 'What is missing still is a synoptic work assessing what overall pattern, if any, can be retrieved from these studies collectively about, for instance, the biogeographic relationships between Indo-Pacific, Pacific and Atlantic Oceans...Despite nearly 25 years of explicit coevolutionary studies based on phylogenetic approaches, we still continue to lack critical information for most host and parasite taxa and in many respects the literature is diverse but fragmented.'

This lack of information is even more obvious if we want to address questions relating to diversity gradients and centres of origin for large groups or even the entire marine fauna. We would need numerous phylogenetic studies of hosts and parasites, which could be superimposed to show an overall pattern. It is important here to emphasize that gradients are recognizable only if large groups are examined. Many small taxa do not conform to the latitudinal gradient in species diversity peaking in the tropics. Nevertheless, such a gradient is well supported for most large groups, for which sufficient data are available, and by overall evidence from many groups. Future phylogenetic studies should address the question of whether basal taxa of parasites and hosts are indeed most common in the tropics, as has been shown, for example, for some plants, and where diversification has occurred.

6.5.3 Autecological studies

Autecological studies deal with the ecology of particular species and populations and have been conducted for a long time. However, Hengeveld and Walter (1999) went further, distinguishing two ecological paradigms, the demographic and autecological, which they consider to be mutually exclusive. In the former, intra- and interspecific competition are important, optimization is frequent, and nature is balanced; emphasis is on quantification and comparison of differences in reproductive outputs between species. In the latter, optimization is impossible because of the variable environment, and emphasis is on survival and reproduction. This approach agrees to some extent with the distinction of nonequilibrium and equilibrium in ecological system. However, as emphasized by Rohde (2005a), both conditions occur, depending on the groups under consideration and the environment. Therefore, it may not be useful to distinguish two 'mutually exclusive' paradigms. Although autecological studies, because of the time involved in making them, cannot be expected to contribute to recognizing diversity patterns and gradients *per se*, they can help in giving causal explanations.

6.6 Future work

Foremost, additional taxonomic surveys of all groups and their parasites in all habitats are necessary. Particular emphasis should be paid to the meio- and deep-sea faunas. The reviews of parasites of various protistans and lower invertebrates in Kinne (1980) show that a rich parasite fauna has been recorded from groups that are also represented in the meio-fauna and deep-sea. There is no reason to assume that they are not parasitized in those systems. Historical (phylogenetic) analyses should be expanded, and models should be developed which realistically reflect natural systems. Autecological studies of some key species from a variety of habitats would be useful in elucidating mechanisms explaining diversity patterns. Finally, attention should be paid to the relative prevalence of cryptic species at different latitudes, using molecular techniques.

References

Allen, A.P., Brown, J.H., and Gillooly, J.F. (2002). Global biodiversity, biochemical kinetics, and the energetic-equivalence rule. *Science*, **297**, 1545–48.

Armonies, W. and Reise, K. (2000). Faunal diversity across a sandy shore. *Marine Ecology Progress Series*, **196**, 49–57.

Belasky, P. (1996). Biogeography of Indo-pacific larger foraminifera and scleractinian corals: a probabilistic approach to estimating taxonomic diversity, faunal similarity, and sampling bias. *Paleogeography, Paleoclimatology, Paleoecology*, **122**, 119–41.

Billari, F.C., Fent, T., Prskawetz, A., and Scheffran, J. (eds.) (2006). *Agent-Based Computational Modelling*. Physica-Verlag, Heidelberg.

Brayard, A., Bucher, H., Escarguel, G., Fluteauc, F., Bourquind, S., and Galfetti, T. (2006). The Early Triassic ammonoid recovery: paleoclimatic significance of diversity gradient. *Paleogeography, Paleoclimatology, Paleoecology*, **239**, 374–95.

Bray, R.A. (2005). Deep-sea parasites. In K Rohde, ed. *Marine Parasitology*, pp. 366–69. CSIRO Publishing, Melbourne and CAB International, Wallingford, Oxfordshire.

Briggs, J.C. (1966). Oceanic islands, endemism, and marine paleotemperatures. *Systematic Zoology*, **15**, 153–63.

Brooks, D.R. (1990). Parsimony analysis in historical biogeography and coevolution: methodological and theoretical update. *Systematic Zoology*, **39**, 14–30.

Brooks, D.R. and McLennan, D.A. (1991). *Phylogeny, Ecology and Behavior. A Research Program in Comparative Biology*. University of Chicago Press, Chicago.

Brown, J.H., Gillooly, J.F., Allen, A.P., Savage, V.M., and West, G.B. (2004). Toward a metabolic theory of ecology. *Ecology*, **85**, 1771–89.

Chowdhury, D. and Stauffer, D. (2005). Evolutionary ecology in silico: does physics help in understanding the "generic" trends? *Journal of Biosciences (India)*, **30**, 277–87.

Copley, J.T.P., Flint, H.C., Ferrero, T.J., and Van Dover, C.L. (2007). Diversity of meiofauna and free-living nematodes in hydrothermal vent mussel beds on the northern and southern East Pacific Rise. *Journal of the Marine Biological Association of the United Kingdom*, **87**, 1141–52.

Cribb, T.H. (2002). The trematodes of groupers (Serranidae: Epinephelidae): knowledge, nature and evolution. *Parasitology*, **124**, S3–S42.

de Buron, I. and Morand, S. (2002). Deep-sea hydrothermal vent parasites: where do we stand? *Cahiers de Biologie Marine*, **43**, 245–6.

de Mesel, I., Lee, H.J., Vanhove, S., Vincx, M., and Vanreusel, A. (2006). Species diversity and distribution within the deep-sea nematode genus *Acantholaimus* on the continental shelf and slope in Antarctica. *Polar Biology*, **29**, 860–971.

Dobson, A., Lafferty, K.D., Kuris, A.M., Hechinger, R.F., and Jetz, W. (2008). Homage to Linnaeus. How many parasites? How many hosts? *Proceedings of the National Academy of Sciences of the U.S.A.*, **105**, 11482–29.

Gaston, K.J. (2000). Global patterns in biodiversity. *Nature*, **405**, 220–27.

Geets, A., Coene, H., and Ollevier, F. (1997). Ectoparasites of the whitespotted rabbitfish, *Siganus sutor* (Valenciennes, 1835) off the Kenyan coast: distribution within the host population and site selection on the gills. *Parasitology*, **115**, 69–79.

George-Nascimento, M., Muñoz, G., Marquet, P.A., and Poulin, R. (2004). Testing the energetic equivalence rule with helminth endoparasites of vertebrates. *Ecology Letters*, **7**, 527–31.

Gillooly, J.F., Charnov, E.L., West, G.B., Savage, V. M., and Brown, J.H. (2002). Effects of size and temperature on developmental time. *Nature*, **417**, 70–73.

Glennon, V., Perkins, E.M., Chisholm, L.A., and Whittington, I.D. (2008). Comparative phylogeography reveals host generalists, specialists and cryptic diversity: hexabothriid, microbothriid and monocotylid monogeneans from rhinobatid rays in southern Australia. *International Journal for Parasitology*, **38**, 1599–612.

Gorelick, R. (2008). Species richness and the analytic geometry of latitudinal and altitudinal gradients. *Acta Biotheoretica*, **56**, 197–203.

Grimm, V. and Railsback, S.F. (2005). *Individual-Based Modeling and Ecology*. Princeton University Press, Princeton.

Hengeveld, R. and Walter, G.H. (1999). The two coexisting ecological paradigms. *Acta Biotheooretica*, **47**, 141–70.

Hennig, W. (1966). *Phylogenetic Systematics*. University of Illinois Press, Urbana.

Hoberg, E. (1986). Evolution and historical biogeography of parasite-host assemblage: *Alcataenia* spp. (Cyclophyllidea: Dilepididae) in Alcidae (Charadriiformes). *Canadian Journal of Zoology*, **64**, 2576–89.

Hoberg, E.P. (1996). Faunal diversity among avian parasite assemblages: the interaction of history ecology and biogeography in marine systems. *Bulletin of the Scandinavian Society for Parasitology*, **6**, 65–89.

Hoberg, E.P. (2005). Marine birds and their helminth parasites. In K Rohde, ed. *Marine Parasitology*, pp. 414–420. CSIRO Publishing, Melbourne and CAB International, Wallingford, Oxfordshire.

Hoberg, E. and Adams, A.M. (1992). Phylogeny, historical biogeography, and ecology of *Anophryocephalus* spp. (Eucestoda: Tetrabothriidae) among pinnipeds of the Holarctic during the late Tertiary and Pleistocene. *Canadian Journal of Zoology*, **70**, 703–19.

Hoberg, E. and Adams, A. (2000). Phylogeny, history and biodiversity: understanding faunal structure and biogeography in the marine realm. *Bulletin of the Scandinavian Society of Parasitology*, **10**, 19–37.

Hoberg, E. and Klassen, G.J. (2002). Revealing the faunal tapestry: coevolution and historical biogeography of hosts and parasites in marine systems. *Parasitology*, **124**, S3–S22.

Kinne, O. (ed.) (1980). *Diseases of Marine Animals, Vol. 1.* John Wiley, Chichester, N.Y., Brisbane, Toronto.

Klassen, G.J. (1994a). Phylogeny of *Haliotrema* species (Monogenea: Ancyrocephalidae) from boxfishes (Tetraodontiformes: Ostraciidae): are *Haliotrema* species from boxfishes monophyletic? *Journal of Parasitology*, **80**, 596–610.

Klassen, G.J. (1994b). On the monophyly of *Haliotrema* species (Monogenea: Ancyrocephalidae) from boxfishes (Tetraodontiformes: Ostraciidae): relationships within the *bodiani* group. *Journal of Parasitology*, **80**, 611–19.

Kleeman, S. (2001). *The Development of the Community Structure of the Ecto- and Endoparasites of* Siganus doliatus, *a Tropical Marine Fish*. BSc Honours thesis, University of New England, Armidale, Australia.

Lambshead, P.J.D. (1993). Recent developments in marine benthic biodiversity research. *Océanis*, **19**, 5–24.

Lambshead, P.J.D., Elge, B.M., Thistle, E., Eckman, J.E., and Barnett, P.R.O. (1994). A comparison of the biodiversity of deep-sea marine nematodes from three stations in the Rockall Trough, Northeast Atlantic, and one station in the San Diego Trough, Northeast Pacific. *Biodiversity Research*, **2**, 95–107.

Lambshead, P. J., Brown, D., Ferrero, C.J. *et al.* (2002). Latitudinal diversity patterns of deep-sea marine nematodes and organic fluxes: a test from the central equatorial Pacific. *Marine Ecology Progress Series*, **236**, 129–35.

Lebedev, B.I. (1969). Basic regularities in the distribution of monogeneans and trematodes of marine fishes in the world ocean. *Zoologicheskij Journal* **48**, 41–50 (in Russian).

Levin, L.A., Etter, R.J., Rex, M.A. *et al.* (2001). Environmental influences on regional deep-sea species diversity. *Annual Review of Ecology and Systematics*, **32**, 51–93.

Luque, J.L. and Poulin, R. (2008). Linking ecology with parasite diversity in Neotropical fishes. *Journal of Fish Biology*, **72**, 189–204.

Manter, H.W. (1967). Some aspects of the geographical distribution of parasites. *Journal of Parasitology*, **53**, 1–9.

Martens, E. and Moens, J. (1995). The metazoan ecto- amd emdoparasites of the rabbitfish, *Siganus sutor* (Cuvier and Valenciennes, 1835) of the Kenyan coast. I. *African Journal of Ecology*, **33**, 405–16.

Martin, P.R. and Mckay, J.K. (2004). Latitudinal variation in genetic divergence of populations and the potential for future speciation. *Evolution*, **58**, 938–45.

Martin, P.R. and Tewksbury, J.J. (2008). Latitudinal variation in subspecies diversification of birds. *Evolution*, **62**, 2775–88.

Mileikovsky, S.A. (1971). Types of larval development in marine bottom invertebrates, their distribution and ecological significance: a reevaluation. *Marine Biology*, **19**, 193–213.

Mittelbach, G.G., Schemske, D.W., Cornell, H.V. *et al.* (2007). Evolution and the latitudinal diversity gradient: speciation, extinction and biogeography. *Ecology Letters*, **10**, 315–31.

Morand, S. (1996). Biodiversity of parasites in relation to their life cycles. In M.E. Hochberg, J. Clobert, and R. Barbault, eds. *Aspects of the Genesis and Maintenance of Biological Diversity*, pp. 243–60. Oxford University Press, Oxford.

Munoz, G., Grutter, A.S., and Cribb, T.H. (2007). Structure of the parasite communities of a coral reef fish assemblage (Labridae): Testing ecological and phylogenetic host factors. *Journal of Parasitology*, **93**, 17–30.

Page, R.D.M. (ed.) (2003). *Tangled Trees: Phylogeny, Cospeciation and Coevolution*. University of Chicago Press, Chicago.

Parukhin, A.M. (1975). On the distribution in the world ocean of Nematoda found in fish from the southern seas. *Vestnik Zoologii*, **1**, 33–88 (in Russian).

Pascual, S., Gonzalez, A.F., and Guerra, A. (2007). Parasite recruitment and oceanographic regime: evidence suggesting a relationship on a global scale. *Biological Reviews*, **82**, 257–63.

Paterson, A.M. and Banks, J. (2001). Analytical approaches to measuring cospeciation of host and parasites: through the glass darkly. *International Journal for Parasitology*, **31**, 1012–22.

Pekalski, A. (2004). A short guide to predator and prey lattice models by physicists. *Computing in Science and Engineering*, **6**, 62–6.

Phillimore, A.B., Orme, C.D.L., Davies, R.G. *et al.* (2007). Biogeographical basis of recent phenotypic divergence among birds: a global study of subspecies richness. *Evolution*, **61**, 942–57.

Poulin, R. (1997). Species richness of parasite assemblages: evolution and patterns. *Annual Review of Ecology and Systematics*, **28**, 341–58.

Poulin, R. and Morand, S. (2000). The diversity of parasites. *Quarterly Review of Biology*, **75**, 277–93.

Poulin, R. and Morand, S. (2004). *Parasite Biodiversity*. Smithsonian Institution Press, Washington.

Poulin, R. and Rohde, K. (1997). Comparing the richness of metazoan ectoparasite communities of marine fishes: controlling for host phylogeny. *Oecologia*, **110**, 278–83.

Powell, M.G. (2007). Latitudinal diversity gradients for brachiopod genera during late Paleozoic time: links between climate, biogeography and evolutionary rates. *Global Ecology and Biogeography*, **16**, 519–28.

Rex, M.A, McClain, C.R., Johnson. N.A. *et al.* (2005). A source-sink hypothesis for abyssal biodiversity. *American Naturalist*, **165**, 163–78.

Rohde, K. (1978). Latitudinal differences in host specificity of marine Monogenea and Digenea. *Marine Biology*, **47**, 125–34.

Rohde, K. (1982). *Ecology of Marine Parasites*. University of Queensland Press, St. Lucia, Australia.

Rohde, K. (1985). Increased viviparity of marine parasites at high latitudes. *Hydrobiologia*, **127**, 197–201.

Rohde, K. (1992). Latitudinal gradients in species diversity: the search for the primary cause. *Oikos*, **65**, 514–27.

Rohde, K. (1993). *Ecology of Marine Parasites*, 2nd Edn. CAB International, Wallingford, Oxfordshire.

Rohde, K. (1997). The larger area of the tropics does not explain latitudinal gradients in species diversity. *Oikos*, **79**, 169–72.

Rohde, K. (1998). Latitudinal gradients in species diversity: area matters, but how much? *Oikos*, **82**, 184–90.

Rohde, K. (1999). Latitudinal gradients in species diversity and Rapoport's rule revisited: a review of recent work, and what can parasites teach us about the causes of the gradients? *Ecography*, **22**, 593–613.

Rohde, K. (2002). Ecology and biogeography of marine parasites. *Advances in Marine Biology*, **43**, 1–86.

Rohde, K. (2005a). *Nonequilibrium Ecology*. Cambridge University Press, Cambridge.

Rohde, K. (2005b). Latitudinal, longitudinal and depth gradients. In K. Rohde, ed. *Marine Parasitology*, pp. 348–51. CSIRO Publishing, Melbourne and CAB International, Wallingford, Oxfordshire.

Rohde, K, and Hayward, C.J. (2000). Oceanic barriers as indicated by scombrid fishes and their parasites. *International Journal for Parasitology*, **30**, 579–83.

Rohde, K. and Heap, M. (1998). Latitudinal differences in species and community richness and in community structure of metazoan endo- and ectoparasites of marine teleost fish. *International Journal for Parasitology*, **28**, 461–74.

Rohde, K. and Rohde, P.P. (2005). The ecological niches of parasites. In K. Rohde, ed. *Marine Parasitology*, pp. 286–93. CSIRO Publishing, Melbourne and CAB International, Wallingford, Oxfordshire.

Rohde, K. and Stauffer, D. (2005). Simulations of geographical trends in the Chowdhury ecosystem model. *Advances in Complex Systems*, **8**, 451–64.

Rohde, K., Heap, M., and Heap, D. (1993). Rapoport's rule does not apply to marine teleosts and cannot explain latitudinal gradients in species diversity. *American Naturalist*, **142**, 1–16.

Rosenzweig, M. (1995). *Species Diversity in Space and Time*. Cambridge University Press, Cambridge.

Sogandares-Bernal, F. (1959). Digenetic trematodes of marine fishes from the Gulf of Panama and Bimini, British West Indies. *Tulane Studies in Zoology*, 7, 70–117.

Stauffer, D. and Rohde, K. (2006). Simulation of Rapoport's rule for latitudinal species spread. *Theory in Biosciences*, 125, 55–65.

Stauffer, D., Kunwar, A., and Chowdhury, D. (2005). Evolutionary ecology in-silico: evolving foodwebs, migrating population and speciation. *Physica A*, **352**, 202–15.

Stauffer, D., Schulze, C., and Rohde, K. (2007). Habitat width along a latitudinal gradient. *Vie et Milieu (Life and Environment)*, **57**, 181–7.

Storer, R.M. (2000). *The Metazoan Parasite Fauna of Grebes (Aves: Podicipediformes) and its Relationship to the Birds' Biology. Miscellaneous Publications Museum of Zoology, No.188*. Museum of Zoology, University of Michigan, Ann Arbor.

Storer, R.M. (2002). *The Metazoan Parasite Fauna of Loons (Aves: Gaviiformes) and its Relationship to the Birds' Evolutionary History and a Comparison with the Parasite Fauna of Grebes. Miscellaneous Publications Museum of Zoology, No. 191*. Museum of Zoology, University of Michigan, Ann Arbor.

Strelkov, Y.A. (1960). Endoparasitic worms of marine fish of eastern Kamchatka. *Proceedings of the Zoological Institute of the USSR Academy of Sciences*, **28**, 147–96 (in Russian).

Stuart, C.T., Rex, M.A., and Etter, R.J. (2003). Large-scale spatial and temporal patterns of deep-sea benthic species diversity. In P.A. Tyler, ed. *Ecosystems of the World. Ecosystems of Deep Oceans*, pp. 295–311. Elsevier, Amsterdam.

Terborgh, J. (1973). On the notion of favorableness in plant ecology. *American Naturalist*, **107**, 481–501.

Vázquez, D.P. and Stevens, R.D. (2004). The latitudinal gradient in niche breadth: concepts and evidence. *American Naturalist*, **164**, E1–E19.

Willig, M.R., Kaufman, D.M., and Stevens, R.D. (2003). Latitudinal gradients in biodiversity: pattern, process, scale, and synthesis. *Annual Review of Ecology and Systematics*, **34**, 273–309.

Wojcicki, M. and Brooks, D.R. (2005). PACT: a simple powerful algorithm for deriving general area cladograms. *Journal of Biogeography*, **32**, 755–74.

CHAPTER 7

Parasite diversity and latitudinal gradients in terrestrial mammals

Frédéric Bordes, Serge Morand, Boris R. Krasnov, and Robert Poulin

7.1 Introduction

Spatial scale can provide important clues to our understanding of general patterns in the abundance and distribution of free-living species (Gaston 2000; Rahbeck 2005). From this perspective, latitudinal variation in species richness has been thoroughly documented for a wide range of taxonomic groups with a relatively high number of potential explanations, including climatic and temperature factors or increased productivity near the equator (Gaston 2000). Unfortunately, global patterns of parasite species richness have received considerably less attention than those concerning free-living species (but see Guernier et al. (2004) for parasites of humans and Rohde and Heap (1998) for ecto- and endoparasites of marine fishes). The search for the existence of latitudinal variation in parasite diversity is thus an open field and a fundamental part of macroecological investigations focusing on potential determinants of parasite species richness.

In this chapter, restricted to parasites of mammals, we first review existing knowledge of patterns of parasite diversity. Second, we summarize the classical predictions likely to explain latitudinal gradients in parasite diversity and the contradictory results of the few studies that have investigated this pattern. We then stress the validity of these classical predictions. Finally, using a large dataset, we present new results on latitudinal variation in helminth species diversity in mammals.

7.2 The search for determinants of parasite diversity

Most animals in natural environments are host to a large diversity of parasites (Poulin and Morand 2004). However, some species harbour more parasites than others. In an attempt to understand the reasons for this inequality, searches for the main determinants of parasite diversity in vertebrates (mainly fish, birds, and mammals) have been intensively carried out over the last decade (Poulin 1995; Morand 2000; Arneberg 2002; Nunn et al. 2003; Krasnov et al. 2005; Ezenwa et al. 2006; Lindenfors et al. 2007; Hughes and Page 2007; Bordes et al. 2007, 2008; Arriereo and Møller 2008). Broadly speaking, these studies have all postulated and put forward that parasite diversity is strongly shaped by host exposure to parasites, which is mainly due to host related factors such as morphological or life-history traits, and to biotic and abiotic environmental factors (Krasnov et al. 2005; Ezenwa et al. 2006; Luque and Poulin 2008). Moreover, hosts are not submissive victims of parasites. The level of host resistance is expected to evolve to match selective pressures from parasites in order to limit damages caused by parasites (Arriereo and Møller 2008). It should then also be considered as a correlate of parasite diversity (Krasnov et al. 2005; Bordes and Morand 2008; Arriereo and Møller 2008).

However, the results of recent studies on determinants of parasite diversity have demonstrated little consistency. Instead, the accumulation of information has lead to strong doubts about the very existence of patterns in parasite communities at the level of host populations or species (Poulin 2007; Luque and Poulin 2008). For instance, many potential determinants of parasite diversity in mammals have been studied. On the one hand, some of them, such as host behaviour or host densities for ectoparasites (Stanko *et al.* 2002; Felso and Rozsa 2006, 2007) or host densities and geographical range for helminths (Feliu *et al.* 1997; Nunn *et al.* 2003; Torres *et al.* 2006), have been supported, while on the other hand many others are still controversial. For example, larger body size has often been predicted to promote higher parasite species richness as larger hosts are supposed to represent larger habitats and greater opportunities for colonization (according to the theory of island biogeography; see MacArthur and Wilson (1967) and Kuris (1980)). However, empirical studies that have tested this prediction have yielded contradictory results. Indeed, Ezenwa *et al.* (2006) and Bordes *et al.* (2008) reported positive correlation between host body size and parasite species richness, whereas Feliu *et al.* (1997), Nunn *et al.* (2003), Krasnov *et al.* (2004) and Korallo *et al.* (2007) found no relationship between these two variables. Another example is the link between host density or life-span and parasite diversity. Although classical epidemiological theory postulates the positive effect of these traits on parasite transmission (Anderson and May 1979), host group size has never, and host longevity has rarely, been found to influence parasite diversity (Morand and Harvey 2000; Ezenwa *et al.* 2006; Torres *et al.* 2006; Lindenfors *et al.* 2007).

7.2.1 Why are so few determinants identified?

Poulin (2007) and Luque and Poulin (2008) have noted that the lack of general patterns emerging from various studies may be related to:

(1) the absence of corrections for sampling effort and phylogeny (in earlier studies, mainly);
(2) a bias toward Holarctic hosts; and
(3) the use of simple measures of parasite diversity, namely species richness, that do not capture taxonomic differences among parasite taxa.

In addition, most studies seeking the main determinants of parasite diversity have often copied previous studies when choosing potential traits to test. For example, we can easily identify some 'historical' or 'classical' determinants for mammalian parasites. These are host body size, diet, density, group size, geographical range, or longevity (but see Ezenwa 2004; Nunn and Donkey 2005; Felso and Rozsa 2008 for new recent perspectives related to host behaviour). Far from stating that parasite species richness is an inappropriate measure of diversity, we would like to emphasize that the lack of clear and strong emerging patterns may be related to the still narrow and rather naive predictions concerning these potential determinants. In other words, it is necessary to reconsider critically these historical predictions and/or look at other traits for new predictions (Felso and Rozsa 2007; Bordes *et al.* 2007, 2008).

7.2.2 New predictions: when host exposure is not always the key factor

The number of parasite species harboured by a host species must result from more than merely the degree of host exposure to parasite infection. For example, it is integral to the classical approach to emphasize that an increase in host group size promotes more frequent intimate contacts among hosts, which leads to higher parasite transmission. However, the roles of classical determinants of parasite diversity such as group size or density *per se* have been recently challenged as they do not always reflect the complexity of host social interactions (Ezenwa *et al.* 2006; Bordes *et al.* 2007). More importantly, they totally ignore host defences. For instance, social species may have evolved sophisticated behavioural defences to limit parasite attacks (Bordes *et al.* 2007). Consequently, the prediction that living in large groups should promote high parasite transmission and thus high parasite species richness may be rather naive, as parasite infection may be impaired in host species with high levels of social complexity (Bordes *et al.* 2007). Moreover, and independently from host defences, an example of the intricate interactions between sociality and parasitism comes from a recent study of tuberculosis in badgers (*Meles meles*) (McDonald *et al.* 2008). Badgers live in socially-structured and territorial groups. A reduction of their densities by culling did not lead to the expected decrease in tuberculosis transmission among groups or between badgers and cattle, but rather to the spread of the disease due to strong destabilization of social groups after culling. Under normal circumstances, the stable social organization of badgers may help to contain the spread of the disease despite very high densities (McDonald *et al.* 2008).

Host longevity is also often believed to be accompanied by high parasite diversity because of the higher chances of transmission and accumulation of parasites over time (Nunn *et al.* 2003; Poulin and Morand 2004). However, the lack of consistent results among numerous tests of this prediction hints that, in reality, it is merely wrong. Indeed, higher exposure to parasites could have promoted stronger investment in anti-parasite defences in natural populations of long-lived hosts, because the investment in resistance is supposed to have evolved and adjusted not only to parasite pressures but also to residual reproductive value (i.e. current and future host reproduction). If so, host species (or conspecific host populations) with greater longevity should invest more in anti-parasite defences to limit deleterious effects (Arrierero and Møller 2008). Consequently, and contrary to the classical prediction related to host exposure, the complicated interplay between exposure and defence could lead to lower parasite species richness in host species with higher adult survival rate.

7.2.3 New predictions: when parasite lifestyle matters

Parasite life-history traits or lifestyles may also strongly interfere with the predictions based on the role of the classical determinants of parasite diversity. For example, if the main habitat of parasites is not the body of a host but rather its habitat, like burrows or nests (see Krasnov *et al.* 2004 and Korallo *et al.* 2007 for fleas and gamasid mites in rodents), then this may explain why host body size has often been found to be a poor predictor of parasite species richness. The complexity of parasite life-cycles and the time spent outside the host could then be more important factors affecting the occurrence of a parasite on or in a host than host body size *per se*. The same criticisms may apply to other host-related

determinants of parasite diversity, such as group or colony size. For example, various studies on bat flies (Streblidae and Nycterobiidae; obligate and specialized ectoparasites of bats) have established that bat fly species richness and even abundance are not positively linked to host colony size (Bordes *et al.* 2008; Reckardt and Kerth 2009). These ectoparasites spend approximately 30 per cent of their life as pupae in bat roosts and only infest a host upon emergence. Roost fidelity seems to be crucial for bat flies to complete their life-cycle, whereas the transmission between hosts does not depend on direct contacts among hosts (Reckardt and Kerth 2009). Moreover, Dittmar *et al.* (2008) have recently shown that the pupae of the bat fly *Trichobius sp.* are localized near bat flyway passages at considerable distances from roosting bats, thus supporting the idea that colony size is less important in the infection process. All this may explain the absence of strong link between colony size of bats and bat fly species richness (Bordes *et al.* 2008).

Thus, parasite species richness is the result of the interplay between parasite lifestyles, host ecology, and host defences. The frameworks based on island biogeography theory and theoretical epidemiology have both led to considerable progress in our understanding of determinants of parasite diversity over the past few decades. However, they have to be considered only as providing some pieces of the puzzle rather than as sole predictors of parasite acquisition and disease transmission.

7.3 Latitudinal gradients of parasite diversity in terrestrial mammals

7.3.1 Why should parasite communities of mammals be richer in the tropics?

Latitude has been considered as a major biogeographical factor that might correlate with parasite diversity, with parasite species richness expected to follow the latitudinal gradient observed in free-living species. Given that mammal species richness is higher in the tropics (Kaufman 1995), we may expect parasite species richness to follow a similar pattern. However, latitude is a proxy variable for a wide range of bioclimatic factors, so latitude *per se* has no meaning as an explanatory factor. This is the reason why more than 25 different mechanisms have been proposed to explain latitudinal gradients of free-living species diversity (Gaston 2000).

The main mechanisms may be, unsurprisingly, related to greater exposure to parasites at lower latitudes due to the particular biotic and abiotic conditions of tropical environments. First, parasite mortality as a result of harsh conditions in winter far from the equator is expected to be a strong constraint for parasite transmission, as many infective stages live off their definitive hosts. This constraint can clearly operate in various major parasite taxa such as helminths, intestinal protozoans, some viruses, and bacteria. Second, invertebrate intermediate hosts (i.e. molluscs or insects), as well as parasitic arthropods that themselves serve as vectors for various parasites, may show latitudinal gradients in abundance and diversity due to the strong effects of temperature or variability and abundance of rainfall on their life-cycles (Guernier *et al*. 2004). This may then affect both ectoparasite species richness and transmission of many intermediate host-dependent and vector-borne parasites such as blood protozoans (e.g. *Plasmodium*), helminths (e.g. filarial nematodes) and viruses (e.g. arboviruses). Third, and more generally, an increase in free-living species diversity near the equator may promote cross-species transmission of generalist parasites and may lead to an increase in parasite species richness among mammals.

7.3.2 Is there any strong emerging pattern linking latitude and parasite diversity in mammals?

Despite expectations of a relation between the latitude at which a host population has been sampled and its parasite species richness, there are still only a few studies investigating this pattern in parasites of mammals (but see Poulin 1995; Krasnov *et al*. 2004 for rodents; Guernier *et al*. 2004 for humans; Nunn *et al*. 2005 for primates; Lindenfors *et al*. 2007 for carnivores) (Fig. 7.1). Moreover, researchers have opted for different methodological approaches for studying the potential effect of latitude on parasite diversity. Some have tested parasite diversity at various latitudinal levels (Guernier *et al*. 2004) (Fig. 7.1a), whereas others have investigated links

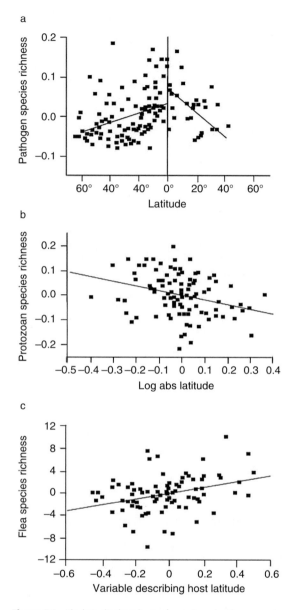

Figure 7.1 The latitudinal gradients of parasite and pathogen species richness. (a) Relationship between pathogen species richness of humans and latitude of human populations sampled for pathogen infectious diseases across the two hemispheres (modified after Guernier *et al.* 2004); (b) relationship between protozoan species richness of primates, and latitude (modified after Nunn *et al.* 2005); (c) relationship between flea species richness and the latitude of centre of geographic range (modified after Krasnov *et al.* 2004).

between the median latitude of the host range and the regional parasite species richness (Krasnov *et al.* 2004) (Fig. 7.1b) or the richness of parasite faunas (Nunn *et al.* 2005; Lindenfors *et al.* 2007) (Fig. 7.1b).

It is thus not especially surprising that the results of these studies are contradictory. The only study that succeeded in finding a strong positive correlation between proximity to the equator and parasite species richness was that of Guernier *et al.* (2004) for humans. Moreover, many studies have established that climatic factors are the main factor behind the link between latitude and parasite species richness. Interestingly, the climatic factor found by Guernier *et al.* (2004) to be best correlated with the diversity of helminths, protozoa, fungi, and indirectly transmitted viruses was the maximum range of precipitation rather than the mean annual temperature or mean annual precipitation. The other studies cited above did not find any positive correlation between proximity to the equator and parasite diversity. Lindenfors *et al.* (2007) and Krasnov *et al.* (2004) found just the opposite for helminths and carnivores and fleas and rodents, respectively. Nunn *et al.* (2005) concluded that parasite species richness increases toward lower latitudes only for protozoan parasites. This last study also suggested that this trend may be a consequence of a greater abundance and diversity of biting arthropods in the tropics, as a high proportion of protozoan parasites in non-human primates are transmitted by arthropods.

7.3.3   Are classical predictions valid?

The above-mentioned contradictory results force us to question the validity of classical predictions. There are at least two pitfalls with the predictions linking positively parasite species richness and proximity to the equator:

(1) the complex interactions between host life-history traits, latitude, and parasite diversity; and
(2) the persistence of parasite transmission despite harsh climatic conditions.

Many empirical studies have highlighted latitudinal variation in body size, geographical range, reproductive effort, and survival in vertebrates (Sand *et al.* 1995; Peach *et al.* 2001; Cardillo 2002; Morrison and Hero 2003; Heibo *et al.* 2005; Hayssen

2008). Interestingly, most of these life-history traits have also been proposed to affect the acquisition of parasite species (Morand 2000; Nunn *et al.* 2003), and several researchers have investigated these relationships (see above). Hence, if body size is positively linked with higher parasite species richness (Kuris 1980; Morand 2000) and if larger body sizes are observed at higher latitudes in endothermic vertebrates (i.e. Bergmann's rule), we could expect that northern species should harbour more parasites. Still from this perspective, it has been proposed that geographical range should decrease towards the equator (Rapoport's rule). If a decrease of host geographical range is often associated with a loss of parasite diversity (Torres *et al.* 2006), we can predict that species living at higher latitudes should, again, harbour more parasite species. Finally, reproductive investment should also not be forgotten as reproductive effort in birds and mammals has been shown to be higher at higher latitudes (Ferguson and McLoughin 2000; Cardillo 2002; Hayssen 2008). Consequently, higher parasite transmission and higher parasite diversity are expected at higher latitudes in these two host taxa due to a trade-off between reproduction and antiparasite defences (see Richner *et al.* (1995) for birds and Pelletier *et al.* (2005) for mammals). Counter-intuitively, complex interactions between latitudinal variation in host life-history traits and parasite acquisition could then promote greater transmission and higher parasite species richness far from the equator.

Parasite mortality, especially of infective stages living off their hosts, is often assumed to be higher at higher latitudes as a result of harsh conditions in winter (Nunn *et al.* 2005). This may negatively affect parasite transmission and result thus in a loss of parasitic diversity. In addition, intermediate hosts are less numerous at northern latitudes, reinforcing this pattern. Some recent studies, however, have cast doubt on these assumptions. In particular, it is well established now that parasites can survive and be transmitted even in very cold conditions (Woodhams *et al.* 2000; Kutz *et al.* 2001; Hrabock *et al.* 2006). For example, Hrabock *et al.* (2006) have clearly demonstrated that in northern Finland nematodes of reindeer are transmitted continuously throughout the year. In addition, some parasites have a very large latitudinal distribution. Jenkins *et al.* (2005) have demonstrated that the nematode *Parelaphostrongylus odocoilei* has an unexpectedly large geographical distribution in North America, ranging from California to central Alaska. These recent results have stressed that parasites can establish at very high latitudes. From this perspective, some authors have suggested that parasite biodiversity in arctic areas could be largely underestimated (Jenkins *et al.* 2005).

7.4   Reinvestigating helminth species diversity and latitudinal gradients in mammals

We reinvestigated helminth species diversity in mammals in relation to latitude using a large dataset that includes only complete parasitic surveys of wild mammals in their native geographical range in three major biogeographical realms (Nearctic, Palaearctic, and Ethiopian). We considered only surveys in which at least 10 host individuals from a given population have been examined. Our dataset for this study included helminths found in 239 mammal species (artiodactyls, perrisodactyls, carnivores, rodents, lagomorphs, and bats).

The link between helminth species richness and sampling latitude was examined at two hierarchical levels. First, we correlated latitude and helminth species richness among the 239 host species. Second, we correlated latitude and helminth species richness within each of 35 host species that have broad latitudinal geographical distribution and for which at least two samples were obtained. We controlled for host phylogeny using independent contrasts in the interspecific analysis and for host sample size in both analyses. We performed multiple regressions and also controlled for the size of the host geographical range.

7.4.1   No latitudinal gradients in helminth diversity among mammal populations and species

In the interspecific analysis, we did not find any latitudinal effect on helminth species richness. Interestingly, however, a positive correlation between the size of the geographical range of a mammal and its helminth species richness was found (Table 7.1), thus supporting earlier studies (Krasnov *et al.* 2004; Torres *et al.* 2006; Hughes and Page 2007). It appeared also that the Ethiopian realm was characterized by a higher diversity of helminths than either the Palaearctic or Nearctic realms independently of latitude (Table 7.1; Fig. 7.2). We also failed to find any positive correlation between latitudinal gradients and helminth species richness in the intraspecific analyses (Table 7.2).

Our results, taken together with previous ones, clearly reinforce the doubt of the existence of any latitudinal gradients in helminth diversity in mammals. At a larger scale, it also supports the earlier idea that latitude *perse* is hardly a determinant of species diversity (Brown 1995). Nonetheless, the latitude gradient of species richness has held an enduring fascination for biologists (Gaston 2000)

Table 7.1   General regression modelling on the following potential determinants of helminth species richness among mammal species: host sample size, latitude, host area size using independent contrasts and incorporating region realms using raw data.

Method	Independent variable	Degrees of freedom	F	p	F model	r² (p)
Independent contrasts	Host sample size	1	53.76	<0.001		
	Latitude	1	0.05	0.83		
	Area	1	4.91	0.02		
		3,157			22.37	0.54 (<0.001)
Raw data	Origin	1	15.92	<0.001		
	Host sample size	1	45.16	<0.001		
	Latitude	1	0.001	0.99		
	Area	1	3.10	0.07		
	Region	2	7.94	<0.001		
		5,198			11.94	0.48 (<0.001)

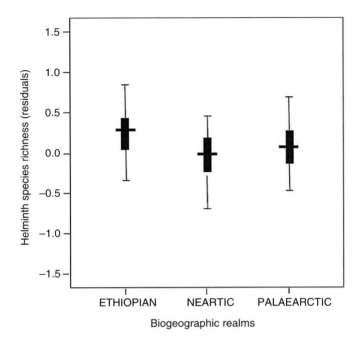

Figure 7.2 Box and whiskers plot showing variation in helminth species in richness of mammals in relation to biogeographic realms (controlled for host sample size). The Ethiopian realm is characterized by a higher diversity of helminths found in mammals than either the Palaearctic or Nearctic realms (see Table 7.1). Each box represents the interquatile range of helminth species richness; the ends of the box are respectively the upper and lower quartiles with the median of helminth species richness marked by the horizontal line; the whiskers show the range of the highest to the lowest observations.

Table 7.2 General regression modelling on the following potential determinants of helminth species richness among mammal populations: host sample size, latitude, host species using raw data.

Independent variable	Degree of freedom	*F*	*p*	*F* model	*r²* (*p*)
Host sample size	1	7.96	0.006		
Latitude	1	0.0001	0.99		
Host species	34	2.17	0.002		
	4,87			2.16	0.69 (0.002)

and parasitologists that have followed this 'classical prediction'.

Nevertheless, we cannot definitively rule out any biogeographical patterns in parasite species richness. The results of Guernier *et al.* (2005) have not only established that tropical areas harbour more human parasite species, but also showed a strong link between parasite diversity and the maximum range of precipitation (instead of the mean annual temperature; see Allen *et al.* 2002). Guernier *et al.* (2005) emphasized that this abiotic factor strongly contributes to the success of para-

site transmission, at least for parasites that have free-living stages (i.e. many helminths, protozoa, and viruses). If the Ethiopian biogeographical realm really harbours a richer helminth diversity in mammals, this reinforces the idea that tropical areas may effectively sustain higher parasite species richness. Rainfall, hygrometry, and soil moisture may be the abiotic determinants of great parasitic biodiversity. Moreover, as parasite diversity has been demonstrated to increase with host diversity (Krasnov *et al.* 2004, 2007), a higher parasite diversity in tropical areas could also be related to a higher mammal species richness in those areas (Ceballos and Ehrlich 2006).

7.5 Conclusion

Many host- or parasite-related factors (i.e. host life-history traits or parasite transmission despite harsh climatic conditions) can act to invalidate the classical expectation that the latitude at which a host population has been sampled may affect its parasite species richness.

Tropical latitudes, however, seem to favour higher parasite transmission and to promote higher parasite diversity in humans (Guernier *et al.* 2005) and in wild mammals (here, helminths in the Ethiopian biogeographical realm). This pattern could be related to specific abiotic factors in tropical areas (i.e. hygrometry and/or mean range of precipitation, for example), but also potentially to biotic factors such as a higher diversity of free-living species in most areas of the tropics. It is necessary to explore this pattern and these predictions by focusing now on the Neotropical or Oriental regions. Unfortunately, the problem will be to gather sufficient and valid parasitic data for these areas which are still largely unexplored (Jimenez *et al.* 2008).

References

Allen, A.P., Brown, J.H., and Gillooly, J.F. (2002). Global biodiversity, biochemical kinetics, and the energetic-equivalence rule. *Science*, **297**, 1545–48.

Anderson, R.M. and May, R.M. (1979). Population biology of infectious diseases. *Nature*, **280**, 361–67.

Arrierero, E. and Møller, A.P. (2008). Host ecology and life-history traits associated with blood parasite species richness in birds. *Journal of Evolutionary Biology*, **21**, 1504–13.

Arneberg, P. (2002). Host population density and body mass as determinants of species richness in parasites communities: comparative analyses of directly transmitted nematodes of mammals. *Ecography*, **25**, 88–94.

Blanck, A. and Lamouroux, N. (2006). Large scale intraspecific variation in life-history traits of European freshwater fish. *Journal of Biogeography*, **34**, 862–75.

Bordes, F., Blumstein, D.T., and Morand, S. (2007). Rodent sociality and parasite diversity. *Biology Letters*, **3**, 692–94.

Bordes, F., Morand, S., and Guerrero R. (2008). Bat fly species richness in Neotropical bats: correlations with host ecology and host brain. *Oecologia*, **158**, 109–16.

Brown, J.H. (1995). *Macroecology*. University of Chicago Press, Chicago.

Cardillo, M. (2002). The life history basis of latitudinal diversity gradients: how do species traits vary from poles to the equator? *Journal of Animal Ecology*, **71**, 79–87.

Ceballos, G. and Ehrlich, P.R. (2006). Global mammal distributions, biodiversity hotspots, and conservation. *Proceedings of the National Academy of Sciences of the USA*, **103**, 19374–79.

Dobson, A., Lafferty, K.D., Kuris, A.M., Hechinger, R.E., and Jetz, W. (2008). Homage to Linneaus: How many parasites? How many hosts? *Proceedings of the National Academy of Sciences of the USA*, **105**, 11482–89.

Ezenwa, V.O. (2004). Host social behavior and parasitic infection: a multifactorial approach. *Behavioral Ecology*, **15**, 446–54.

Ezenwa, V., Price, S.A., Altizer, S., Vitone, N.D., and Cook, C. (2006). Host traits and parasite species richness in even and odd-toed hoofed mammals, Artiodactyla and Perissodactyla. *Oikos*, **115**, 526–37.

Feliu, C., Renaud, F., Catzeflis, F., Durand, P., Hugot J.-P., and Morand, S. (1997). A comparative analysis of parasites species richness of Iberian rodents. *Parasitology*, **115**, 453–66.

Felso, B. and Rozsa, L. (2006). Reduced taxonomic richness of Lice (Insecta: Phthiraptera) in diving birds. *Journal of Parasitology*, **92**, 867–69.

Felso, B. and Rozsa, L. (2007). Diving behavior reduces genera richness of Lice (Insecta: Phthiraptera) of mammals. *Acta Parasitologica*, **52**, 82–85.

Ferguson, S.H. and McLoughlin, P.D. (2000). Effect of energy availability, seasonality and geographical range on brown bear life history. *Ecography*, **23**, 193–200.

Gaston, K.J. (2000). Global patterns in biodiversity. *Nature*, **405**, 220–27.

Guégan, J.F., Morand, S., and Poulin, R. (2005). Are there general laws in parasite community ecology? The emergence of spatial ecology and epidemiology. In F. Thomas, F. Renaud, and J.F. Gueguan, eds. *Parasitism and Ecosystems*, pp. 22–42. Oxford University Press, Oxford.

Guernier, V., Hochberg, M.E., and Guégan, J.F. (2004). Ecology drives the worldwide distribution of human infectious diseases. *PloS Biology*, **2**, 740–46.

Hayssen, V. (2008). Reproductive effort in squirrels: ecological, phylogenetic, allometric and latitudinal patterns. *Journal of Mammalogy*, **89**, 582–606.

Heibo, E., Magnhagen, C. and Vollestad, A. (2005). Latitudinal variation in life history traits in Eurasian perch. *Ecology*, **86**, 3377–86.

Hrabock, J.T., Oksanen, A., Nieminen, M., and Waller, P.J. (2006). Population dynamics of nematode parasites of reindeer in the sub-artic. *Veterinary Parasitology*, **142**, 301–11.

Hughes, J. and Page, R.D.M. (2007). Comparative tests of ectoparasite species richness in seabirds. *BMC Evolutionary Biology*, **7**, 227–48.

Jenkins, E.J., Appleyard, G.D., Hoberg, E.P. et al. (2005). Geographic distribution of the muscle-dwelling nematode *Parelaphostrongylus odocoilei* in North America, using molecular identification of first-stage larvae. *Journal of Parasitology*, **91**, 574–84.

Jimenez, F.A., Braun, J.K., Campbell, M.L., and Gardner, S. (2008). Endoparasites of fat- tailed mouse opossums (*Thylamys*: didelphidae) from Northwestern Argentina and Southern Bolivia, with the description of a new species of tapeworm. *Journal of Parasitology*, **94**, 1098–102.

Kaufman, D.M. (1995). Diversity of new world mammals: universality of the latitudinal gradients of species and bauplans. *Journal of Mammalogy*, **76**, 322–34.

Kutz, S.J., Hoberg, E.P., and Polley, L. (2001). A new lungworm in muskoxen: an exploration in Artic parasitology. *Trends in Parasitology*, **17**, 276–80.

Korallo, N.P, Vinarski, M.V, Krasnov, B.R, Shenbrot, G.I, Mouillot, D., and Poulin, R. (2007). Are there general rules governing parasite diversity? Small mammalian hosts and gamasid mite assemblages. *Diversity and Distribution*, **13**, 353–60.

Krasnov, B.R, Shenbrot, G.I., Khokhlova, I., and Degen, A.A. (2004). Flea species richness and parameters of host body, host geography and host "milieu". *Journal of Animal Ecology*, **73**, 1121–28.

Krasnov, B.R, Shenbrot, G.I., Khokhlova, I., and Poulin, R. (2007). Geographical variation in the "bottom-up" control of diversity: fleas and their small mammalian hosts. *Global Ecology and Biogeography*, **16**, 179–86.

Krasnov, B.R, Korallo-Vinarskaya, N.P., Vinarsky, M.V., Shenbrot G.I, Mouillot, D., and Poulin, R. (2008). Searching for general patterns in parasite ecology: host identity versus environmental influence on gamasid mite assemblages in small mammals. *Parasitology*, **135**, 229–42.

Lindenfors, P., Nunn, C.L., Jones, K.E., Cunningham, A.A., Sechrest, W., and Gittleman, J.L. (2007). Parasite species richness in carnivores: effects of host body mass, latitude, geographical range and population density. *Global Ecology and Biogeography*, **1**, 1–14.

Luque, J.L. and Poulin, R. (2008). Linking ecology with parasite diversity in Neotropical fishes. *Journal of Fish Biology*, **72**, 189–204.

MacArthur, R.H. and Wilson, E.O. (1967). *The Theory of Island Biogeography*. Princeton University Press, Princeton.

Mc Donald, R.A., Delahay, R.J., Carter, S.P., Smith, G.C., and Cheeseman, C.L. (2008). Perturbing implications of wildlife ecology for disease control. *Trends in Ecology and Evolution*, **23**, 53–56.

Morand, S. (2000). Wormy world: comparative tests of theorical hypotheses on parasite species richness. In R. Poulin, S. Morand, and A. Skorping, eds. *Evolutionary Biology of Host-Parasite Relationships: Theory Meets Reality*, pp. 63–79. Elsevier, Amsterdam.

Morand, S. and Harvey, P.H. (2000). Mammalian metabolism, longevity and parasite species richness. *Proceeding of the Royal Society of London B*, **267**, 1999–2003.

Morrison, H. and Hero, J.M. (2003). Geographic variation in life history characteristics of amphibians: a review. *Journal of Animal Ecology*, **72**, 270–79.

Nunn, C.L., Altizer, S. Jones, K.E., and Sechrest, W. (2003). Comparative tests of parasites species richness in primates. *American Naturalist*, **162**, 597–614.

Nunn, C., Altizer, S., Jones, K.E., Sechrest, W., and Cunningham, A.A. (2005). Latitudinal gradients of parasite species richness in primates. *Diversity and Distributions*, **11**, 249–56.

Nunn, C.L. and Dokey, A.T.W. (2006). Ranging patterns and parasitism in primates. *Biology Letters*, **2**, 351–54.

Peach, W.J., Hammer, D.B., and Oatley, T. (2001). Do southern African songbirds live longer than their European counterparts? *Oikos*, **93**, 235–49.

Pelletier, F., Page, K.A., Ostiguy, T., and Festa-Bianchet, M. (2005). Fecal counts of lungworm larvae and reproductive effort in Bighorn Sheep, *Ovis canadensis. Oikos*, **110**, 473–80.

Poulin, R. (1995). Phylogeny, ecology and the richness of parasite communities in vertebrates. *Ecological Monographs*, **65**, 283–302.

Poulin, R. and Morand, S. (2004). *Parasite Biodiversity*. Smithsonian Institution Press, Washington.

Poulin, R. (2007). Are there general laws in parasite ecology? *Parasitology*, **134**, 763–76.

Rahbeck, C. (2005). The role of spatial scale and the perception of large-scale species richness patterns. *Ecology Letters*, **7**, 1–15.

Reckardt, K. and Kerth, G. (2008). Does the mode of transmission between hosts affect the host choice strategies of parasites? Implications from a field study on bat fly and wing infestation of Bechstein's bats. *Oikos*, **118**, 183–90.

Richner, H., Christe, P., and Oppliger, A. (1995). Paternal investment affects prevalence of malaria. *Proceedings National Academy of Sciences of the USA*, **92**, 1192–94.

Rohde, K. and Heap, M. (1998). Latitudinal differences in species and in community structure of metazoan endo- and ectoparasites of marine teleost fish. *International Journal of Parasitology*, **28**, 461–74.

Sand, H., Cederlund, G., and Danell, K. (1995). Geographical and latitudinal variation in growth patterns and adult body size of Swedish moose (*Alces alces*). *Oecologia*, **102**, 433–42.

Stanko, M.D., Miklisova, D., Goüy de Bellocq, J., and Morand, S. (2002). Mammal density and patterns of ectoparasite species richness and abundance. *Oecologia*, **131**, 289–95.

Torres, J., Miquel, J., Casanova, J.C., Ribas, A., Feliu, C., and Morand, S. (2006). Parasite species richness of Iberian carnivores: influences of host density and range distribution. *Biodiversity and Conservation*, **15**, 4619–32.

Woodhams, D.C., Costanzo, J.P., Kelty, J.D., and Richard, E.L. (2000). Cold hardiness in two helminth parasites of the freeze-tolerant wood frogs, *Rana sylvatica*. *Canadian Journal of Zoology*, **78**, 1085–91.

CHAPTER 8

Ecological properties of a parasite: species-specific stability and geographical variation

Boris R. Krasnov and Robert Poulin

8.1 Introduction

The ability of a species in which individuals with a given genotype can manifest this genotype via different phenotypes in different environments is a well-known phenomenon called phenotypic plasticity. In this context, the expression of any trait depends, on the one hand, on species-specific limits of variation and, on the other hand, on the external environment. Indeed, the interplay between intrinsic species-specific boundaries of variation and extrinsic environmental influences has been reported repeatedly for morphological traits (e.g. body size; Peters 1983) or physiological traits (e.g. metabolic rate; Degen 1997). In contrast, the plasticity of ecological traits is much less understood. Any population of any species is characterized by ecological properties such as abundance and niche breadth. However, these properties may vary greatly among populations of the same species. Consequently, a question arises: Is the observed manifestation of an ecological trait a species attribute subjected to natural selection, or is it merely a reflection of the local effects exerted by a variety of both biotic and abiotic ecological factors (e.g. Fox and Morrow 1981)? Obviously, this question applies equally well to both free-living and parasitic species. However, parasites are very convenient models for studying geographic patterns of stability and variation in the ecological parameters of a species for at least two reasons. First, it is relatively easy to obtain replicated samples of parasites (e.g. host individuals, host populations, or host species), making it possible to examine geographical variation in ecological parameters on a range of spatial scales. Second, the ecological niche of a parasite is often easier to define than that of a free-living organism (Timms and Read 1999) because the main living environment or habitat of a parasite and its food are represented by its host. In this chapter, we will consider species-specific stability and spatial variation in abundance and host specificity (niche breadth) of several parasite taxa. We will show that, on the one hand, these properties show some geographical stability and represent genuine species attributes, but, on the other hand, they also parallel to some extent the responses of parasites to their external environment.

8.2 Species properties, sources of variation, and methodological approach

Among the many properties that a species possesses, abundance and niche breadth are possibly the most fundamental. In most cases, the level of abundance of a species is measured as the mean number of individuals of this species per unit area. However, this measure is not generally applicable for parasite species. This is because parasites are almost universally aggregated among their hosts. In other words, most parasite individuals occur in a few host individuals, while most host individuals have only a few, if any, parasites (Anderson and May 1978; Poulin 1993; Shaw and Dobson 1995; Wilson *et al.* 2001). This pattern of distribution prevents us from characterizing the abundance of a parasite by a single value. Consequently, the three common measures of parasite abundance are mean abundance (mean number of parasites per host individual, including both infested and uninfested hosts), intensity of infestation (mean number of parasites per infested host individual), and prevalence (proportion of infested hosts in the population). Obviously, intensity of infestation is the product of mean abundance and prevalence. These measures are straightforward, simple to understand, and easily calculated. However, in many old literature sources, only a single measure of parasite abundance is reported.

The niche breadth of a parasite is usually considered in terms of its host specificity, which is traditionally defined as the number of host species exploited by a given parasite species. A parasite species using a single host species is usually considered as having a narrow niche, whereas if it exploits multiple host species it is thought to have a broad niche. However, from an evolutionary perspective, the host specificity of a parasite is not merely a function of how many host species it can exploit, but also of how closely related these host species are to each other (Poulin and Mouillot 2003). For example, consider two parasite species each exploiting the same number of host species; if one of these parasite species exploits only host species belonging to the same genus whereas the other exploits hosts belonging to different families or orders, then the host specificity of the former should be considered higher than that of the latter. Therefore, the study of host specificity should take into account phylogenetic (taxonomic) relationships among all host species used by a parasite species. This can be achieved using the specificity index, S_{TD}, and its variance, $VarS_{TD}$, proposed by Poulin and Mouillot (2003). The index S_{TD} measures the average taxonomic distinctness of all host species used by a parasite species. When these host species are placed within a taxonomic hierarchy, the average taxonomic distinctness is simply the mean number of steps up the hierarchy that must be taken to reach a taxon common to two host species, computed across all possible pairs of host species. The greater the taxonomic distinctness between host species, the higher the number of steps needed, and the higher the value of the index S_{TD}: thus it is actually inversely proportional to specificity. A high index value means that on average the hosts of a parasite species are not closely related. Since the

index cannot be computed for parasites exploiting a single host species, a S_{TD} value of 0 is usually assigned to these parasite species, to reflect their strict host specificity. The variance in S_{TD}, $VarS_{TD}$, provides information on any asymmetries in the taxonomic distribution of host species (Poulin and Mouillot 2003); it can only be computed when a parasite exploits three or more host species (it always equals zero with two host species).

If the abundance and host specificity of a parasite are genuine species properties, then they should vary among different populations of the same parasite species only within some species-specific boundaries. However, the level of abundance and host specificity manifested by a given parasite population may also depend on given extrinsic conditions. In contrast to free-living species, these extrinsic conditions for a parasite population comprise both the host species exploited by this parasite population and the environment where this population occurs. In other words, the observed level of abundance and host specificity demonstrated by a parasite population is expected to depend on (a) parasite identity; (b) host identity; and (c) geographic location. Obviously, the following questions arise:

1. What are the relative effects of parasite identity, host identity, and geographic location on abundance and niche breadth of parasites?
2. Do patterns in these effects vary among different parasite taxa?

A simple, but convincing, methodological approach to evaluate the effect of different factors on the spatial stability or variation of parasite abundance and host specificity is repeatability analysis, a method first proposed by Arneberg et al. (1997). Essentially, in the repeatability analyses, the variation in parasite abundance or host specificity is analysed using a one-way ANOVA in which either parasite species, host species, or locality is the independent factor. Obviously, this can be done only for parasite species for which at least two samples (from different hosts or localities) are available. A significant effect of parasite species, host species, or locality would indicate that values of abundance/host specificity are repeatable within parasite species, host species, or locality, respectively. For example, the significant effect of parasite species would demonstrate that abundance or host specificity values for the same parasite species exploiting different hosts or occurring in different geographic regions are more similar to each other than to values from other parasite species. Then, the proportion of the total variance originating from differences among parasite species, host species, or localities, as opposed to within parasite species, host species, or localities, respectively, can be evaluated following procedure described in statistical textbooks (e.g. Sokal and Rohlf 1995).

8.3 Mean abundance, intensity of infestation, and prevalence

To date, several studies of intraspecific variability versus stability of parasite population parameters have been carried out. In general, these studies suggest that abundance, intensity of infestation,

and prevalence, albeit varying to some degree, are true species attributes of parasites; this is true for parasites belonging to a wide range of taxa (Table 8.1; see Fig. 8.1 for illustrative example with fleas). Furthermore, when repeatability analyses are run for taxonomic levels higher than species (e.g. genera, tribes, subfamilies, and families; see Krasnov *et al.* 2006), the results demonstrate that parasite abundance appears to be a genuine property of at least some of these higher taxonomic units, although the percentage of the variation among samples accounted for by differences among taxa, as opposed to within-taxa, is lower for senior taxonomic units (e.g. tribes) than junior taxonomic units (e.g. genera). This suggests that limits of abundance as well as intrinsic properties of parasites that determine these limits are phylogenetically constrained. In other words, a relatively narrow, species-specific range of inter-population variation in abundance is

Table 8.1 Proportion of the total variance in mean abundance (A), intensity of infestation (I) or prevalence (P) originating from differences among parasite species as opposed to within parasite species.

Parasites	Hosts	Measure of abundance	% variance	Source
Monogeneans	Freshwater fish	A	84	Poulin 2006
		I	88	
		P	57	
Digeneans	Freshwater fish	A	69	Poulin 2006
		I	87	
		P	11	
Cestodes	Freshwater fish	A	12	Poulin 2006
		I	18	
		P	32	
Nematodes	Freshwater fish	A	52	Poulin 2006
		I	46	
		P	29	
	Mammals	A	36	Arneberg *et al.* 1997
		I	52	
Acanthocephalans	Freshwater fish	A	84	Poulin 2006
		I	92	
		P	51	
Copepods	Freshwater fish	A	48	Poulin 2006
		I	64	
		P	42	
Gamasid mites	Small mammals	A	47	Korallo-Vinarskaya *et al.* 2009
Larval and nymphal ixodid ticks	Small mammals	A	21	Krasnov *et al.* 2007
		P	4	
Fleas	Small mammals	A	46	Krasnov *et al.* 2006

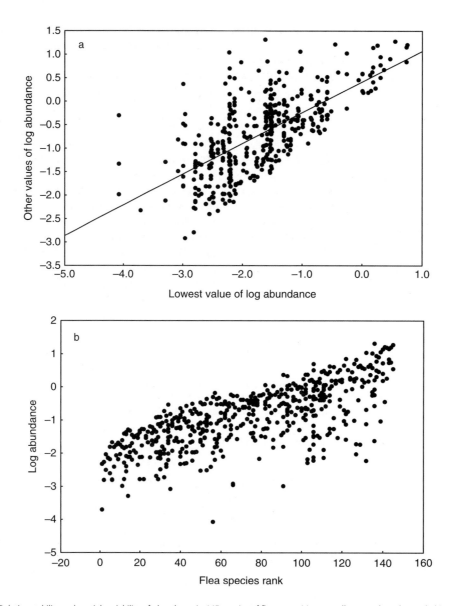

Figure 8.1 Relative stability and spatial variability of abundance in 145 species of fleas parasitic on small mammals and recorded in at least two geographic locations. (a) Relationship between the lowest abundance recorded for a given flea species and all other abundance values recorded in other regions for this species on the same host species. A positive correlation between the lowest value and other values of abundance indicates that abundances of the same flea species are consistent with each other across different geographic localities. (b) Rank plot of flea abundance. The 145 flea species are ranked according to their mean log-transformed abundance values, with rank 1 given to the species with the lowest mean abundance; all sample estimates are plotted for each species. The points falling along a narrow band stretching from the lower left to the upper right corner, with few or no points in either the upper left or lower right corner, indicate that variation is small within compared to between flea species. After Krasnov *et al.* (2006). Reproduced with permission of Springer Verlag.

the result, at least in part, of natural selection. Consequently, the observed range of abundance for a given parasite species has not only evolved for some particular ecological reason, but also seems to be nested within the phylogeny and, thus, it can be inherited by descent.

Repeatability of population parameters within parasite species implies that some species-specific life-history traits determine the limits of abundance. Lower limits of abundance can be affected by species-specific mating systems, whereas upper limits of abundance can be determined by species-specific reproductive outputs, generation times, ability to withstand crowding, or the action of density-dependent processes. For example, in fleas, some species can mate without a bloodmeal (Iqbal and Humphries 1974), whereas in other species, an unfed male is unable to inseminate a female (Dean and Meola 1997). The need to feed prior to mating may possibly increase the number of fleas on a host body at any given time. Fecundity varies drastically among parasite species of the same taxon (Marshall 1981; Krasnov 2008) with difference in fecundity often being paralleled by a difference in mean abundance (Kosminsky 1965). For example, in nematodes, daily egg production per female varies among species by five orders of magnitude (Morand 1996). The literature contains a huge variety of examples illustrating interspecific differences in life-history traits that may underpin intraspecific differences in the population parameters of parasites.

Interestingly, when the repeatability analyses are carried out for all three infection measures, it appears that, in general, intensity of infestation is the most conservative character within a species. In contrast, mean abundance demonstrates more inter-population variability but is, nevertheless, intraspecifically stable, whereas prevalence varies greatly. Moreover, the variability of prevalence within species is sometimes so high that it cannot be considered as a species attribute (e.g. Poulin 2006 for digeneans in fish). The reason for this may be that prevalence is determined by the rate at which parasites encounter suitable hosts. This rate undoubtedly depends on a variety of extrinsic factors such as host number, host behaviour, and survivability of infective stage, in other words, on factors strongly affected by local conditions. Consequently, between-locality differ-

ences may overcome the conservatism in the species-specific pattern of distribution of a parasite.

No parasite species may exist without its host. It is therefore logical that infestation parameters can also be seen as properties of the host species. In other words, one may ask whether certain host species are characterized by higher infection levels, by any parasite species, than other host species. The results of several studies suggest that the answer to this question is, in general, positive (Table 8.2). Thus, it appears that mean abundance, intensity of infestation, and/or prevalence of parasites can be considered as host species properties as well. In other words, independently of parasite species, some fish or mammal species have many parasites per individual, whereas other fish or mammals have only a few parasites per individual. As a result, some host species always harbour more parasites than other host species (e.g. Krasnov *et al*. 2008a for fleas). Nevertheless, the repeatability of parasite population parameters among host species is weaker than that among parasite species (compare Tables 8.1 and 8.2). Interestingly, the results of different studies provide a very similar ratio between percentages of the variation in parasite abundance associated with differences among host species and among parasite species (compare Arneberg *et al*. 1997 and Krasnov *et al*. 2006).

The repeatability of parasite abundance within host species suggests that some host properties constrain to some extent the number of parasites harboured by an individual. In most cases, these constraints may be related to processes that act in or on the host body. For example, the abundance of adult parasites can be limited by host body features, host immune defence, and host anti-parasitic behaviour. Host body size can limit the number of parasites per host, as more space is available for multiple parasite individuals on larger hosts (e.g. Morand and Guégan 2000). Host species also differ in the level of their immune and behavioural defence against parasites (Klein and Nelson 1998; Mooring *et al*. 2000). A host species with lower immunocompetence and/or a more limited ability to self-groom can be exploited by a higher number of parasites than a host species with higher immunocompetence and/or higher self-grooming efficiency, all else being equal. In addition, the fecundity of

Table 8.2 Proportion of the total variance in mean abundance (A), intensity of infestation (I) or prevalence (P) originating from differences among host species as opposed to within host species.

Hosts	Parasites	Measure of abundance	% variance	Source
Freshwater fish	Monogeneans, digeneans, cestodes, nematodes, acanthocephalans, copepods	A	13	Poulin 2006
		I	6	
		P	19	
Marine gastropods	Larval digeneans	P	23	Poulin and Mouritsen 2003
Mammals	Nematodes	A	22	Arneberg et al. 1997
		I	18	
	Gamasid mites	A	29	Korallo-Vinarskaya et al. 2009
	Larval and nymphal ixodid ticks	A	10–11	Krasnov et al. 2007
		P	0.2–0.9	
	Fleas	A	24	Krasnov et al. 2006

parasites depends on which host species they exploit (e.g. Krasnov et al. 2004a for fleas; Riquelme et al. 2006 for acanthocephalans) which might, in turn, determine host-specific levels of parasite abundance.

Notably, the repeatability of prevalence within host species proved to be stronger than within parasite species. This supports the aforementioned explanation that the main mechanism determining prevalence level is the rate of encounters between hosts and parasites, which are likely to depend on host species traits such as mobility.

The effect of geographic locality on parasite abundance levels is much less obvious than that of parasite and host identities. This is because the vast majority of parasites live inside the host's body, so the influence of a complex of external environmental factors on them is likely limited. However, ectoparasites as well as the free-living stages of many endoparasites may be strongly influenced by the off-host environment. Moreover, environmental factors such as temperature, relative humidity, salinity, and solar radiation may have similar effects on parasites belonging to the same taxon as they likely have similar biochemical and/or physiological requirements. Consequently, if key population parameters of both ecto- and endoparasites are also determined by the local environment, then the abundance and distribution of parasites may to some extent be a property of a given geographic location. To date, two studies only have examined

the repeatability of parasite abundance within geographic locations. These studies have focused on the abundance of two taxa of ectoparasitic arthropods, fleas (Krasnov et al. 2006), and gamasid mites (Korallo-Vinarskaya et al. 2009), collected from the bodies of small mammalian hosts. Both these studies demonstrate that parasite abundance is repeatable within locations, although the proportion of the variance accounted for by differences among locations as opposed to within-location is consistently low, being 13 per cent for fleas and 10 per cent for mites. In other words, some geographic locations are characterized by higher flea or mite abundance than other locations, whatever the identity of the host and parasite species involved. The reason behind the repeatability of flea or mite abundance within a geographic location is likely related to the fact that both taxa are ectoparasitic and are thus subjected to the influence of environmental factors. Indeed, the effect of factors such as climate and soil structure on abundance is well known for both fleas (e.g. Krasnov 2008) and mites (e.g. Crystal 1986). Furthermore, both fleas and mites spend a substantial part of their life in the burrow and/or nest of the host. Burrow structure within one host species can differ among different locations (e.g. Shenbrot et al. 2002). In addition, the burrows of different host species in the same location can be characterized by similar microclimate conditions due to the same macroclimate (Kucheruk 1983). This can lead to among-location differences (as opposed to within-location similarities among host species) in ectoparasite abundance and, together with climatic differences between locations, can, at least partly, explain the repeatability of within-location parasite abundances.

In addition, from the point of view of each parasite species, a geographic location is characterized not only by certain abiotic conditions, but also by a biotic component that includes the co-occurring assemblage of other parasite species. In fleas, for example, the abundance of a given species (a) correlates positively with the total abundance of all other co-occurring species in the community and (b) correlates negatively with the diversity of the flea community (Krasnov *et al.* 2005). Consequently, the repeatability of parasite abundance within localities as opposed to among localities can be explained by differences between localities in a range of biotic and abiotic factors.

8.4   Host specificity

The relative stability and geographic variability of host specificity have been studied to date in two parasite taxa only. These again were fleas (Krasnov

et al. 2004b) and gamasid mites (Korallo-Vinarskaya *et al.* 2009). In fleas, host specificity, expressed as either of the three measures previously mentioned, appears to be repeatable among different populations of the same species, i.e. host specificity varies significantly more among flea species than within flea species with 30, 14, and 17.4 per cent (for the number of host species exploited, the taxonomic distinctness, and taxonomic asymmetry of host assemblages, respectively) of the variation among samples accounted for by differences between flea species (see Fig. 8.2 for an example with the number of exploited hosts). The number of host species exploited and the taxonomic asymmetry of host assemblages were repeatable also across flea genera with 21 and 15 per cent, respectively, of the variation among samples accounted for by among-genera differences. However, the taxonomic distinctness of host assemblages was not repeatable within a flea genus. The effects of either host species identity or geographic location were not considered in this study.

In mites, host specificity, taken as the number of host species exploited, was highly repeatable within mite species and less, albeit significantly, so within

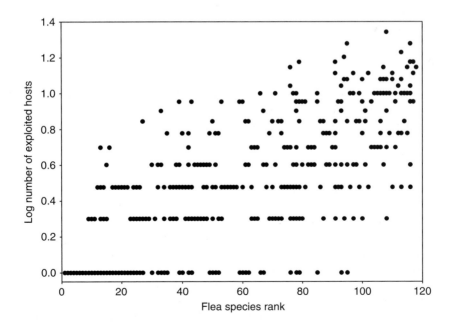

Figure 8.2   Rank plots of number of host species used for 118 flea species parasitic on small mammals and recorded in at least two geographic locations. See Fig. 8.1b for explanations. After Krasnov *et al.* (2004b). Reproduced with permission of Wiley Blackwell.

host species or geographic localities (with 33, 14, and 8 per cent, respectively, of the variation among samples accounted for by among species differences). However, host specificity measured as the taxonomic diversity of the host spectrum was highly variable within mite or host species. It was not repeatable within the same mite species and cannot be considered as a mite or host species attribute. In contrast, the taxonomic diversity of the mite's host spectrum was repeatable across mite species within a locality, although the amount of variation associated with differences between regions was rather low (9 per cent).

The level of host specificity of a parasite species is determined by the range of conditions to which this species is adapted. These conditions are related to the ecological, behavioural, physiological, and biochemical traits of a particular host species or group of species (Ward 1992; Poulin 2007). If we adopt a Hutchinsonian representation of the ecological niche of a parasite species as an n-dimensional hypervolume, the axes of which are host traits, then a parasite species would demonstrate either broad or narrow tolerance along each of these axes. The repeatability of the degree of host specificity among populations of the same parasite species suggests that host specificity is a true species attribute that can be envisaged as the entire set of parasite's responses along all axes representing host traits. Natural selection can therefore act on this set of responses as a whole. On the other hand, the lack of within-genus repeatability in S_{TD} supports, albeit indirectly, the hypothesis that specificity, at least for the taxonomic diversity of the hosts, is not an evolutionary blind alley and that the evolution of specialization has no intrinsic direction (Desdevises *et al.* 2002; Poulin *et al.* 2006).

Nevertheless, host specificity values from different populations of the same parasite species are still somewhat variable. To some extent, this may reflect geographic differences in host availability. For example, an increase in regional host species richness correlates with an increase in the number of hosts exploited by a mite species (Korallo-Vinarskaya *et al.* 2009), suggesting that whenever a new host species appears in a regional pool, at least some species of mites add it to their host spectrum. However, this is not the case for fleas (Krasnov *et al.* 2004b). These contrasting results for the two parasite taxa reflect our limited knowledge of geographic patterns of host specificity, and suggest that more studies on other parasite–host associations are required prior to the elucidation of any general trend.

8.5 Geographic patterns of variation in parasite abundance and host specificity

As we already know, despite the relative stability of parasite abundance and niche breadth over geographical space, both these properties vary spatially to some extent. Logically, the next question should be: Do they vary in some predictable way or, alternatively, is their spatial variation following a mosaic pattern? Not much is known about this. However, a few recent studies have attempted to test whether spatial variation of abundance and/or host specificity of parasites follows some well-known geographic patterns, such as the 'abundance optimum' model and latitudinal gradients.

8.5.1 'Abundance optimum' pattern

Almost universally, a species is abundant in only a few of the localities where it occurs, whereas it is rare in most other localities (Gaston 2003). One of the explanations for this pattern is that a species' abundance decreases from the centre to the periphery of its geographic range (Hengeveld 1990). This explanation is based on the assumptions that:

(1) abundance represents the response of a local population to local conditions;
(2) local abundance reflects the relationship between local environmental conditions and the Hutchinsonian niche requirements of a species; and
(3) environmental variables that affect abundance are spatially autocorrelated (Brown 1984).

However, the 'centre of abundance' hypothesis appears not to be especially well supported by empirical data and, therefore, has been recently strongly criticized (e.g. Sagarin *et al.* 2006). Nevertheless, if this hypothesis is reformulated as the 'abundance optimum' hypothesis, that is, that species abundance peaks in the locality with the most favourable conditions and decreases with increasing distance from that locality, then many arguments against the 'centre of abundance' model can be ruled out (see Sagarin *et al.* 2006 versus Krasnov *et al.* 2008b for details).

The first attempt to test the 'abundance optimum' pattern in parasites has been carried out by Poulin and Dick (2007) using data on eight helminth endoparasites of the fish *Perca flavescens* in continental North America. Assuming that body condition and anti-parasite defence systems do not vary substantially among different populations of conspecific hosts, the general lack of an 'abundance optimum' pattern in helminths (except for prevalence of a cestode *Proteocephalus pearsei*) reported by Poulin and Dick (2007) is not particularly surprising. In other words, the high predictability of a parasite's immediate environment (host body) could be the main reason behind the lack of significant spatial patterns in abundance. However, this predictability does not apply for ectoparasites. Indeed, Seifertová *et al.* (2008) have reported a significant decrease of abundance with an increase of distance from the most favourable locality in monogeneans parasitic on the European fish *Leuciscus cephalus*, but not for cestodes, nematodes, and acanthocephalans. Intriguingly, Seifertová *et al.* (2008) also found this pattern for larval trematodes of the genus *Diplostomum*, while no significant pattern was found for parasitic crustaceans. Krasnov *et al.* (2008b) tested the 'abundance optimum' hypothesis using data on ectoparasites of terrestrial hosts, namely 9 flea and 13 gamasid mite species. The results of this study supported an existence of the tested pattern, although it appeared not to be particularly strong. Admittedly, a significant negative effect of the distance from the locality of maximum abundance on relative abundance in a particular locality was revealed only when a meta-analysis of coefficients of correlations between these two variables for individual species was performed across all species. This was despite the fact that the relationships between relative abundance in a region and the distance from that region to the region of maximum abundance, though generally negative, were mostly statistically non-significant. Nevertheless, a cross-species comparison between relative abundances in the localities closest to and furthest from the locality

of maximum abundance demonstrated that the former were significantly higher than the latter. The lack of significance of the correlation between abundance and distance to the optimal locality in most species, coupled with a significant difference between relative abundances in the localities closest to and furthest from the locality of maximum abundance in cross-species analysis, suggest that the shape of the 'abundance optimum' pattern may be not merely a monotonic decline of abundance with an increasing distance from the optimal locality, but also involve a sharp step-wise drop of abundance in the most remote locality. Furthermore, the results of the study of Seifertová *et al.* (2008) suggest that the existence of the 'abundance optimum' pattern may be associated with the level of host specificity of the parasite species under consideration. For example, the pattern may hold for highly specific monogeneans because, in contrast to generalist endoparasites, their abundance values are not affected by the distribution of intermediate hosts or other potential definitive hosts. In addition, between-localities differences in parasite abundance may be related to differences between populations of their hosts. Seifertová *et al.* (2008) found that geographical distances between localities correlated positively with phylogenetic distances between populations of *L. cephalus* occurring in these localities, while the correlation between abundance or prevalence of parasites and the phylogenetic distance of a host population from the locality with maximum abundance or prevalence of these parasite species was negative, albeit non-significant. These results allowed Seifertová *et al.* (2008) to conclude that the 'abundance optimum' pattern could be driven more by phylogenetic distances between host populations than by geographical distances.

Another species trait that may vary with the distance between a given locality and the optimal locality is niche breadth, or host specificity if we consider parasites. As the optimal locality often (albeit not always) coincides with the centre of the geographic range of a species (Hengeveld 1990, but see Sagarin *et al.* 2006), the level of host specificity in a parasite species can be distorted in its peripheral populations because the latter often live under conditions sharply different from those of core populations, for example highly unpredictable and/or suboptimal conditions (in terms of both host populations and off-host environment), relative to core areas. Furthermore, due to marginal ecological conditions at the periphery, populations there may be small and isolated and adapted to a narrower range of ecological conditions (Carson 1959). To the best of our knowledge, the only study that compared host specificity of parasites in core versus peripheral populations examined fleas exploiting small mammals (Krasnov *et al.* 2004b). Contrary to expectation, no correlation between any of the host specificity measures and the relative distance of a location from the centre of the geographic range was found in any of the 23 studied flea species. However, in 17 of these species at least one measure of host specificity correlated positively or negatively with at least one of the parameters that describe the departure of environmental conditions in a location from the mean value for that environmental factor calculated across the entire geographic range of a flea species. In other words, the geographic variation in the host specificity of a flea species is not related to differences between core and peripheral flea populations; rather it is associated with environmental variation, suggesting therefore the existence of a causal association between specificity and local ecological factors. Environmental factors thus play an important role in the geographic variation of flea host specificity, although they are not the only factors involved.

8.5.2 Latitudinal patterns

The most pervasive patterns in biogeography are those related to latitude. A variety of these patterns, such as latitudinal gradients in species richness and geographic range size, have been extensively studied and documented (Gaston 2003; Ruggiero and Werenkraut 2007; Rohde in this volume for recent reviews). However, despite the fact that the manifestation of latitudinal patterns is generally similar (linear positive or negative correlations), the mechanisms underlying these patterns may be profoundly different.

One group of latitudinal patterns is undoubtedly associated not with the latitude per se, but with climatic correlates of latitude. Nobody would deny that, for example, air, surface, and water temperatures monotonically decrease with the distance to the north or to the south of the equator. Consequently, a correlation between the abundance of a species and latitude is expected in any taxon because all living organisms are affected by temperature. The relationships between parasite abundance and latitude have been studied in various parasite taxa exploiting various host taxa and occurring in various geographic regions (Table 8.3). In many, albeit not all, parasites, an increase or decrease in parasite abundance with changes in latitude has been reported. The contrasting directions of correlations between abundance and latitude in different

Table 8.3 Pattern of correlation (P—positive, N—negative, NC—no correlation) between latitude and some measure of parasite abundance (A—mean abundance, P—prevalence, I—intensity of infestation, L—for blood parasites, the proportion of red blood cells infected by parasites per host).

Parasite	Host	Measure of abundance	Geographic region	Pattern of correlation	Source
Protozoan *Hepatozoon hinuliae*	Lizard *Eulamprus quoyii*	L	Eastern Australia	P	Salkel *et al.* 2008
Protozoan *Leucocytozoon*	Forest birds	P	Chile	P	Merino *et al.* 2008
Protozoan *Plasmodium*				N	
Protozoan *Haemoproteus*				N	
Cestode *Proteocephalus pearsei*	Fish *Perca flavescens*	A, P	North America	P	Poulin and Dick 2007
Cestode *Bothriocephalus cuspidatus*		A, P		NC	
Acanthocephalan *Leptorhynchoides thecatus*		P		P	
Nematode *Capillaria gruweli*	Bird *Alectoris rufa*	A	Iberian Peninsula	N	Calvete 2003
Trematode *Discrocoelium* sp.		A		P	
Flea *Hystrichopsylla talpae*	Shrew *Neomys fodiens*	A	Holarctic	P	Krasnov *et al.* 2006
Flea *Megabothris turbidus*	Rodent *Myodes glareolus*	A		N	
Flea *Amalaraeus penicilliger*		A		NC	

parasites support the idea that this pattern is related to climatic correlations of latitude that depend on species-specific environmental preferences.

Another group of latitudinal patterns seems to be driven by global processes, such as climatic changes due to periodical changes in the orbit of the Earth (Dynesius and Jansson 2000). If this is indeed the case, then these patterns are expected to be universal (Hillebrand 2004). For example, one of the most famous latitudinal patterns is Rapoport's rule, which postulates an increase in a species' geographic range size with increasing latitude (Stevens 1989; Hawkins and Diniz-Filho 2006). From an ecological perspective, Rapoport's pattern may exist because of one of three alternative mechanisms. First, it may represent a particular case of the more general relationship between a species' niche breadth and the size of its geographic range (Brown 1984; Ruggiero and Hawkins 2006). Second, it may exist because the level of specialization is determined by population stability (MacArthur 1955) which, in turn, is determined by environmental stability (MacArthur 1972). The latter is assumed to be lower in temperate regions compared to the tropics (MacArthur 1972). Third, the negative relationship between the level of specialization and latitude would occur if there are (a) a latitudinal gradient in species richness and (b) an effect of richness on niche breadth (Vázquez and Stevens 2004). This mechanism seems more complex than the previous ones, thus deserving a more detailed explanation. The effect of richness on niche breadth can occur if the species are involved in an asymmetrically specialized interaction network in which specialists tend to interact with generalists (Vázquez and Aizen 2003, 2004). An important characteristic of asymmetrically specialized interaction networks is that asymmetric specialization among species increases with an increase in the number of species in the network (Vázquez and Aizen 2004). In particular, this tendency is manifested by an increase in the number of extreme specialists with an increase in community size (Vázquez and Stevens 2004), that is, higher species richness may lead to more extreme specialization. Therefore, a decrease in niche breadth (increase in specialization) at low latitudes may be a by-product of an increase in species richness. In other words, if species richness affects niche breadth, this should result in a link between niche breadth and latitude.

Whatever the mechanism, a decrease in the level of specialization with an increase of latitude is expected. Furthermore, if the hypothesis of Vázquez and Stevens (2004) is correct, then the negative relationship between the level of specialization and latitude may not be universal, but rather applicable mainly in those taxa that (a) demonstrate latitudinal gradients in species richness and (b) are involved in asymmetric interaction networks. This is exactly the case for parasites (Poulin and Rohde 1997; Calvete et al. 2004; Vázquez et al. 2005).

To the best of our knowledge, there is only a single study that tested latitudinal patterns of niche breadth (i.e. host specificity) in parasites using data on the host specificity of 120 flea species parasitic on small mammals in the Palaearctic (Krasnov et al. 2008a). This study demonstrated that the number of species in a flea's host spectrum did not correlate with latitude, but the taxonomic distinctness and taxonomic asymmetry of these hosts increased significantly with increasing distance of the centre of a flea's geographic range from the equator (Fig. 8.3), although the former relationship was significant only when the data were controlled for the confounding effect of flea phylogeny. Interestingly, the distribution of data points in Fig. 8.3b is roughly triangular. This suggests that although 'northern' fleas can have both taxonomically symmetric and taxonomically asymmetric host spectra, no main host branch can be distinguished in the host spectrum of a 'southern' flea. This trend certainly requires further investigation. Moreover, the fact that a latitudinal gradient in host specificity, measured as the taxonomic distinctness of the host spectrum, was detected only when phylogenetic relationships among fleas were controlled for, suggests that there can be an adaptive component to this geographic pattern (i.e. resulting from species-level responses) that may occur independently of phylogeny.

Nevertheless, the existence of latitudinal patterns of niche breadth in parasites should be further validated by studies on other parasite–host associations in both terrestrial and aquatic realms.

8.6   Concluding remarks

The geographic variation in abundance and niche breadth of parasites results from the interplay

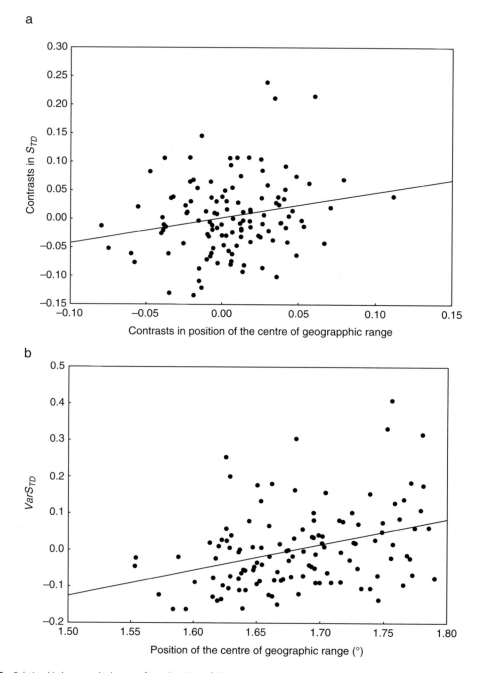

Figure 8.3 Relationship between the log-transformed position of the geographic range and (a) the taxonomic distinctness of a flea's host spectrum, S_{TD}, using the method of phylogenetically independent contrasts, and (b) the taxonomic asymmetry of a flea's host spectrum, $VarS_{TD}$, across 120 Palaearctic flea species. The index S_{TD} was corrected for the number of host species in a flea's host spectrum, whereas the index $VarS_{TD}$ was corrected for the S_{TD} of that spectrum. After Krasnov *et al.* (2008a). Reproduced with permission of Wiley Blackwell.

between the intrinsic features of a species that tend to maintain the stability of these parameters across all populations of the same species, and extrinsic factors that differ among hosts and/or localities and thus tend to produce spatial variability in these parameters. The patterns of stability and variability of two major properties of parasites suggest that they, although variable within some limits, (a) are true species characters, (b) are determined to some extent by host species identity, and (c) are affected to a small extent by local biotic and abiotic conditions. This supports the idea that the biological attributes of parasite species are primary determinants of parasite dynamics compared with characteristics of hosts and of the local environment (Arneberg et al. 1997; Poulin 2006).

Furthermore, as we have seen, the manifestation of some geographic patterns depends on parasite life-history. For instance, the 'abundance optimum' pattern held (albeit weakly) for ectoparasites such as fleas and mites, but did not hold for intestinal helminths. Different sensitivity to environmental factors stemming from the life-history strategy is a likely reason for this difference. However, many other geographic patterns of abundance and host specificity are strikingly similar in vastly different parasite taxa. For example, the repeatability of parasite abundance was found in both endo- and ectoparasites that belong to taxa of different origin, with different life-cycles, exploiting different (phylogenetically, morphologically, and ecologically) hosts, and from both terrestrial and aquatic ecosystems. This suggests that some fundamental rules in parasite ecology exist and that populations of apparently different species may be governed by some common principles.

References

Anderson, R.M. and May, R.M. (1978). Regulation and stability of host-parasite population interactions. I. Regulatory processes. *Journal of Animal Ecology*, **47**, 219–47.

Arneberg, P., Skorping, A., and Read, A.F. (1997). Is population density a species character? Comparative analyses of the nematode parasites of mammals. *Oikos*, **80**, 28–300.

Brown, J.H. (1984). On the relationship between abundance and distribution of species. *American Naturalist*, **124**, 255–79.

Calvete, C. (2003). Correlates of helminth community in the red-legged partridge (*Alectoris rufa* L.) in Spain. *Journal of Parasitology*, **89**, 445–51.

Calvete, C., Blanco-Aguiar, J.A., Virgós, E., Cabezas-Díaz, S., and Villafuerte, R. (2004). Spatial variation in helminth community structure in the red-legged partridge (*Alectoris rufa* L.): effects of definitive host density. *Parasitology*, **129**, 101–13.

Carson, H.L. (1959). Genetic conditions that promote or retard the formation of species. *Cold Spring Harbor Symposia in Quantitative Biology*, **24**, 87–103.

Crystal, M.M. (1986). Artificial feeding of northern fowl mites, *Ornithonyssus sylviarum* (Canestrini and Fanzago) (Acari: Macronyssidae), through membranes. *Journal of Parasitology*, **72**, 550–54.

Dean, S.R. and Meola, R.W. (1997). Effect of juvenile hormone and juvenile hormone mimics on sperm transfer from the testes of the male cat flea (Siphonaptera: Pulicidae). *Journal of Medical Entomology*, **34**, 485–88.

Degen, A.A. (1997). *Ecophysiology of Small Desert Mammals*. Springer Verlag, Berlin.

Desdevises, I., Morand, S., and Legendre, P. (2002). Evolution and determinants of host specificity in the genus *Lamellodiscus* (Monogenea). *Biological Journal of the Linnean Society*, **77**, 431–43.

Dynesius, M. and Jansson, R. (2000). Evolutionary consequences of changes in species' geographical distributions driven by Milankovitch climate oscillations. *Proceedings of the National Academy of Sciences of the USA*, **97**, 9115–20.

Fox, L.R. and Morrow, P.A. (1981). Specialization: species property or local phenomenon? *Science*, **211**, 887–93.

Iqbal, Q.J. and Humphries, D.A. (1974). The mating behaviour of the rat flea *Nosopsyllus fasciatus* Bosc. *Pakistan Journal of Zoology*, **8**, 39–41.

Gaston, K.J. (2003). *The Structure and Dynamics of Geographic Ranges*. Oxford University Press, Oxford.

Hawkins, B.A. and Diniz-Filho, J.A.F. (2006). Beyond Rapoport's rule: evaluating range size patterns of New World birds in a two-dimensional framework. *Global Ecology and Biogeography*, **15**, 461–69.

Hengeveld, R. (1990). *Dynamic Biogeography*. Cambridge University Press, Cambridge.

Hillebrand, H. (2004). On the generality of the latitudinal diversity gradient. *American Naturalist*, **163**, 192–211.

Klein, S.L. and Nelson, R.J. (1998). Adaptive immune responses are linked to the mating system of arvicoline rodents. *American Naturalist*, **151**, 59–67.

Korallo-Vinarskaya, N.P., Krasnov, B.R., Vinarski, M.V., Shenbrot, G.I., Mouillot, D., and Poulin, R. (2009). Stability in abundance and niche breadth of gamasid mites across environmental conditions, parasite identity and host pools. *Evolutionary Ecology*, **23**, 329–45.

Kosminsky, R.B. (1965). Feeding and reproduction of fleas of house mice under natural and experimental conditions. *Zoologicheskyi Zhurnal*, **44**, 1372–75 (in Russian).

Krasnov, B.R. (2008). *Functional and Evolutionary Ecology of Fleas*. Cambridge University Press, Cambridge.

Krasnov, B.R., Khokhlova, I.S., Burdelova, N.V., Mirzoyan, N.S., and Degen, A.A. (2004a). Fitness consequences of density-dependent host selection in ectoparasites: testing reproductive patterns predicted by isodar theory in fleas parasitizing rodents. *Journal of Animal Ecology*, **73**, 815–20.

Krasnov, B.R., Mouillot, D., Shenbrot, G.I., Khokhlova, I.S., and Poulin, R. (2004b). Geographical variation in host specificity of fleas (Siphonaptera): the influence of phylogeny and local environmental conditions. *Ecography*, **27**, 787–97.

Krasnov, B.R., Mouillot, D., Shenbrot, G.I., Khokhlova, I.S., and Poulin R. (2005). Abundance patterns and coexistence processes in communities of fleas parasitic on small mammals. *Ecography*, **28**, 453–64.

Krasnov, B.R., Shenbrot, G.I. Khokhlova, I.S., and Poulin, R. (2006). Is abundance a species attribute of haematophagous ectoparasites? *Oecologia*, **150**, 132–40.

Krasnov, B.R., Stanko, M., and Morand, S. (2007). Host community structure and infestation by ixodid ticks: repeatability, dilution effect and ecological specialization. *Oecologia*, **154**, 185–94.

Krasnov, B.R., Shenbrot, G.I., Khokhlova, I.S., Mouillot, D., and Poulin, R. (2008a). Latitudinal gradients in niche breadth: empirical evidence from haematophagous ectoparasites. *Journal of Biogeography*, **35**, 592–601.

Krasnov, B.R., Shenbrot, G.I., Khokhlova, I.S., Vinarski, M.V., Korallo-Vinarskaya, N.P., and Poulin, R. (2008b). Geographical patterns of abundance: testing expectations of the "abundance optimum" model in two taxa of ectoparasitic arthropods. *Journal of Biogeography*, **35**, 2187–94.

Kucheruk, V.V. (1983), Mammal burrows: their structure, topology and use. *Fauna and Ecology of Rodents*, **15**, 5–54 (in Russian).

MacArthur, R.H. (1955). Fluctuations of animal populations and a measure of community stability. *Ecology*, **36**, 533–36.

MacArthur, R.H. (1972). *Geographical Ecology*. Princeton, Princeton University Press.

Marshall, A.G. (1981). *The Ecology of Ectoparasite Insects*. Academic Press, London.

Merino, S., Moreno, J., Vasquez, R.A. *et al*. (2008). Haematozoa in forest birds from southern Chile: latitudinal gradients in prevalence and parasite lineage richness. *Austral Ecology*, **33**, 329–40.

Mooring, M.S., Benjamin, J.E., Harte, C.R., and Herzog, N.B. (2000). Testing the interspecific body size principle in ungulates: the smaller they come, the harder they groom. *Animal Behaviour*, **60**, 35–45.

Morand, S. (1996). Life-history traits in parasitic nematodes: a comparative approach for the search of invariants. *Functional Ecology*, **10**, 210–18.

Morand, S. and Guégan, J.-F. (2000). Distribution and abundance of parasite nematodes: ecological specialization, phylogenetic constraints or simply epidemiology? *Oikos*, **88**, 563–73.

Peters, R.H. (1983). *The Ecological Implications of Body Size*. Cambridge University Press, Cambridge.

Poulin, R. (1993). The disparity between observed and uniform distibutions: a new look at parasite aggregation. *International Journal for Parasitology*, **23**, 937–44.

Poulin, R. (2006). Variation in infection parameters among populations within parasite species: intrinsic properties versus local factors. *International Journal for Parasitology*, **36**, 877–85.

Poulin, R. (2007). *Evolutionary Ecology of Parasites: From Individuals to Communities*, 2nd Edn. Princeton University Press, Princeton.

Poulin, R. and Dick, T.A. (2007). Spatial variation in population density across the geographical range in helminth parasites of yellow perch, *Perca flavescens*. *Ecography*, **30**, 629–36.

Poulin, R. and Mouillot, D. (2003). Parasite specialization from a phylogenetic perspective: a new index of host specificity. *Parasitology*, **126**, 473–80.

Poulin, R. and Mouritsen, K.N. (2003). Large-scale determinants of trematode infections in intertidal gastropods. *Marine Ecology Progress Series*, **254**, 187–98.

Poulin, P. and Rohde, K. (1997). Comparing the richness of metazoan ectoparasite communities of marine fishes: controlling for host phylogeny. *Oecologia*, **110**, 278–83.

Poulin, R., Krasnov, B.R., Shenbrot, G.I., Mouillot, D., and Khokhlova, I.S. (2006). Evolution of host specificity in fleas: is it directional and irreversible? *International Journal for Parasitology*, **36**, 185–91.

Riquelme, C., George-Nascimento, M., and Balboa, L. (2006). Morphometry and fecundity of *Profilicollis bullocki* Mateo, Cordova & Guzman 1982 (Acanthocephala: Polymorphidae) in sympatric coastal bird species of Chile. *Revista Chilena de Historia Natural*, **79**, 465–74.

Ruggiero, A. and Hawkins, B.A. (2006). Mapping macroecology. *Global Ecology and Biogeography*, **15**, 433–37.

Salkeld, D.J, Trivedi, M., and Schwarzkopf, L. (2008). Parasite loads are higher in the tropics: temperate to tropical variation in a single host-parasite system. *Ecography*, **31**, 538–44.

Seifertová, M., Vyskočilová, M., Morand, S., and Šimková, A. (2008). Metazoan parasites of freshwater cyprinid fish (*Leuciscus cephalus*): testing biogeographical hypotheses of species diversity. *Parasitology*, **135**, 1417–35.

Shaw, D.J. and Dobson, A.P. (1995). Patterns of macroparasite abundance and aggregation in wildlife populations: a quantitative review. *Parasitology*, **111**, S111–S127.

Shenbrot, G.I., Krasnov, B.R., Khokhlova, I.S., Demidova, T., and Fielden, L.J. (2002). Habitat-dependent differences in architecture and microclimate of the Sundevall's jird (*Meriones crassus*) burrows in the Negev Desert, Israel. *Journal of Arid Environments*, **51**, 265–79.

Sokal, R.R. and Rohlf, F.J. (1995). *Biometry*, 3rd Edn. W. H. Freeman and Co., New York.

Stevens, G.C. (1989). The latitudinal gradient in geographical range: how so many species coexist in the tropics. *American Naturalist*, **133**, 240–56.

Timms, R. and Read, A.F. (1999). What makes a specialist special? *Trends in Ecology and Evolution*, **14**, 333–34.

Vázquez, D.P. and Aizen, M.A. (2003). Null model analyses of specialization in plant-pollinator interactions. *Ecology*, **84**, 2493–501.

Vázquez, D.P. and Aizen, M.A. (2004). Asymmetric specialization: a pervasive feature of plant-pollinator interactions. *Ecology*, **85**, 1251–57.

Vázquez, D. and Stevens, R.D. (2004). The latitudinal gradient in niche breadth: concepts and evidence. *American Naturalist*, **164**, E1–E19.

Vázquez, D.P., Poulin, R., Krasnov, B.R., and Shenbrot, G.I. (2005). Species abundance patterns and the distribution of specialization in host-parasite interaction networks. *Journal of Animal Ecology*, **74**, 946–55.

Ward, S.A. (1992). Assessing functional explanations of host specificity. *American Naturalist*, **139**, 883–91.

Wilson, K., Bjørnstad, O.N., Dobson, A.P. *et al.* (2001). Heterogeneities in macroparasite infections: Patterns and processes. In P.J. Hudson, A. Rizzoli, B.T. Grenfell, H. Heesterbeek, and A.P. Dobson, eds. *The Ecology of Wildlife Diseases*, pp. 6–44. Oxford University Press, Oxford.

CHAPTER 9

Similarity and variability of parasite assemblages across geographical space

Robert Poulin and Boris R. Krasnov

9.1 Introduction

Saying that the similarity between two localities decreases with increasing distance between them is an almost trivial observation. Everyone would agree that, either within terrestrial or marine realms, both the abiotic characteristics of a locality and the composition of its biota are generally very similar to those of any nearby locality within a radius of, say, a few kilometres. And everyone would agree that the similarity between these two localities would be much lower if they were instead separated by hundreds or thousands of kilometres. Yet this obvious and well-known pattern underpins much of the large-scale habitat transitions and species turnover that account for the biogeography and diversity of living organisms (Nekola and White 1999; Morlon *et al.* 2008). The distance decay of similarity, that is, the rate at which similarity in species composition decreases with distance between localities, can also shed light on the relative importance of processes such as dispersal or niche-based community structure.

This chapter will explore the importance of this phenomenon for our understanding of the biogeography of parasitic organisms. First, we will briefly discuss the mechanisms that can generate decreases in similarity with increasing geographical distance, and the various ways of quantifying this decrease. We will then examine the available empirical evidence that similarity between parasite communities decreases with distance, and attempt to identify the main processes responsible for the observed patterns. Our goal is to establish decrease of similarity with distance as a cornerstone in parasite biogeography. To this end, we propose that the role of distance *per se* must be taken into account in investigations of other large-scale biogeographical patterns, such as latitudinal gradients or correlations of parasite diversity with environmental variables.

9.2 Basic processes

In terms of species composition, a decrease of similarity with distance corresponds to a decrease in the proportion of species shared by two communities as a function of the geographical distance between them (Fig. 9.1). Thus, nearby communities tend to have many species in common, whereas distant communities share very few, if any, species.

Several mechanisms can act, either solely or in combination, to produce a decrease of the similarity in species composition between two communities with an increase of the distance between them. Soininen *et al.* (2007) have grouped these into three general classes of mechanisms. First, community similarity may decrease with distance simply because the similarity of climatic and physicochemical variables also decreases with distance (e.g. Steinitz *et al.* 2006). Most climatic or environmental factors vary in space along some gradient (Fig. 9.1). Different species have different tolerances for abiotic factors, and each performs best under only a specific range of abiotic conditions. The free-living infective stages of parasites, for instance, can only survive and successfully infect a host within narrow

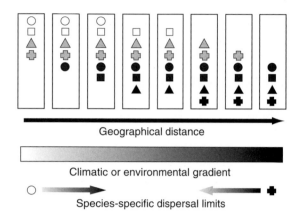

Geographical distance

Climatic or environmental gradient

Species-specific dispersal limits

Figure 9.1 Schematic representation of 8 communities (large rectangles) arranged in order of increasing distance. The species present in each community are indicated by different symbols. Nearby communities share many species, whereas distant communities have few species in common. This pattern can result from several processes, including associations between the presence/absence of species and existing climatic or environmental gradients, or differences in the dispersal potential of the different species.

and species-specific sets of environmental conditions (Pietrock and Marcogliese 2003). The geographical range of a species is therefore determined in large part by its ability to survive environmental conditions, and the ranges of different species overlap to the extent that these species share the same preferences for abiotic conditions, that is, to the extent that their niches overlap (Gilbert and Lechowicz 2004). This can lead to species-sorting in geographical space, with the sets of species occurring in different localities reflecting the match between local conditions and the species' tolerances of those conditions.

Second, the topography of the landscape can either facilitate or impede the dispersal of organisms among localities. Assuming that different species have similar rates of dispersal, we would expect that similarity would decrease at a low and even rate in an open landscape presenting no major physical barriers to dispersal. In contrast, community similarity should decrease more abruptly and rapidly in fragmented landscapes with major physical discontinuities. For example, all else being equal, the dispersal of fish parasites, and therefore the homogenization of parasite communities over space, should be less likely in freshwater fish than in marine coastal fish species. Freshwater habitats

such as lakes and rivers are physically isolated from each other, whereas localities along continental coasts are more or less connected along a continuous shoreline.

Third, even in homogeneous and continuous settings, with no environmental gradients or barriers to dispersal, community similarity would also decrease with distance because of the limited dispersal of organisms (Fig. 9.1). This is even true under neutral scenarios that assume that all species have the same dispersal abilities: random dispersal and ecological drift can on their own generate species distributions that yield decreasing similarity with distance (Hubbell 2001). In reality, different species do not have identical dispersal capabilities. The free-living infective stages of parasites generally have extremely limited mobility, and on a geographical scale, their direct dispersal abilities are essentially non-existent. However, it is the indirect dispersal of parasites via host movements that vary greatly among species. Using again the example of fish parasites in freshwater systems, phylogeographic studies have shown this convincingly. Based on estimates of gene flow, parasite species using birds as definitive hosts achieve more frequent and more extensive dispersal than those using fish as final hosts (Criscione and Blouin 2004). The use of birds as both hosts and dispersal agents frees parasites from the physical isolation normally associated with lake organisms. We might thus expect more pronounced decrease in community similarity with increasing distance among parasite assemblages using only hosts with limited dispersal, than among parasite assemblages using highly vagile hosts such as birds.

The three general processes described above (species-sorting along environmental gradients, dispersal constraints imposed by topography, and species differences in dispersal) are not mutually exclusive, and it is not easy to distinguish between them or assess their relative importance. For instance, observing stronger decreases in similarity among parasite communities utilizing freshwater fish hosts than among those in bird hosts would tell us that dispersal mediated by host movements matters, but it does not tell us whether environmental gradients or topography also play important roles. Similarly, differences in decreases of similarity between parasite communities of freshwater and marine fish hosts points towards a role for dispersal barriers, but does not rule out other mechanisms acting jointly with topographical structure. Soininen *et al.* (2007) have conducted a meta-analysis of the results of studies on distance decay relationships available to date, in an attempt to identify key properties of organisms or environments associated with the rates of decrease of similarity with distance. Most published studies focus on vascular plants in terrestrial systems; nevertheless, Soininen *et al.* (2007) managed to show that the observed rate of decay depends on the spatial scale of the study, the latitude at which communities are sampled, and the ecosystem studied (terrestrial versus marine). Most interestingly, they also found that the rate of decrease in community similarity was higher among organisms that are highly mobile than among sessile organisms capable only of passive dispersal. The few studies now available on parasite communities are summarized later in this chapter; perhaps when more of them become available, a meta-analysis similar to that of Soininen *et al.* (2007) will be possible, and we may then be able to pinpoint which mechanism(s) is/ are most important in generating distance decay of similarity.

9.3   The measurement of distance decay in similarity

The standard approach to quantify distance decay in similarity consists of two simple steps. First, one needs to calculate, for each possible pair of communities in a dataset, both the similarity in species composition between the communities and the distance (in km) between them. Second, the pairwise similarity measures are regressed against inter-community distance across all pairs of communities. Similarity in species composition is usually based on presence/absence data only, and is computed using either the Jaccard or Sorensen indices; both these measures range from 0 to 1, and essentially represent the proportion of species shared between two communities. When plotted against distance in normal (untransformed) parameter space, similarity usually decreases exponentially with increasing distance. When similarity values are log-transformed, or when both similarity and distance values are log-transformed, the relationship becomes linear (Fig. 9.2). The slope of the decay relationship, that is, that of the straight line in the log-log plot, was the first measure used for comparative purposes (Nekola and White 1999). Steeper slopes indicate a more rapid decrease in similarity with distance, and comparing the slopes obtained for different types of communities is one way to test

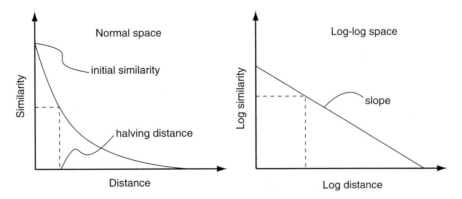

Figure 9.2 Schematic representation of distance decay of community similarity as a function of distance. The exponential curve visible in normal space becomes a linear relationship in log-normal or log-log space. The initial similarity corresponds to the intercept, and the halving distance to the distance at which similarity is half the maximum (initial) value; both these measures of the rate of decay can be obtained from either the plot in normal space or that in log-log space. The slope of the decay relationship is that of the straight line in the log-log plot.

hypotheses about, say, the role of dispersal in maintaining similarity over geographical space.

Recently, Soininen *et al.* (2007) have proposed two other measures for comparisons of similarity decay rates, both also derived from plots in either normal space, log-normal space, or log-log space (Fig. 9.2). The first one, initial similarity, should in principle correspond to the intercept, although in practice it is taken as the similarity at 1 km distance because it cannot otherwise be estimated from log-log plots. The second one, the halving distance, is the distance at which similarity is half the maximum (initial) value; it is therefore greatly dependent on the slope of the decay relationship. Both these measures are easy to obtain from any type of regressions of similarity versus distance. Initial similarity measures the differences in species composition between nearby communities, and may therefore reflect beta diversity on small scales: high initial similarity thus indicates low beta diversity (Soininen *et al.* 2007). In contrast, the halving distance captures the rate of species turnover per unit distance, and thus provides an estimate of the scale-dependency of beta diversity. Large halving distances suggest that the rate of species turnover does not change much with increasing spatial scale, whereas short halving distances indicate that species turnover is strongly scale-dependent (Soininen *et al.* 2007). The three quantitative measures of distance decay in similarity (slope, initial similarity, halving distance), although related and therefore not entirely independent of each other in a statistical sense, nevertheless each capture a slightly different aspect of the relationship between community similarity and geographical distance.

9.4 Empirical patterns of distance decay

The species richness of parasite communities, that is, the number of parasite species coexisting within one host population, generally appears to be a host species characteristic (Poulin and Mouritsen 2003; Poulin and Mouillot 2004; Bordes and Morand 2008). In other words, the number of parasite species per host population varies less across populations of the same host species than among host species. For a given host species, the number of parasite species supported by a host population

therefore varies within species-specific limits. This relative stability in species richness across conspecific host populations, however, does not mean that the species composition of the parasite assemblages is also stable. Different host populations harbour different subsets of the entire pool of parasite species known to exploit a particular host species. Although it has been shown that pairwise distances between host populations is often the best predictor of the similarity between their parasite communities (Poulin and Morand 1999), attempts to quantify the decay of similarity with geographical distance are relatively recent. Comparative data from all available studies of distance decay in parasite communities are summarized in Table 9.1, and their findings are discussed below.

Significant negative relationships between parasite community similarity and distance between host populations have been found in most, but not all, fish species studied to date (e.g. Poulin 2003; Oliva and González 2005; Pérez-del-Olmo *et al.* 2009). In many of these, the relationship is quite strong, with distance being a good predictor of the variation in pairwise similarity values (Fig. 9.3). In general, values for both initial similarity and halving distance are higher for marine fish species than for freshwater fish species (Table 9.1). This is consistent with the view that marine habitats allow much greater dispersal and homogenization of the fauna than freshwater habitats that are physically isolated in a fragmented landscape. In addition, connections between freshwater habitats are not necessarily determined by spatial proximity, as becomes apparent when comparing adjacent river systems. Barger (2007) sampled creek chub, *Semotilus atromaculatus*, from different sites within rivers, and found that drainage-level differences in parasite species composition provide better explanations for the spatial pattern of community similarity than does physical distance among sites.

This assumes that parasite dispersal across host populations can only occur via fish movements. However, parasites may use other host species as intermediate, definitive, or paratenic hosts, and use them as vehicles for the colonization of other nearby or distant localities. In freshwater fish, for example, parasite faunas comprising several species of helminths using birds as definitive hosts consist of

Table 9.1 Summary results from studies of distance decay of similarity in parasite communities.

Host	Parasites	Total no. parasite species	No. host populations[*]	Maximum distance (km)	Initial similarity	Halving distance (km)	Reference
MOLLUSCS							
Littorina littorea	Trematodes	10	35	1800	0.689	1214	Thieltges et al. 2008
Hydrobia ulvae	Trematodes	44	36	1520	0.383	1047	Thieltges et al. 2008
Cerastoderma edule	Trematodes	15	19	3900	0.614	2381	Thieltges et al. 2008
FRESHWATER FISH							
Perca flavescens	Endohelminths	38	21	2800	0.252	2410	Poulin 2003
Catostomus commersoni	Endohelminths	34	6	2850	—	—	Poulin 2003
Esox lucius	Endohelminths	32	6	4000	0.407	1905	Poulin 2003
Coregonus lavaretus	All metazoans	11	8	245	0.625	155	Karvonen and Valtonen 2004
Lepomis macrochirus	Endohelminths	29	25	1700	0.664	1107	Fellis and Esch 2005b
Leuciscus cephalus	All metazoans	62	15	2510	0.502	330	Seifertova et al. 2008
MARINE FISH							
Trachurus murphyi	All metazoans	21	7	3930	0.691	3455	Oliva and González 2005
Merluccius gayi	All metazoans	30	5	3300	0.804	1340	Oliva and González 2005
Sebastes capensis	Ectoparasites	15	7	4465	0.739	1848	Oliva and González 2005
Hippoglossina macrops	All metazoans	28	6	900	—	—	Oliva and González 2005
MAMMALS							
Ondatra zibethicus	Endohelminths	35	8	5000	—	—	Poulin 2003
Procyon lotor	Endohelminths	47	8	2800	0.449	1100	Poulin 2003
Canis latrans	Endohelminths	37	7	2950	0.549	1540	Poulin 2003
Mastomys natalensis	Nematodes	9	10	70	—	—	Brouat and Duplantier 2007
Mastomys erythroleucus	Nematodes	11	10	70	0.875	90	Brouat and Duplantier 2007
Apodemus agrarius	Fleas	30	8	4503	—	—	Krasnov et al. 2005
Apodemus uralensis	Fleas	37	6	4503	0.596	1860	Krasnov et al. 2005
Mus musculus	Fleas	40	14	6813	—	—	Krasnov et al. 2005
Arvicola terrestris	Fleas	31	8	3230	0.560	1717	Krasnov et al. 2005
Myodes glareolus[†]	Fleas	27	5	4481	—	—	Krasnov et al. 2005

(continued)

Table 9.1 Continued

Host	Parasites	Total no. parasite species	No. host populations[*]	Maximum distance (km)	Initial similarity	Halving distance (km)	Reference
Myodes rutilus	Fleas	44	7	5415	0.355	1380	Krasnov *et al.* 2005
Microtus arvalis	Fleas	40	9	4481	—	—	Krasnov *et al.* 2005
Microtus gregalis	Fleas	27	6	3997	0.487	1600	Krasnov *et al.* 2005
Microtus oeconomus	Fleas	27	7	5415	0.587	1684	Krasnov *et al.* 2005
Cricetulus migratorius	Fleas	50	8	3237	—	—	Krasnov *et al.* 2005
Neomys fodiens	Fleas	15	5	4481	—	—	Krasnov *et al.* 2005
Sorex araneus	Gamasid mites	27	7	4602	—	—	Vinarski *et al.* 2007
Apodemus agrarius	Gamasid mites	33	6	6305	—	—	Vinarski *et al.* 2007
Arvicola terrestris	Gamasid mites	26	9	4237	—	—	Vinarski *et al.* 2007
Clethrionomys glareolus	Gamasid mites	28	9	4602	0.738	4470	Vinarski *et al.* 2007
Clethrionomys rutilus	Gamasid mites	32	16	4980	—	—	Vinarski *et al.* 2007
Clethrionomys rufocanus	Gamasid mites	27	11	4980	—	—	Vinarski *et al.* 2007
Microtus arvalis	Gamasid mites	25	9	4429	—	—	Vinarski *et al.* 2007
Microtus gregalis	Gamasid mites	25	6	3174	—	—	Vinarski *et al.* 2007
Microtus oeconomus	Gamasid mites	25	15	5092	—	—	Vinarski *et al.* 2007
Ondatra zibethicus	Gamasid mites	13	7	6305	—	—	Vinarski *et al.* 2007
Sicista betulina	Gamasid mites	17	4	697	—	—	Vinarski *et al.* 2007

— No statistically significant decay of similarity with distance was observed.

[*] For N populations (or N localities sampled), the number of pairwise comparisons on which the results are based equals $N(N-1)/2$.

[†] Distance decay relationship based on the Morisita–Horn similarity index was significant, but not considered here.

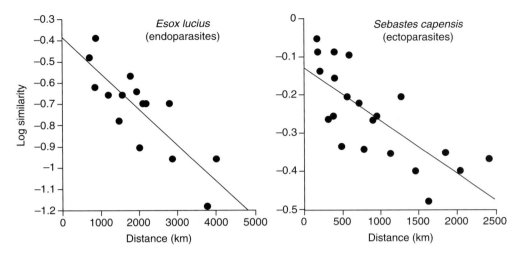

Figure 9.3 Distance decay of similarity among endoparasite communities from populations of a freshwater fish, the pike *Esox lucius*, in North American lakes (data from Poulin 2003), and among ectoparasite communities from populations of a marine fish, the red rockfish *Sebastes capensis*, sampled along the Pacific coast of South America (data from Oliva and González 2005). Each point represents a contrast between two localities, i.e. between two host populations.

more homogeneous and predictable local communities than faunas comprising mostly parasite species incapable of dispersing from one water body to another (Esch *et al.* 1988). Parasites can thus have a colonizing ability independent of the vagility of the fish host, a fact that increases the similarity among communities. Fish parasites using bird definitive hosts, or allogenic parasites, tend to form a more predictable species subset among communities than autogenic species, that is, those completing their life-cycle in water and incapable of crossing land barriers between freshwater bodies. In the whitefish *Coregonus lavaretus*, for example, communities of helminth parasites in different interconnected lakes share allogenic species independently of the distance separating them. In contrast, similarity in the composition of their autogenic species decreases significantly with increasing distance (Karvonen and Valtonen 2004). The same phenomenon has even been observed on a much smaller spatial scale, with the similarity of communities of autogenic parasites decreasing with increasing distance along the shore among fish samples from the same lake (Karvonen *et al.* 2005). A somewhat different pattern has been reported for helminth parasites in a different fish species. On a relatively small geographic scale, Fellis and Esch (2005a) found that the similar-

ity of communities of allogenic parasites in sunfish, *Lepomis macrochirus*, decreased with increasing distance among isolated ponds, whereas similarity in autogenic parasites was independent of inter-pond distances. On a much larger geographical scale, the similarity of both allogenic and autogenic parasites decreased with increasing distance between localities, but at different rates (Fellis and Esch 2005b). Unlike the situation in the interconnected lakes studied by Karvonen and Valtonen (2004), the physical isolation of the lakes and ponds in Fellis and Esch's (2005a, b) studies and the inability of autogenic parasites to disperse readily mean that measuring the distance decay in similarity of these parasites is greatly dependent on study scale.

Helminth communities in terrestrial mammal hosts have also received some attention, with significant distance decay relationships observed in several species, at scales ranging from tens to thousands of kilometres (Poulin 2003; Brouat and Duplantier 2007). In contrast, among the many mammal host species studied, significant decreases in the similarity of ectoparasitic arthropod communities (fleas and mites) as a function of increasing distance are rarely seen (Krasnov *et al.* 2005; Vinarski *et al.* 2007). Values of initial similarity or halving distance are no different from those reported for

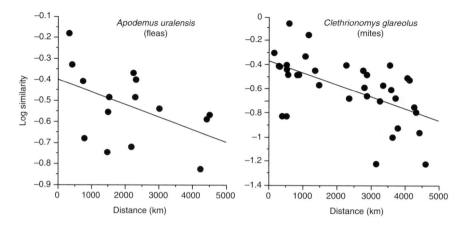

Figure 9.4   Distance decay of similarity among ectoparasitic flea communities from populations of the mouse *Apodemus uralensis* (data from Krasnov *et al.* 2005), and among ectoparasite gamasid mite communities from populations of the bank vole *Clethrionomys glareolus* (data from Vinarski *et al.* 2007); both studies were performed across localities in Eurasia. Each point represents a contrast between two localities, i.e. between two host populations.

helminth endoparasites, and thus the relationships reported are typical ones (Fig. 9.4). However, there are simply fewer host–parasite systems in which distance decay relationships are detected out of the many that have been investigated (Table 9.1). This may be due to fundamental differences between endo- and ectoparasites. The chances that an ectoparasite species becomes established in a locality may depend much more strongly on the local environmental conditions, since ectoparasites are exposed to external conditions throughout their lives. Mere geographical distance between two host populations may therefore be a poor proxy for the probability that they share ectoparasite species (see Section 9.5).

It would be interesting to compare the rates of distance decay in similarity among parasite communities of fish and mammals with those of parasite communities in bird hosts, given that birds are highly vagile and potentially capable of dispersing parasites over greater geographical areas. Unfortunately, there are at present no estimates of distance decay rates for parasite communities in bird hosts. Recently, Gómez-Diaz *et al.* (2008) have reported that the similarity of ectoparasite communities among breeding colonies of shearwaters, *Calonectris* spp., correlated negatively with the geographic distance between colonies. However, their measure of similarity differs from those in all other

studies: it is based on similarity in abundance or prevalence values of the four ectoparasite species that were present at all colonies. It does not examine parasite species composition, since there was no spatial variation in species composition among bird colonies.

Indirect evidence of the dispersing effect of bird hosts on the distance decay of parasite communities can be obtained from studies of parasite assemblages in the intermediate hosts of bird parasites. For instance, shore birds can visit several coastal locations within a short time period, potentially releasing eggs of helminth parasites in each location. It is difficult to assign one 'home' location to such birds, since, unlike fish in a pond, they can potentially travel several hundred kilometres within days. However, the snails serving as intermediate hosts to all of the numerous trematode species using birds as definitive hosts are essentially motionless on a geographic scale. Infected snails are only capable of limited movements, generally limited to an area several metres wide only; although the larval stages of snails can be planktonic, they are not infected by trematodes. Snails are therefore spatially fixed accumulators of trematodes, providing a long-term record of past bird visits to an area (Hechinger and Lafferty 2005). In the two coastal snails species investigated to date, similarity in trematode communities decreases with increasing

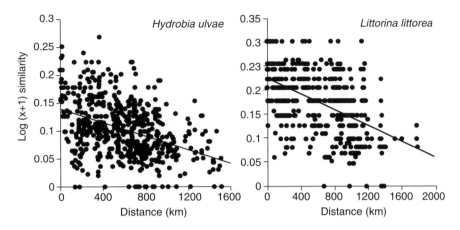

Figure 9.5   Distance decay of similarity among larval trematode communities from populations of two marine snails, the mudsnail *Hydrobia ulvae* and the periwinkle *Littorina littorea* (data from Thieltges *et al.* 2009); both snail species were sampled along the European coast. Each point represents a contrast between two localities, i.e. between two host populations.

distance, though with greater scatter among data points than for most other known relationships (Fig. 9.5). In particular, nearby localities were often found to harbour very different trematode faunas. Values of initial similarity and halving distance for trematode communities in these snails are not noticeably different from those reported for fish and mammal hosts (Table 9.1). The same is true for trematode communities in the cockle *Cerastoderma edule*, which acts as second intermediate host for many trematodes, providing an equally immobile link between snails and birds (Table 9.1). However, a more telling comparison is provided by contrasting the halving distance for trematode species in the snail *Hydrobia ulvae* that use birds as definitive hosts with those that use fish as definitive hosts: the halving distance of the latter is about half that of the former (Thieltges *et al.* 2009). Thus, trematode assemblages relying on fish for dispersal lose their similarity with increasing distance twice as fast as those relying on birds as dispersers. This suggests that, for these parasites at least, dispersal potential is a key determinant of community similarity in geographical space.

Although there are too few available data for a proper or comprehensive meta-analysis, some revealing patterns emerge from some simple statistical analyses of the data in Table 9.1. Firstly, the estimates of initial similarity compiled in Table 9.1 show a strong negative correlation with the number of

parasite species involved per study ($r = -0.638$, $p = 0.0024$). This may indicate that in systems with a large pool of parasite species, beta diversity is generally higher, that is, nearby communities generally differ more in species composition than in systems with a restricted species pool. Secondly, estimates of halving distance are clearly positively correlated with the spatial scale of the study, that is, with the maximum distance between two localities ($r = 0.622$, $p = 0.0034$). This echoes the findings of Soininen *et al.* (2007), and suggests that the choice of an appropriate spatial scale must be considered carefully prior to any study. Finally, the number of host populations included in analyses that have found a significant distance decay relationship (mean = 12.6 populations) is slightly higher than the number of populations in studies that have found no distance decay (mean = 8.5 populations; two-tailed *t*-test, $t = 826$, $p = 0805$); this weak pattern disappears after excluding the three studies on mollusc hosts. Nevertheless, there may still be simple issues of statistical power limiting our ability to detect distance decay relationships, and one should aim to include as many independent communities as possible in these analyses.

9.5   Decay along other distance measures

Geographical distance is a surrogate measure for the real processes that underlie similarity decay relationships, such as dispersal limits or environmental gradients. It is used because it is convenient, not because it is the most ecologically relevant measure. Other approaches used recently have focused on alternative 'distance' measures, and taken together, these provide much more information on the underlying processes than the mere use of geographical distance.

The first of these alternative approaches consists in measuring exchanges or movements of host individuals between localities more directly, by using estimates of gene flow. Frequent exchanges of individual hosts among nearby or distant host populations leave genetic footprints, detectable from molecular data. The 'genetic distance' between two populations of hosts provides a more accurate estimate of host movements than that inferred from geographical distance, and thus a more direct estimate of potential parasite dispersal. Seifertová *et al.* (2008) investigated patterns of similarity among communities of metazoan parasites from fish populations (chub, *Leuciscus cephalus*) across Europe. They looked at both geographical distance and genetic distance between host populations as potential predictor variables, and found that in a multivariate analysis, host genetic distances were the only significant determinant of similarity in parasite species composition: after controlling for the influence of genetic distances, there was no relationship left

between similarity values and geographic distance (Fig. 9.6). Although both geographical and genetic distances correlated negatively with similarity, geographic distance *per se* irrelevant, and only genetic distance truly matters for these parasites. This finding clearly points toward parasite dispersal via host movement as the key determinant of community similarity in geographical space.

In contrast, in their study of ectoparasites of birds (shearwaters, *Calonectris* spp.), Gómez-Diaz *et al.* (2008) found that the similarity in prevalence and abundance of the four ectoparasites shared across all localities depended strictly on geographical distances between them, and not on genetic distances among bird colonies. This result suggests that for these ectoparasites, abundance levels achieved locally may depend solely on local environmental conditions, the latter being more likely to resemble those of nearby localities, and not on dispersal probabilities.

This brings us to the second alternative approach to the study of decay in community similarity: the proportion of shared parasite species between two localities (that is, between two host populations) may depend on the 'environmental distance' between them. This environmental distance can be measured easily as the pairwise difference in local conditions between localities, providing a more

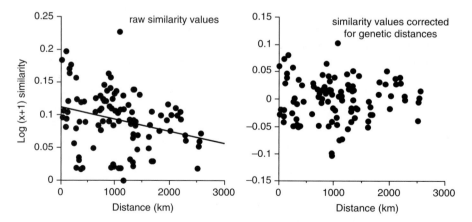

Figure 9.6 Distance decay of similarity among metazoan parasite communities from populations of a freshwater fish, the chub *Leuciscus cephalus*, across Europe (data from Seifertová *et al.* 2008). When 'raw' similarity values are plotted against geographical distance, the usual negative relationship is observed; however, when similarity values are corrected for genetic distances (by taking the residuals of the similarity versus genetic distance regression, where genetic distances are estimated from divergence in DNA sequences), the relationship disappears. Each point represents a contrast between two localities, i.e. between two host populations.

relevant measure of the possible role of environmental gradients in generating species sorting. One can focus on the biotic environment, and calculate for instance the difference between pairs of localities with respect to the taxonomic diversity of host taxa. The persistence of generalist parasites, that is, parasites capable of exploiting several host species, in a given locality may depend on the availability of alternative host species, and thus similarity in parasite communities between localities may mirror the similarity in host communities. In addition, one can also quantify the 'distance' in the abiotic environment between pairs of localities, by measuring the differences in variables such as temperature, rainfall, elevation, etc. (Steinitz *et al.* 2006). These abiotic variables directly affect the survival and infectivity of transmission stages in helminth parasites, and can also modulate population dynamics of ectoparasites.

Krasnov *et al.* (2005) investigated the similarity in flea communities across populations of the same host species, separately for several different host species. They found that similarity in flea communities was equally well, if not better, explained by differences in rodent host faunal composition between localities, than by mere geographical distance. Similarly, Vinarski *et al.* (2007) found that 'environmental distance' (both in terms of local host fauna and abiotic variables) generally explained more of the variance in similarity of ectoparasitic mite communities on rodents than geographic distance between host populations. The use of these more sophisticated measures of 'distance' shines a light on the mechanisms responsible for distance decay patterns: in the case of fleas and mites parasitic on rodents, the presence of particular species in a locality appears to depend much more on the locality's characteristics than on pure dispersal processes.

9.6   Concluding remarks

Distance decay relationships have now been documented for a broad range of host and parasite taxa. The empirical evidence reviewed in this chapter suggests that these relationships may have different underlying causes, depending on the type of parasites. For instance, it appears that similarity in communities of endoparasitic helminths in fish hosts are maintained by host movement (genetic distance between host populations), whereas similarity in ectoparasite communities of terrestrial hosts are more likely to depend on the 'environmental distance' (both biotic and abiotic) between host populations. It is still too early to draw any conclusion from the few available studies, however.

Future studies should go beyond looking merely at similarity in species composition. In particular, it will be crucial to look at both the abundance of different parasite species in different communities, and at their taxonomic identity. First, it should be clear that two parasite communities that consist of the exact same species actually differ considerably if these species occur at vastly different relative abundances. For any given parasite species, the mean number of individual parasites per host (abundance) or the percentage of hosts infected by the parasite (prevalence) appear to be true species traits, since they only vary within species-specific limits (Arneberg *et al.* 1997; Poulin 2006; Krasnov *et al.* 2006). However, abundance and prevalence still vary in space and time; for instance, these values often peak close to the centre of a parasite species' geographical range and decrease toward its edges (Poulin and Dick 2007; Krasnov *et al.* 2008). Future studies of distance decay relationships should try to incorporate data on parasite abundance in measures of similarity between communities, instead of relying solely on presence/absence data and the simple Jaccard index (see Chao *et al.* 2005).

Second, it is important to recognize that even if two communities have no parasite species in common, they may still be very similar. Decay in community similarity with increasing distance may have little impact at the functional level if the turnover of species involves a gradual replacement of certain species over distance with other species that are their ecological equivalent. For instance, two distant communities of parasites may share no species in common, but one of them may consist of species taxonomically close (e.g. congeners) to those found in the other community, playing similar roles in the community, and having similar impacts on the hosts. Similarity between communities should thus take into account the phylogenetic affinities of their species. It is perfectly reasonable to postulate that higher-level taxonomic similarity and possibly also functional similarity between parasite communities may be more resilient to increasing spatial distances. The replacement of species in parasite communities over hundreds or thousands of kilometres may be constrained within taxonomic limits: parasite species of host populations found in one part of a continent or ocean could be replaced by taxonomically (or functionally) equivalent species elsewhere. The net result would be that although species similarity decays rapidly with increasing geographical distance, the higher-level taxonomic and functional composition of parasite communities would show a greater stability across large spatial scales. This certainly needs to be examined using similarity indices that incorporate phylogenetic distances between sets of species in different communities, rather than the all-or-nothing approach based on species presence/absence data.

In spite of the need for further refinements in the study of distance decay relationships, it is clear that the decrease in parasite community similarity with increasing geographical distance is a widespread, if not universal, pattern. We feel that it may underpin a range of other biogeographical patterns, and we propose that distance decay should be incorporated as an underlying phenomenon in other biogeographical studies (see also Morlon *et al.* 2008). Distance decay relationships provide simple null patterns, expected to exist on scales of hundreds or thousands of kilometres, even in homogeneous habitats since they can arise from passive dispersal processes. For example, consider latitudinal gradients in parasite species diversity (Rohde 2002): How much of these are simple by-products of distance decay in similarity stemming from limited dispersal or environmental gradient? Is there any need to invoke more complex mechanisms to account for the existence of latitudinal gradients? Or, more realistically, how much of any given gradient is explained by the mechanisms usually proposed, and how much is the simple product of distance-based processes? Parasite biogeography would gain from a wider recognition that the mere linear distance between two points can explain much of their biotic similarities and differences.

References

Arneberg, P., Skorping, A., and Read, A.F. (1997). Is population density a species character? Comparative analyses of the nematode parasites of mammals. *Oikos*, **80**, 289–300.

Barger, M.A. (2007). Congruence of endohelminth community similarity in creek chub (*Semotilus atromaculatus*) with drainage structure in southeastern Nebraska. *Comparative Parasitology*, **74**, 185–93.

Bordes, F. and Morand, S. (2008). Helminth species diversity of mammals: parasite species richness is a host species attribute. *Parasitology*, **135**, 1701–05.

Brouat, C. and Duplantier, J.M. (2007). Host habitat patchiness and the distance decay of similarity among gastrointestinal nematode communities in two species of *Mastomys* (southeastern Senegal). *Oecologia*, **152**, 715–20.

Chao, A., Chazdon, R.L., Colwell, R.K, and Shen, T.-J. (2005). A new statistical approach for assessing similarity of species composition with incidence and abundance data. *Ecology Letters*, **8**, 148–59.

Criscione, C.D. and Blouin, M.S. (2004). Life cycles shape parasite evolution: comparative population genetics of salmon trematodes. *Evolution*, **58**, 198–202.

Esch, G.W., Kennedy, C.R., Bush, A.O., and Aho, J.M. (1988). Patterns in helminth communities in freshwater fish in Great Britain: alternative strategies for colonization. *Parasitology*, **96**, 519–32.

Fellis, K.J. and Esch, G.W. (2005a). Autogenic-allogenic status affects interpond community similarity and species-area relationship of macroparasites in the bluegill sunfish, *Lepomis macrochirus*, from a series of freshwater ponds in the Piedmont area of North Carolina. *Journal of Parasitology*, **91**, 764–67.

Fellis, K.J. and Esch, G.W. (2005b). Variation in life cycle affects the distance decay of similarity among bluegill sunfish parasite communities. *Journal of Parasitology*, **91**, 1484–86.

Gilbert, B. and Lechowicz, M.J. (2004). Neutrality, niches, and dispersal in a temperate forest understory. *Proceedings of the National Academy of Sciences of the USA*, **101**, 7651–56.

Gómez-Diaz, E., Navarro, J., and González-Solis, J. (2008). Ectoparasite community structure on three closely related seabird hosts: a multiscale approach combining ecological and genetic data. *Ecography*, **31**, 477–89.

Hechinger, R.F. and Lafferty, K.D. (2005). Host diversity begets parasite diversity: bird final hosts and trematodes in snail intermediate hosts. *Proceedings of the Royal Society of London B*, **272**, 1059–66.

Hubbell, S.P. (2001). *The Unified Neutral Theory of Biodiversity and Biogeography*. Princeton University Press, Princeton.

Karvonen, A., Cheng, G.-H., and Valtonen, E.T. (2005). Within-lake dynamics in the similarity of parasite assemblages of perch (*Perca fluviatilis*). *Parasitology*, **131**, 817–23.

Karvonen, A. and Valtonen, E.T. (2004). Helminth assemblages of whitefish (*Coregonus lavaretus*) in interconnected lakes: similarity as a function of species-specific parasites and geographical separation. *Journal of Parasitology*, **90**, 471–76.

Krasnov, B.R., Shenbrot, G.I., Mouillot, D., Khokhlova, I.S., and Poulin, R. (2005). Spatial variation in species diversity and composition of flea assemblages in small mammalian hosts: geographic distance or faunal similarity? *Journal of Biogeography*, **32**, 633–44.

Krasnov, B.R., Shenbrot, G.I., Khokhlova, I.S., and Poulin, R. (2006). Is abundance a species attribute? An example with haematophagous ectoparasites. *Oecologia*, **150**, 132–40.

Krasnov, B.R., Shenbrot, G.I., Khokhlova, I.S., Vinarski, M.V., Korallo-Vinarskaya, N.P., and Poulin, R. (2008). Geographical patterns of abundance: testing expectations of the "abundance optimum" model in two taxa of ectoparasitic arthropods. *Journal of Biogeography*, **35**, 2187–94.

Morlon, H., Chuyong, G., Condit, R. *et al.* (2008). A general framework for the distance-decay of similarity in ecological communities. *Ecology Letters*, **11**, 904–17.

Nekola, J.C. and White, P.S. (1999). The distance decay of similarity in biogeography and ecology. *Journal of Biogeography*, **26**, 867–78.

Oliva, M.E. and González, M.T. (2005). The decay of similarity over geographical distance in parasite communities of marine fishes. *Journal of Biogeography*, **32**, 1327–32.

Pérez-del-Olmo, A., Fernández, M., Raga, J.A., Kostadinova, A., and Morand, S. (2009). Not everything is everywhere: the distance decay of similarity in a marine host-parasite system. *Journal of Biogeography*, **36**, 200–09.

Pietrock, M. and Marcogliese, D.J. (2003). Free-living endohelminth stages: at the mercy of environmental conditions. *Trends in Parasitology*, **19**, 293–99.

Poulin, R. (2003). The decay of similarity with geographical distance in parasite communities of vertebrate hosts. *Journal of Biogeography*, **30**, 1609–15.

Poulin, R. (2006). Variation in infection parameters among populations within parasite species: intrinsic properties versus local factors. *International Journal for Parasitology*, **36**, 877–85.

Poulin, R. and Dick, T.A. (2007). Spatial variation in population density across the geographical range in helminth parasites of yellow perch *Perca flavescens*. *Ecography*, **30**, 629–36.

Poulin, R. and Morand, S. (1999). Geographic distances and the similarity among parasite communities of conspecific host populations. *Parasitology*, **119**, 369–74.

Poulin, R. and Mouillot, D. (2004). The evolution of taxonomic diversity in helminth assemblages of mammalian hosts. *Evolutionary Ecology*, **18**, 231–47.

Poulin, R. and Mouritsen, K.N. (2003). Large-scale determinants of trematode infections in intertidal gastropods. *Marine Ecology Progress Series*, **254**, 187–98.

Rohde, K. (2002). Ecology and biogeography of marine parasites. *Advances in Marine Biology*, **43**, 1–86.

Seifertová, M., Vyskočilová, M., Morand, S. and Šimková, A. (2008). Metazoan parasites of freshwater cyprinid fish (*Leuciscus cephalus*): testing biogeographical hypotheses of species diversity. *Parasitology*, **135**, 1417–35.

Soininen, J., McDonald, R., and Hillebrand, H. (2007). The distance decay of similarity in ecological communities. *Ecography*, **30**, 3–12.

Steinitz, O., Heller, J., Tsoar, A., Rotem, D., and Kadmon, R. (2006). Environment, dispersal and patterns of species similarity. *Journal of Biogeography*, **33**, 1044–54.

Thieltges, D.W., Ferguson, M.A.D., Jones, C.S., Krakau, M., de Montaudouin, X., Noble, L.R., Reise, K., and Poulin, R. (2009). Distance decay of similarity among parasite communities of three marine invertebrate hosts. *Oecologia*, **160**, 163–73.

Vinarski, M.V., Korallo, N.P., Krasnov, B.R., Shenbrot, G.I., and Poulin, R. (2007). Decay of similarity of gamasid mite assemblages parasitic on Palaearctic small mammals: geographic distance, host species composition or environment? *Journal of Biogeography*, **34**, 1691–700.

CHAPTER 10

Gap analysis and the geographical distribution of parasites

Mariah E. Hopkins and Charles L. Nunn

10.1 Introduction

Studies of host–parasite biogeography are influenced greatly by geographically inconsistent sampling patterns, especially if studies combine patterns of local variation to make inferences at regional or global scales. A variety of factors result in heterogeneous sampling across space. For example, parasite sampling is often limited by logistical factors that include a lack of suitable roads or airports, risks arising from unstable political climates, and difficulty in acquiring and preserving samples in remote locations, all of which produce gaps in our knowledge of parasite distributions. In addition, it is often easier to obtain funding to study parasites that have large economic impacts, including the potential for transmission to humans (zoonoses). These funding opportunities produce sampling biases towards particular parasite species, as well as towards hosts that harbour parasites that can also infect humans or domesticated animals. Last, any of these factors can change through time, producing temporal variation that can further complicate studies of parasite biogeography.

In this chapter, we focus on quantifying these biases. Specifically, we demonstrate how global gap analysis—a method used in conservation biology to identify conservation targets—can be applied to identify and quantify bias in geographic sampling for parasites. We review and illustrate the principles of gap analysis by describing its recent application to identify geographic gaps in our knowledge of primate parasites (Hopkins and Nunn 2007). We also provide new analyses to demonstrate how

optimality techniques can be applied within a gap analysis framework to guide future sampling, namely by targeting areas where additional sampling would reduce geographic bias in large-scale datasets. In addition to prioritizing sampling sites to correct biases in the future, the results from gap analyses are important for developing spatially-explicit statistical corrections for sampling bias in host–parasite biogeography.

10.1.1 Goals of spatial analyses of parasite sampling effort

Understanding the evolutionary diversification of parasites and their hosts requires a better understanding of the geographic distribution of parasites in relation to host characteristics or ecological factors that vary at continental or global scales (Gregory 1990; Poulin 1997; Poulin and Morand 2000). At the most basic level, parasite distributions are inextricably linked to the distributions of their hosts (Guegan and Kennedy 1996; Poulin 1997; Lindenfors et al. 2007). Thus, any analysis of geographic sampling patterns for parasites must first take into account host ranges. Detailed analyses of parasite sampling may also choose to take into account whether parasites have been sampled across a diversity of host characteristics, such as body mass, population density, home range size, and diet, as these factors have been suggested to impact parasite species richness (Poulin and Morand 2000; Nunn et al. 2003; Araujo and Guisan 2006). Furthermore, ensuring that parasite sampling incorporates environmental gradients, such as distance from the equator, temperature, precipitation, and habitat variability may also be critical to understanding host–parasite biogeography (Guernier *et al.* 2004; Nunn *et al.* 2005; Nunn and Altizer 2006; Lindenfors *et al.* 2007; Krasnov *et al.* 2008).

When evaluating host–parasite biogeography in relation to host or environmental factors that vary at continental scales, it is essential to ensure that sampling effort has been distributed evenly or in proportion to the expected abundance of parasites, and to take remedial action if sampling biases are present. In other words, when interpreting observed trends in parasite–host biogeography, one has to be careful that the patterns generated do not simply reflect taxonomic or geographic biases in research effort (Poulin and Morand 2000; Burke 2007). The likelihood of such biases is expected to increase with the scale of the analysis. At a global scale, biases are especially likely because it is usually infeasible to achieve complete sampling; some regions, taxonomic groups, or subsets of parasites with different transmission modes are likely to be sampled better than others.

A number of methods to correct for uneven sampling effort have been developed, particularly in the context of studying parasite species richness (Walther *et al.* 1995; Poulin 1998; Walther and Morand 1998; Walther and Martin 2001; Cam *et al.* 2002; Robertson and Barker 2006; Lobo 2008). These methods range in complexity from simply including the number of sampled sites or individuals as independent factors in regression analyses, to more complex adjustments such as developing accumulation curve models and non-parametric estimators of species richness. Research has shown that certain performance estimates do better than others at low sampling effort (Walther and Morand 1998). However, few studies have incorporated *spatially-explicit* examinations of geographic variation in sampling effort when correcting for sampling bias. Here, we take the first step in this regard, namely by identifying methods that can help to quantify biases; once such biases are known, additional methods can be developed to control for the biases, which will ultimately depend on the question at hand.

In order to be most effective, measures of geographic bias must move beyond simply counting the number of sites or ecosystems in which a parasite species has been sampled, to spatially-explicit

analyses that take into account both the characteristics of each sampling site and its actual location. These spatially-explicit explorations of geographic bias can provide critical information that is currently unaccounted for in global studies of parasites, and is the essential first step in the development of effective corrections for geographic sampling effort in studies at these larger regional or global scales. For example, using measures such as the number of publications on a particular host or parasite species, or the number of sampling localities at which a species has been sampled, leaves out critical information such as the following: Were the sites at which sampling occurred extremely close together or far apart? Which hosts were present at that locality, and how many of those hosts were actually sampled? Were the sites at which sampling occurred representative of the range of environmental conditions in which the host/parasite occurs? What percentage of potential microhabitats in which the host occurs was sampled?

Developing an approach to correct for geographic bias that can incorporate answers to these questions requires that geographic bias is first quantified within a spatially-explicit framework that can compare the distribution of sampling localities to the distribution of factors thought to influence host–parasite biogeography (e.g. host ranges and environmental characteristics). Gap analysis provides such a framework (Jennings 2000; Funk and Richardson 2002; Funk *et al.* 2005).

10.1.2 Gap analysis

Gap analysis provides a conceptual, technical, and organizational basis for identifying and quantifying gaps between two or more spatial distributions (Jennings 2000). In evaluating gaps between distributions, gap analysis may incorporate traditional measures of spatial data analysis, such as display mapping and spatial statistics. However, these traditional measures form only a part of the gap analysis framework. Gap analysis encompasses an entire process ranging from the establishment of an 'optimal' distribution to the comparison of the observed and 'optimal' distributions and the subsequent targeting of actions to remedy gaps between these distributions.

Early gap analyses were developed to solve location-allocation (i.e. distance-based) problems in a variety of fields including operations research and transportation engineering (Tansel *et al.* 1983; Brandeau and Chui 1989). For example, urban planners use gap analysis to determine where cities should build fire stations, with the goal of minimizing the distance between all homes and the nearest fire station. Recent extensions of gap analysis have moved beyond purely distance-based location-allocation problems to incorporate attribute data as well. For example, conservation biologists use geographic distributions of attribute values such as habitat type, species richness, and projected levels of environmental change to distribute protected areas in such a way that the highest amount of biodiversity is conserved (Scott *et al.* 1987; Ferrier 2002; Rodrigues *et al.* 2004; Sarkar *et al.* 2006).

Gap analysis has more recently been applied to guide spatial patterns in biological sampling efforts, including sampling for disease (Funk and Richardson 2002; Funk *et al.* 2005; Hortal and Lobo 2005; Hopkins and Nunn 2007). These studies typically use both location and attribute data to determine whether sampling has been distributed along critical host or environmental gradients. They also use optimization techniques to identify future sampling sites that have the highest probability of yielding additional biodiversity (Funk *et al.* 2005; Hortal and Lobo 2005).

10.2 Gap analysis methods: a case study examining geographic patterns of sampling for primate parasites

In the following sections, we illustrate the most common steps of a gap analysis, providing illustrative examples based on our recent study of parasite sampling across primates (Hopkins and Nunn 2007).

10.2.1 Selection of an 'optimal distribution'

Gap analyses of parasite sampling effort rest on the assumption that while uniform data sampling may be suitable for studies of parasites at local or regional scales, uniform sampling is neither feasible nor the most representative sampling technique at continental or global scales. Instead, proportional sampling techniques may better inform studies of global biogeography (Schoereder *et al.* 2004; Hortal and Lobo 2005). For example, since parasite distributions inevitably follow host distributions, one could argue that parasite sampling intensity should be allocated geographically in proportion to host diversity; in such a case, we would want to sample most heavily where the most primate species occur. Alternatively, since ecological characteristics (such as population density or the diversity of habitat types) have been hypothesized to increase parasite richness in a host species, it could be argued that comparatively more parasite sampling effort should be devoted to hosts in areas with higher host abundance and/or greater ecological diversity. We conducted a gap analysis based on the assumption that global patterns of parasite sampling should be allocated proportionally to host diversity (i.e. species richness; Hopkins and Nunn 2007). In the following sections, we provide examples from this research to illustrate typical gap analysis methods.

10.2.2 Data acquisition

Once the theoretical optimal distribution has been selected, the next step in any global gap analysis involves data acquisition. While this step may seem self-evident, we include discussion of it here because in studies of the global biogeography of host–parasite interactions, the acquisition of spatially-explicit sampling data may perhaps be the greatest challenge. Global GIS clearinghouses increasingly distribute information on environmental characteristics (e.g. Earth Resources Observation Systems Data Center, The Geography Network, National Geospatial Intelligence Agency Products and Services, and Tropical Rain Forest Information Center). However, data on parasite sampling is often derived from literature searches where authors frequently fail to geo-reference their sampling sites. For example, we obtained data for our analyses on primate parasite sampling from the Global Mammal Parasite Database (GMPD; Nunn and Altizer 2005). Although this represents the most comprehensive database of parasites in wild primates, sufficient spatial information (coordinates or a unique locality name) to geo-reference parasite sampling sites was lacking in approximately one half of all primate parasite studies. This finding is consistent with previous studies of biodiversity sampling, which have urged scientists to adopt systematic geo-referencing methods when collecting samples (Araujo and Guisan 2006; Guralnick *et al.* 2007). This lack of geo-referenced sampling localities considerably limits the amount of data that can be used in spatially-explicit analyses of host–parasite biogeography. However, while spatially-explicit analyses of sampling gaps cannot incorporate all previous studies, they do illustrate geographic sampling trends and can guide future data collection efforts.

10.2.3 Display mapping

Increasingly widespread use of Geographic Information Systems (GIS) allows for the illumination of patterns that would be more difficult to discern if the data were analysed in tabular form. For example, mapping the distribution of primate parasite sampling using data from Hopkins and Nunn (2007) indicates that primates have been over-sampled in East Africa, and have been comparatively under-sampled in Southeast Asia (Fig. 10.1a). Gradient or proportional symbol maps are commonly used to further illustrate numeric discrepancies between localities (Rodrigues *et al.* 2004; Rinaldi *et al.* 2006). In our example, applying a proportional circle mapping technique to the number of primate parasite sampling records at each locality further emphasizes the higher abundance of studies on primate parasites in Africa, as compared to Asia and South America (Fig. 10.1b).

10.2.4 Quantifying spatial distributions

While display maps yield general trends, further statistical analysis is needed to quantitatively compare two spatial distributions. Spatial statistics can be applied to continuous data, point data, or tessellated/polygon data, in order to measure both first and second order spatial effects. First order effects refer to spatial variation in the mean value of a process (i.e. a global trend). Second order effects refer to the spatial correlation structure or spatial

dependence within the dataset, which may cause deviations from the global trend in specific smaller regions. A number of spatial statistics software modules have recently become available, some of which are present within GIS frameworks (e.g. ESRI's Geostatistical Analyst, Spatial Analyst, and Spatial Statistics toolboxes, GEODA, SpaceStat, mapping and spatial statistics modules for Matlab, S-Plus, and R; see Anselin 2005 for a discussion of available spatial statistics GIS modules). We discuss several of the most common spatial statistical measures here that can be used to quantify geographic patterns of parasite sampling by measuring the degree of spatial clumping or correlation present in the dataset(s):

(1) *Summary statistics:* When analysing geographic datasets, it may be useful to identify anomalous or highly variable regions by calculating regional summary statistics. In this case, the entire dataset is usually divided into local regions (also called 'windows' or 'neighbourhoods') of a size specified by the researcher. Rectangular windows are used for ease of calculation and the size of the window depends on the overall dimensions of the area being studied, as well as the average distances between data locations (Isaaks and Srivastava 1989). Windows can be overlapped for small datasets or irregularly spaced data, which produces a smoothing effect that can easily quantify regional trends and isolate outliers. For example, Fig. 10.2a quantifies the intensity of primate parasite sampling points in dense areas such as East Africa relative to sparsely populated areas such as East Asia, by applying a moving mean statistic with a 5 x 5 decimal degree window.

(2) *Measures of clumping and dispersion:* When summary statistics indicate clumping of data points as in the primate parasite example, spatial statistics can be applied to quantitatively measure the degree of clumping and to test whether the data are significantly more clumped than expected (i.e. based on random or uniform distributions). Measures can be applied both globally and to local neighbourhoods or regions. Examples include: Kernel estimation, nearest neighbour distances, K-function approaches, the Clark–Evans test, the Cuzick and Edwards method, the GAM/K method, and spatial scan statistics (Bailey and Gatrell 1995; Kulldorff and

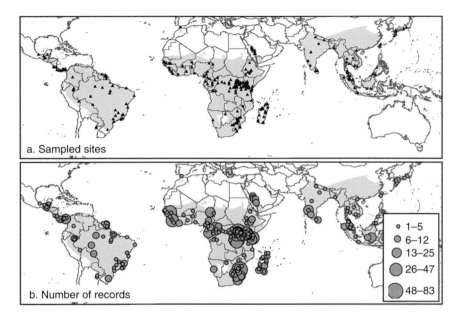

Figure 10.1   Geographic distribution of sampling for primate parasites. (a) Distribution of primate parasite sampling points. Redrawn from Hopkins and Nunn (2007), using primate parasite records added to the Global Mammal Parasite Database (Nunn and Altizer 2005) prior to 2009. (b) Distribution of primate parasite sampling points, weighed by the number of records in the GMPD at each locality.

Nagarwalla 1995; Cuzick and Edwards 1996; Openshaw *et al.* 1999; Ward and Carpenter 2000; Anselin 2005). These estimates have been critical in identifying clusters of disease in human and animal populations (Kulldorff and Nagarwalla 1995; Cuzick and Edwards 1996; Rinaldi *et al.* 2006). In field research, however, the researcher decides on the distribution of sample points. Hence, the question is often not whether sample points are more clumped than random or uniform distributions. Rather, the question of interest becomes whether attribute values at some points are more similar to the values at closer points than the values at farther points. When attribute values are spatially clustered, this phenomenon is termed *spatial autocorrelation*.

(3) *Spatial autocorrelation:* Measures of spatial autocorrelation can not only inform the researcher as to whether clustering in attribute values exists, but also to the scale and direction of that effect. In positive associations, attribute values increase in similarity the closer the sampling points. In negative associations, dissimilar values are found in close spatial association. Two of the most common meas-

ures of global spatial autocorrelation are Moran's I and Geary's C (Moran 1950; Geary 1954). The most common measures of local spatial autocorrelation are collectively known as the LISA statistics (Local Indicators of Spatial Association), and include both local Moran and Geary statistics as well as the Getis–Ord statistics (Getis and Ord 1992; Anselin 1995). These measures of spatial autocorrelation have been central to many epidemiological efforts seeking to identify spatial clumping in disease prevalence or intensity (Guernier *et al.* 2004; Zhang and Lin 2007; Crighton *et al.* 2008). In studies of parasite sampling effort, measures of spatial autocorrelation may be most useful as a means to identify violation of standard statistical assumptions when applying non-spatial statistical models to spatial data (see Section 10.2.5).

10.2.5   Data modeling: comparing two or more distributions

Measuring spatial correlation

Studies of host–parasite biogeography often seek to correlate geographic attributes (e.g. host diversity or environmental characteristics) with parasite species richness, prevalence, intensity, or abundance. Understanding patterns of spatial correlation is also important in studies that aim to identify the factors that drive spatial patterns of parasite sampling, or to investigate whether observed sampling distributions correlate with optimal sampling distributions. Researchers often attempt to address correlation in spatial variables with standard methods that do not incorporate a spatial component. However, spatial data often violate the central assumption of many standard statistical procedures, namely the independence of data points. Non-independent data can yield faulty statistical results, and they can result in lower statistical power than models that incorporate spatial information.

Thus, statistical analyses of spatially-distributed data should allow for the possibility that two data points may have similar values not due to one or more of the explanatory variables, but because the distance between these two data points is very small (i.e. spatial autocorrelation in the response variable is present). When using spatial data in statistical tests that are not spatially-explicit (e.g. OLS regression), tests for autocorrelation of standard model residuals should be conducted. These tests can include the Moran's I or Geary's C methods discussed above, as well as Lagrange Multipliers (Anselin 1988; Anselin et al. 1996).

If spatial autocorrelation is present, spatial models often result in better model fits (e.g. for regression: simultaneous autoregressive models (SAR) or conditional autoregressive models (CAR)). For example, when conducting a regression analysis of the relationship between species richness ('the optimal distribution') and the intensity of primate parasite sampling ('the observed distribution'), we found that the density of primate parasite sampling localities was correlated with host species richness, but that residuals from this regression model were spatially autocorrelated (Hopkins and Nunn 2007). When we applied an appropriate spatial regression model, the explanatory power of the model increased from $r^2 = 0.008$ to $r^2 = 0.05$, but still was relatively poor, indicating that 95 per cent of the variation in sampling effort could not be explained by host distributions. If the researchers are confident of model specifications, the residuals from spatial models can themselves serve as a

quantification of the relationship between an optimal distribution and observed sampling distribution. However, traditional gap analyses also commonly use spatial layer manipulations within a GIS to quantify and illustrate differences between distributions.

Spatial layer manipulations

Spatial data within a GIS can be represented in two forms: vector data and raster data. Vector data are represented as the intersection of points, lines, and/ or polygons. Raster data are represented in a grid format. When data are in raster format, two or more rasters can be combined by applying mathematical operations to each grid cell. In what follows, we provide two examples of spatial raster manipulations in which the distribution of primate parasite sampling points is compared to host distributions (from Hopkins and Nunn 2007).

Example 1: Quantile subtraction

Quantile subtraction provides a straightforward way to quantify dissimilarities between distributions, based on the concept of proportional sampling. Data from each layer are distributed evenly into bins ('quantiles') and the differences between layer quantile values are displayed according to standard deviations from the mean. Resulting maps provide a geographic quantification of over- and under-sampling. Fig. 10.2 demonstrates this process for a comparison of primate parasite sampling distributions and primate host species richness. The resulting values clearly point to Central Africa, portions of the Amazon, and Borneo as the regions most in need of sampling for primate parasites.

Example 2: Sampling factor estimation

While quantile subtraction provides a straightforward means of quantifying dissimilarities in distributions, it fails to take into account pertinent factors other than geography, such as historical patterns of taxonomic sampling. The sampling factor approach is based on conventional biodiversity gap analyses, which attempt to maximize species representation or complementarity in reserve networks. It combines geographic distributions of host species with historical sampling patterns to identify the sites that are the most under-sampled. Specifically, the sampling factor approach uses the percentage of hosts

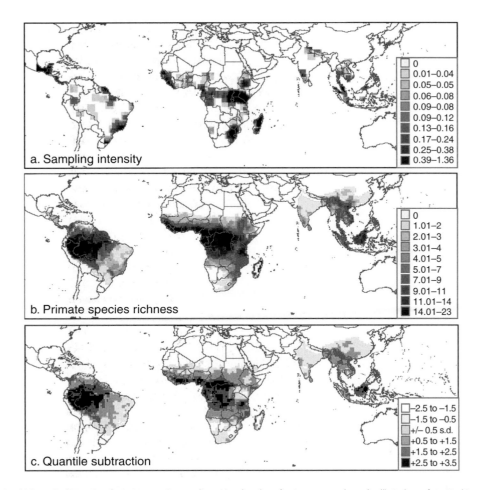

Figure 10.2   (a) Smoothed intensity of primate parasite sampling points (number of points per one degree² cell). Redrawn from Hopkins and Nunn (2007), using an updated version of the Global Mammal Parasite Database (Nunn and Altizer 2005). A moving mean was calculated using overlapping windows of 5 x 5 decimal degrees. Resulting values are displayed in 10 quantiles. (b) Distribution of primate species richness, displayed in 10 quantiles. (c) Quantile subtraction of parasite sampling intensity from primate species richness, displayed as standard deviations from the mean. Positive values indicate under-sampling.

within each cell that have not been sampled at *any* geo-referenced location (Fig. 10.3a) to determine the number of host species within a particular cell that need to be sampled in order to reach mean sampling levels (Fig. 10.3b). Thus, it prioritizes areas both with high host species richness and high numbers of previously unsampled species, and pinpoints the geographic areas that would be most complementary to the current suite of sampled species and localities.

Sampling factor analyses applied to the distribution of primate parasite sampling revealed that while the overall mean percentage of unsampled

primates at any given site is low (14 per cent), this pattern varies extensively across regions. For example, up to 90 per cent of the primates in large portions of Southeast Asia have not been sampled at a geo-referenced location in the GMPD, and to reach mean sampling levels, up to eight primate species would need to be sampled at some sites. Thus, by allocating sampling points according to species complementarity instead of species richness, a different optimal distribution was created, and results differed from the quantile subtraction method. Where quantile subtraction highlighted large portions of Africa as the most in need of sampling, this

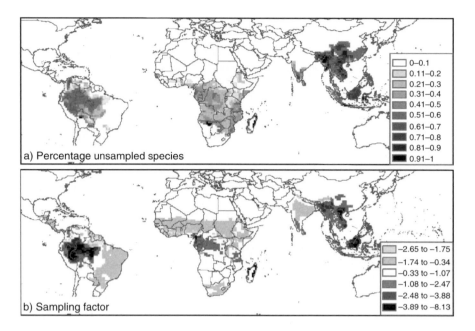

Figure 10.3 (a) Percentage of unsampled primate species per 1x1 decimal degree cell. Redrawn from Hopkins and Nunn (2007), using an updated version of the Global Mammal Parasite Database (Nunn and Altizer 2005). (b) The number of species that need to be sampled in each cell in order to reach mean sampling levels. Values are distributed in 1 SD bins (mean value = 0.36) to allow for visual comparison to the quantile subtraction approach (Fig. 10.2c). Positive values indicate under-sampling.

analysis indicated that Africa is comparatively over-sampled and instead allocated most research effort towards Asia.

10.2.6 Remedying sampling gaps through new data collection

The final step in most gap analyses is to provide enough information on discrepancies between the observed and optimal distributions such that future sampling efforts can be targeted to remedy these discrepancies. Both measures of spatial correlation and spatial layer manipulations, such as the quantile subtraction and sampling factor methods listed above, provide quantitative measures of how geographic patterns of parasite sampling differ from geographic patterns of host species distributions. These methods are useful for quantifying geographic trends. However, researchers seeking to target just a few of the *most* under-sampled sites for future sampling efforts might benefit from incorporating optimization techniques with traditional gap analysis techniques. For example, a number of sites in

Southeast Asia and the Central Amazon have up to eight primate species that need to be sampled in order to reach global mean levels of parasite sampling. Optimization techniques can prioritize the sites to visit and even the order in which to visit them.

10.3 Guiding future research efforts: targeting the most under-sampled sites

10.3.1 Prioritizing sites for future sampling efforts

Optimization approaches derived from fields such as operations research and transportation engineering are increasingly being used in conservation biology to predict patterns of biodiversity (e.g. 'covering' problems, '*p*-dispersion' problems, '*p*-center problems', '*p*-median problems', cluster analysis, compositional dissimilarity) (Tansel *et al.* 1983; Brandeau and Chui 1989; Faith and Walker 1996; Snelder *et al.* 2006; Arponen *et al.* 2008). With only minor modifications, these methods can be used to prioritize future sampling for parasites.

Optimization techniques can be distance-based and/or attribute-based. Distance methods use Euclidean distances between sites in order to place a site in an under-sampled area. Distance-based methods are most frequently applied to select sampling sites in epidemiological analyses conducted at smaller regional scales, where regular sampling is often a pre-requisite for statistical methods that create continuous disease-risk surfaces (e.g. kriging or bayesian surface estimation; see Best *et al.* 2005; Rinaldi *et al.* 2006). In larger scale analyses, attribute values may have equal or greater weight than distance values. In these cases, non-Euclidean distances can be incorporated into analyses. Imagine, for example, that a researcher wishes to prioritize one of two sites that each contains one unsampled host species and an equal number of sampled species (the sites may or may not have host species in common). In such a case, it might be valuable to increase the phylogenetic breadth of sampling. In this example, phylogenetic distance between the unsampled species and its closest sampled relative at the site could serve as a 'non-Euclidean' distance.

In the next section, we use both Euclidean distances and non-Euclidean distances (phylogenetic relationships between unsampled and sampled host species) to illustrate two of the most common optimality approaches used currently for site selection for biological sampling. The first method follows a traditional gap analysis approach in which the site that is most different from the current suite of sites is selected. This approach has been used in a variety of contexts within gap analyses (Faith and Norris 1989; Belbin 1993; Faith and Walker 1996; Jennings 2000). Here, we refer to this approach as the F–N criterion after Faith and Norris (1989). The second optimality approach—'the *p*-median problem'—differs from the F–N criterion in that it seeks to identify the site (or a suite of *p* sites) that, if sampled, would reduce the overall mean distance between unsampled sites and the most similar sampled site (Tansel *et al.* 1983; Faith and Walker 1996). Thus, the *p*-median approach identifies the suite of sites that is most representative of all remaining target sites, whereas the F–N criterion selects the suite of sites that are most dissimilar from currently sampled sites. Both approaches allow for the incorporation of Euclidean and non-Euclidean

(attribute-based) distances. In this section, we apply both approaches to the dataset from Hopkins and Nunn (2007) on primate parasite sampling to illustrate how these approaches differ.

10.3.2 Analysis

We used both the F–N criterion and the *p*-median methods to calculate the top five sites most in need of future sampling, according to two variables: Euclidean distance and phylogenetic distance (Bininda-Emonds *et al.* 2007, 2008). These distances were chosen because a negative relationship has been observed between parasite community similarity and both distance between sampled habitats and phylogenetic distance between hosts (Poulin 2003; Davies and Pedersen 2008). While a number of other factors could be incorporated in the analysis (e.g. ecosystem type, temperature, and levels of precipitation), we feel that using just these two values illustrates the differences between the F–N and *p*-median approaches reasonably well, while allowing for comparison to spatial layer manipulations conducted in the previous section.

Analyses were conducted in a similar grid format to Hopkins and Nunn (2007), with 5017 one degree^2 grid cells, to allow for appropriate comparisons. All unsampled grid cells were considered as target sites in a discrete analysis. Primate geographic range data (Sechrest *et al.* 2002; Nunn *et al.* 2003, 2005) were used to identify the species composition of grid cells, and the cells were identified as sampled or unsampled based on the data from the Global Mammal Parasite Database (Nunn and Altizer 2005).

The two distance metrics were calculated for each cell as follows:

(1) *Euclidean distance*: Distance (km) between a potential future sampling site (i.e. an unsampled one degree^2 grid cell) and the nearest already sampled site. Locations of primate parasite sampling were obtained from the Global Mammal Parasite Database, and include sampling through 2008.
(2) *Phylogenetic distance*: Phylogenetic distance (millions of years) between an unsampled species and its closest sampled relative, summed for all unsampled species present at a site. Phylogenetic

distances between pairs of primate species were calculated as time to last common ancestor using the mammalian supertree from Bininda-Emonds *et al.* (2007, 2008).

Each cell's attribute value was normalized by the maximum value prior to calculations, and both calculations were executed using an iterative greedy algorithm (i.e. only one site was selected at a time). An iterative process was chosen, as this approach reduces the necessary computational power required for analyses. Since the total number of future sites is unknown, an iterative process also demonstrates how regional priorities change as additional sites are sampling. Prioritization of sampling sites during each iteration proceeded as follows:

F–N Criterion:

$$k_{t+1} = \max \prod_{i=1}^{n} \left| x_{ij} - x_{ik} \right|_t w_i \qquad \text{(Eq. 10.1)}$$

In the next time step ($t+1$), sample the cell that has the maximum overall distance to its nearest cell (j) for all distance metrics (i), in the current time step (t). Relative impacts of distance metrics can be specified by giving each metric (i) different weights (w_i).

p-Median:

$$k_{t+1} = \min \sum_{k=1}^{m} \prod_{i=1}^{n} \left| x_{ij} - x_{ik} \right|_t w_i \qquad \text{(Eq. 10.2)}$$

In the next time step ($t+1$), sample the cell that would minimize the sum of the distances between all 5017 grid cells (k) and their nearest sampled neighbour (j) for all distance metrics (i). Relative impacts of distance metrics can be specified by giving each metric (i) different weights (w_i).

F–N and *p*-median values were generated for both distance metrics separately and together. Analyses were conducted with equal weights for analyses of both Euclidean and phylogenetic distances.

10.3.3 Site placement

The placement and sequence of selected sites differed significantly depending upon how the opti-

mal cell was calculated (F–N criterion or *p*-median), and which distance metrics were used (Euclidean distance, phylogenetic distance, or both). The initial target site fell in the same general area (East and Southeast Asia), regardless of the method or distance metric used (Figs. 10.4 and 10.5). However, substantial differences occurred between the remaining four sites. For example, the F–N criterion using only Euclidean distance prioritized the sites that are farthest from existing sampling sites, resulting in the selection of sites at the most northern and southern tips of Africa. Sites in these regions were not prioritized by any of the other methods. The least amount of variation between the F–N criterion and p-median methods is evident when considering phylogenetic distances alone (Fig. 10.4). Using this criterion, three out of five sites remained the same, regardless of selection method. Most methods placed only one site in the Americas, although they differed somewhat in regional placement of this site. When both criteria (Euclidean and phylogenetic distances) were given equal weights, the *p*-median and F–N approaches only converged at two localities. The *p*-median method using both criteria placed the most undersampled site in Myanmar, whereas the F–N method placed it in Laos.

By selecting a series of five complementary sites, these optimization approaches set different sampling priorities than were evident in the previous spatial layer manipulations. Through the incorporation of phylogenetic distances, these approaches can prioritize sites with similar levels of sampling effort, as reflected by sampling factor calculations. For example, although the sampling factor approach identified a large area in South America that requires between three and eight primate species to be sampled in order to reach mean sampling levels (Fig. 10.3b), the Americas were less commonly prioritized using either the *p*-median or F–N methods. This might be expected because species in the Americas are, on average, more closely related, whereas unsampled species in East and Southeast Asia are comparatively unique from an evolutionary perspective. In addition, by using an iterative approach to select a complementary set of sites, we obtained an indication of how regional priorities demonstrated by the static maps in the

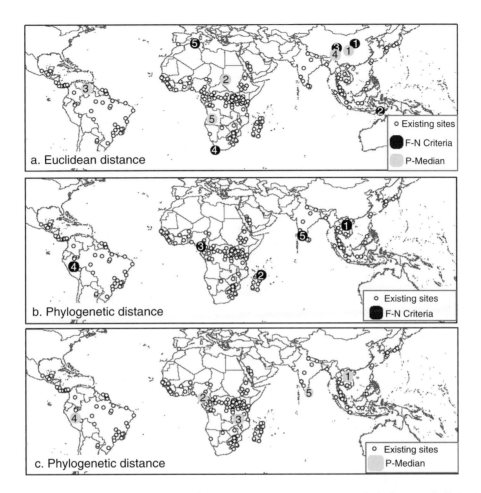

Figure 10.4 Prioritization of five geographic areas in need of future sampling using two optimization methods: the F-N method (prioritization of the most dissimilar site) and the *p*-median method (prioritization of the site that reduces the mean dissimilarity of unsampled and sampled sites the most). Values are selected according to (a) Euclidean distance (F-N and *p*-median methods); (b) phylogenetic distance (F-N method); and (c) phylogenetic distance (*p*-median method).

previous sections are likely to change with the addition of each future sampling point.

10.4 Conclusions

In the previous sections, we used data on the geographic sampling patterns of primate parasites to illustrate the potential of gap analysis techniques to quantify geographic patterns of sampling effort. These methods are widely accessible due to their extensive use in fields ranging from conservation biology to transportation engineering, and have user-friendly implementations within GIS frameworks. As a result, they have great potential to inform

studies of host–parasite biogeography. At the most basic level, these studies can aid in illuminating geographic sampling biases. Such gaps provide a means to prioritize future data collection. By quantifying geographic processes in a spatially-explicit way, these studies also provide the first necessary step towards developing quantitative statistical approaches to account for spatial biases in sampling effort.

Nevertheless, some qualification in the interpretation of results is necessary. Gap analyses ultimately rely on the optimal distribution selected. This selection invariably results from a subjective process that depends largely on the goals of the

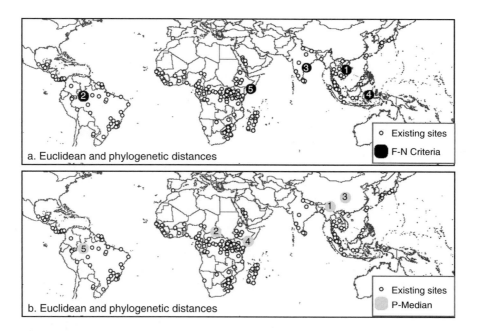

Figure 10.5 Prioritization of five geographic areas in need of future sampling by giving equal weights to Euclidean and phylogenetic distances. Calculations made using (a) the F-N method and (b) the *p*-median method.

researcher. Thus, two gap analyses on the same dataset may differ in conclusions depending upon which host or environmental criteria are prioritized. In addition, due to a widespread lack of geo-referenced sample sites, any spatially explicit analysis of sampling patterns relies upon a subset of all historical sampling for parasites. Thus, results can only yield relative trends, and no absolute conclusions regarding historical sampling patterns. Yet, even with these limitations, quantitative measures of sampling have enormous potential to aid those seeking to better understand how patterns of sampling effort impact our knowledge of host–parasite biogeography.

References

Anselin, L. (1988). Lagrange multiplier test diagnostics for spatial dependence and spatial heterogeneity. *Geographical Analysis*, **20**, 1–17.

Anselin, L. (1995). Local indicators of spatial association-LISA. *Geographical Analysis*, **27**, 93–115.

Anselin, L. (2005). Spatial statistical modeling in a GIS environment. In D.J. Maguire, M. Batty and M.F. Goodchild, (eds.) *GIS, Spatial Analysis, and Modeling*, pp. 93–107. ESRI Press, Redlands.

Anselin, L., Bera, A.K., Florax, R., and Yoon, M.J. (1996). Simple diagnostic tests for spatial dependence. *Regional Science and Urban Economics*, **26**, 77–104.

Araujo, M.B. and Guisan, A. (2006). Five (or so) challenges for species distribution modeling. *Journal of Biogeography*, **33**, 1677–88.

Arponen, A., Moilanen, A., and Ferrier, S. (2008). A successful community-level strategy for conservation prioritization. *Journal of Applied Ecology*, **45**, 1436–45.

Bailey, T.C. and Gatrell, A.C. (1995). *Interactive Spatial Data Analysis*. Pearson Education, Edinburgh.

Belbin, L. (1993). Environmental representativeness: regional partitioning and reserve selection. *Biological Conservation*, **66**, 223–30.

Best, N., Richardson, S., and Thomson, A. (2005). A comparison of bayesian spatial models for disease mapping. *Statistical Methods in Medical Research*, **14**, 35–59.

Bininda-Emonds, O.R.P., Cardillo, M., Jones, K.E., Macphee, D.E., Beck, R.M.D., Grenyer, R., and Price, R.D. (2007). The delayed rise of present-day mammals. *Nature*, **446**, 507–12.

Bininda-Emonds, O.R.P., Cardillo, M., Jones, K.E., Macphee, D.E., Beck, R.M.D., Grenyer, R., and Price, R.D. (2008). Corrigendum. The delayed rise of present-day mammals. *Nature*, **456**, 274.

Brandeau, M.L. and Chui, S.S. (1989). An overview of representative problems in location research. *Management Science*, **35**, 654–74.

Burke, A. (2007). How sampling effort affects biodiversity measures in an arid succulent karoo biodiversity hotspot. *African Journal of Ecology*, **46**, 488–99.

Cam, E., Nichols, J.D., Hines, J.E., Sauer, J.R., Alpizar-Jara, R., and Flather, C.H. (2002). Disentangling sampling and ecological explanations underlying species-area relationships. *Ecology*, **83**, 1118–30.

Crighton, E.J., Elliott, S.J., Kanaroglou, P., Moineddin, R., and Upshur, R.E.G. (2008). Spatio-temporal analysis of pneumonia and influenza hospitalizations in ontario, canada. *Geospatial Health*, **2**, 191–202.

Cuzick, J. and Edwards, R. (1996). Methods for investigating localized clustering of disease. Clustering methods based on k nearest neighbour distributions. *International Agency for Research on Cancer Scientific Publications*, **135**, 53–67.

Davies, T.J. and Pedersen, A.B. (2008). Phylogeny and geography predict pathogen community similarity in wild primates and humans. *Proceedings of the Royal Society of London B*, **275**, 1695–701.

Faith, D.P. and Norris, R.H. (1989). Correlation of environmental variables with patterns of distribution and abundance of common and rare fresh-water macroinvertebrates. *Biological Conservation*, **50**, 77–98.

Faith, D.P. and Walker, P.A. (1996). Environmental diversity: on the best-possible use of surrogate data for assessing the relative biodiversity of sets of areas. *Biodiversity and Conservation*, **5**, 399–415.

Ferrier, S. (2002). Mapping spatial pattern in biodiversity for regional conservation planning: where to from here? *Systematic Biology*, **51**, 331–63.

Funk, V.A. and Richardson, K.S. (2002). Systematic data in biodiversity studies: Use it or lose it. *Systematic Biology*, **51**, 303–16.

Funk, V.A., Richardson, K., and Ferrier, S. (2005). Survey-gap analysis in expeditionary research: where do we go from here? *Biological Journal of the Linnean Society*, **85**, 549–67.

Geary, R.C. (1954). The contiguity ratio and statistical mapping. *Incorporated statistician*, **5**, 115–45.

Getis, A. and Ord, J.K. (1992). The analysis of spatial association by use of distance statistics. *Geographical Analysis*, **24**, 189–206.

Gregory, R.D. (1990). Parasites and host geographic range as illustrated by water fowl. *Functional Ecology*, **4**, 645–54.

Guegan, J.-F. and Kennedy, C.R. (1996). Parasite richness/sampling effort/host range: the fancy three-piece jigsaw puzzle. *Parasitology Today*, **12**, 367–69.

Guernier, V., Hochberg, M.E., and Guegan, J.F. (2004). Ecology drives the worldwide distribution of human diseases. *PLoS Biology*, **2**, 740–46.

Guralnick, R.P., Hill, A.W., and Lane, M. (2007). Towards a collaborative, global infrastructure for biodiversity assessment. *Ecology Letters*, **10**, 663–72.

Hopkins, M.E. and Nunn, C.L. (2007) A global gap analysis of infectious agents in wild primates. *Diversity and Distributions*, **13**, 561–72.

Hortal, J. and Lobo, J.M. (2005). An ED-based protocol for optimal sampling of biodiversity. *Biodiversity and Conservation*, **14**, 2913–47.

Isaaks, E.H. and Srivastava, R.M. (1989). *An Introduction to Applied Geostatistics*. Oxford University Press, Oxford.

Jennings, M.D. (2000). Gap analysis: concepts, methods, and recent results. *Landscape Ecology*, **15**, 5–20.

Krasnov, B.R., Shenbrot, G.I., Khokhlova, I.S., Mouillot, D., and Poulin, R. (2008). Latitudinal gradients in niche breadth: empirical evidence from haematophagous ectoparasites. *Journal of Biogeography*, **35**, 592–601.

Kulldorff, M. and Nagarwalla, N. (1995). Spatial disease clusters: detection and inference. *Statistics in Medicine*, **14**, 799–810.

Lindenfors, P., Nunn, C.L., Jones, K.E., Cunningham, A.A., Sechrest, W., and Gittleman, J.L. (2007). Parasite species richness in carnivores: effects of host body mass, latitude, geographical range and population density. *Global Ecology and Biogeography*, **16**, 496–509.

Lobo, J.M. (2008). Database records as a surrogate for sampling effort provide higher species richness estimations. *Biodiversity and Conservation*, **17**, 873–81.

Moran, P.A.P. (1950). Notes on continuous stochastic phenomena. *Biometrika*, **37**, 17–23.

Nunn, C.L. and Altizer, S.M. (2005). The global mammal parasite database: an online resource for infectious disease records in wild primates. *Evolutionary Anthropology*, **14**, 1–2.

Nunn, C.L. and Altizer, S.A. (2006). *Infectious Diseases in Primates: Behaviour, Ecology, and Evolution*. Oxford University Press, Oxford.

Nunn, C.L., Altizer, S., Jones, K.E., and Sechrest, W. (2003). Comparative tests of parasite species richness in primates. *American Naturalist*, **162**, 597–614.

Nunn, C.L., Altizer, S.M., Sechrest, W., and Cunningham, A.A. (2005). Latitudinal gradients of disease risk in primates. *Diversity and Distributions*, **11**, 249–56.

Openshaw, S., Turton, E., and Macgill, J. (1999). Using the geographical analysis machine to analyze limiting long-term illness census data. *Geographical and Environmental Modeling*, **3**, 83–99.

Poulin, R. (1997). Species richness of parasite assemblages: evolution and patterns. *Annual Review of Ecology and Systematics*, **28**, 341–58.

Poulin, R. (1998). Comparison of three estimators of species richness in parasite component communities. *Journal of Parasitology*, **84**, 485–90.

Poulin, R. (2003). The decay of similarity with geographical distance in parasite communities of vertebrate hosts. *Journal of Biogeography*, **30**, 1609–15.

Poulin, R. and Morand, S. (2000). The diversity of parasites. *Quarterly Review of Biology*, **75**, 277–93.

Rinaldi, L., Musella, V., Biggeri, A., and Cringoli, G. (2006). New insights into the application of geographical information systems and remote sensing in veterinary parasitology. *Geospatial Health*, **1**, 33–47.

Robertson, M.P. and Barker, N.P. (2006). A technique for evaluating species richness maps generated from collections data. *South African Journal of Science*, **102**, 77–85.

Rodrigues, A.S.L., Akcakaya, H.R., Andelman, S.J. *et al.* (2004). Global gap analysis: priority regions for expanding the global protected-area network. *BioScience*, **54**, 1092–100.

Sarkar, S., Pressey, R., Faith, D. *et al.* (2006). Biodiversity conservation planning tools: present status and challenges for the future. *Annual Review of Environmental Resources*, **31**, 123–59.

Schoereder, J.H., Glabiati, C., Ribas, C., Sobrinho, T., Sperber, C., Desouza, O., and Lopes-Andrade, C. (2004). Should we use proportional sampling for species-area studies? *Journal of Biogeography*, **31**, 1219–26.

Scott, J.M., Csuti, B., Jacobi, J.D., and Estes, J.E. (1987). Species richness. *BioScience*, **37**, 782–28.

Sechrest, W., Brooks, T.M., Da Fonseca, G.A.B. *et al.* (2002). Hotspots and the conservation of evolutionary history. *Proceedings of the National Academy of Sciences of the USA*, **99**, 2067–71.

Snelder, T., Dey, K., and Leathwick, J. (2006). A procedure for making optimal selection of input variables for multivariate environmental classifications. *Conservation Biology*, **21**, 365–75.

Tansel, B.C., Francis, R.L., and Lowe, T.J. (1983). Location on networks: A survey. Part I: The p-center and p-median problems. *Management Science*, **29**, 482–97.

Walther, B.A. and Morand, S. (1998). Comparative performance of species richness estimation methods. *Parasitology*, **116**, 395–405.

Walther, B.A. and Martin, J.-L. (2001). Species richness estimation of bird communities: how to control for sampling effort. *Ibis*, **143**, 413–9.

Walther, B.A., Cotgreave, P., Price, R.D., Gregory, R.D., and Clayton, D.H. (1995). Sampling effort and parasite species richness. *Parasitology Today*, **11**, 306–10.

Ward, M.P. and Carpenter, T.E. (2000). Techniques for analysis of disease clustering in space and in time in veterinary epidemiology. *Preventative Veterinary Medicine*, **45**, 257–84.

Zhang, T.L. and Lin, G. (2007). A decomposition of Moran's I for clustering detection. *Computational Statistics and Data Analysis*, **51**, 6123–37.

PART III

Geography of Interactive Populations

In the hosts' footsteps? Ecological niche modelling and its utility in predicting parasite distributions

Eric Waltari and Susan L. Perkins

11.1 Introduction

Ecological niche modelling (ENM) is a rapidly emerging development in biogeographical analysis. ENM predicts species' geographic distributions based on the attributes of chosen parameters, primarily environmental conditions, at locations of known occurrence (Guisan and Thuiller 2005; Pearson 2007). Due to the development of a number of factors, ENM is now a powerful multidisciplinary tool. Computational advances have led to desktop ENM programs such as GARP (Stockwell and Peters 1999) and Maxent (Phillips *et al*. 2006; Phillips and Dudik 2008) that can rapidly create ENMs from large datasets. GIS advances themselves have produced large online databases of environmental data, including estimates of future and historical climates that are available using standard (e.g. ArcGIS) and biologically specific (e.g. DIVA-GIS, see Hijmans *et al*. 2008) GIS software. Lastly, informatics advances have led to collaborative online databases of museum specimens, such as the online Global Biodiversity Information Facility (http://www.gbif.org). As a result, ENM is increasingly used in the fields of ecology, evolution, conservation biology, and biogeography (see reviews in Kozak *et al*. 2008 and Swenson 2008).

ENM has many potential uses beyond mapping current geographic distributions of organisms, including the prediction of potential range expansion of invasive species (Fitzpatrick *et al*. 2007), inferring species ranges in past or future climates (Pearson and Dawson 2003; Waltari *et al*. 2007), and assessment of biodiversity hotspots and the prediction of sites likely to have undocumented species (Araújo and Williams 2000; Raxworthy *et al*. 2003). New studies have begun to incorporate biotic interactions into the models and one such study has shown that this can substantially improve the predictive ability of distribution modelling (Heikkinen *et al*. 2007). While some notable studies have used ENM to examine the distributions of parasites (Peterson *et al*. 2002; Cruz-Reyes and Pickering-Lopez 2006) and to examine disease ecology and evolution (Peterson 2006; Peterson *et al*. 2007a), the uses of ENM clearly can be applied to other important questions regarding parasites, such as the prediction of current or potentially invasive parasites and emerging diseases.

One of the greatest benefits of ENM is its spatially explicit results, which can directly confirm or refute existing biogeographical, evolutionary, and ecological hypotheses. Because parasites clearly have additional dimensions of their ecological niche due to their (sometimes multiple) hosts, it is as yet unclear if current ENM methods can accurately predict parasite distributions, but researchers can make specific predictions to test using ENM. First, the common null hypothesis that a parasite is coevolving with one host can be tested using ENM; under this scenario one could predict the niche of the parasite to be equal to or completely contained within that of its host. Parasites with complex life-histories might be expected to have different distribution patterns compared to their host(s). For example, a parasite with intermediate and final hosts might have a distribution smaller than either host's range. In contrast, a parasite with low host specificity could be expected to have a distribution larger than any one host's range. Comparison of host and parasite ranges can be used to make further assessments, for example inferences about a parasite's ability to invade new regions and/or hosts or hosts' potential to escape parasites by inhabiting areas that are not suitable for the transmission of the parasite.

11.2 Some 'nuts and bolts' of ENM

Niche modelling requires three things:

(1) occurrence data as a listing of coordinates (e.g. latitude/longitude, UTM) of localities where the species of interest is known to exist;
(2) raster GIS layers that are often environmental or climatalogical variables; and
(3) an ENM algorithm, such as those provided by the desktop programs GARP (Stockwell and Peters 1999; http://www.nhm.ku.edu/desktopgarp/) or Maxent (Phillips *et al.* 2006; http://www.cs.princeton.edu/~schapire/maxent).

Researchers may provide their own occurrence data, or use collaborative databases of specimen data, many of which have associated geo-references. We caution researchers who use these resources, however. While some online databases are well vetted to remove obvious errors such as terrestrial geo-references in watery locations or missing "-" signs for degrees west or south (Guralnick *et al.* 2006), some are not. Similarly, the potential uncertainty or lack of precision of the geo-references for samples should be well understood. Finally, taxonomic revisions or simple taxonomic database errors can also lower the quality of inputted occurrence data. Thus, we encourage all ENM users to re-vet their input occurrences. One way to do this is to independently map occurrences using a GIS platform such as ArcGIS or DIVA-GIS (Hijmans *et al.* 2008).

The geo-referenced occurrence data are then matched with one or more layers that are examined to make the niche model. While nearly any type of layer can be used, most ENMs are made using environmental variables, particularly temperature and precipitation-based variables. A commonly used set of variables is the WorldClim dataset that uses 19 bioclimatic variables (Hijmans *et al.* 2005; http://www.worldclim.org), which are derived from monthly temperature and precipitation values (e.g. maximum temperature of warmest month, precipitation of warmest quarter); this derivation allows them to be used uncorrected in studies across Southern and Northern Hemispheres. Due to these characteristics, they are considered most relevant to determining species distributions (Waltari and Guralnick 2009) and so in the following analyses, we use these 19 bioclimatic variables to make our ENMs.

Inputted raster layers can be at any resolution but, in general, coarse resolution results in less biologically meaningful niche models. Given the proper environmental layers and very precise occurrence data, niche models even at the scale of microclimates can be made (e.g. Eisen and Wright 2001). In cases where the study species has a limited distribution or where it is expected to be impacted by climatic variables at smaller scales, higher resolutions are likely warranted. Current Worldclim layers are available in 30 arc-second resolution, or roughly 1 square kilometre pixels. We caution users to more closely examine their occurrence data when using layers at such fine resolution; geo-referencing errors can be greater than the resolution of the environmental layers, sometimes much more so. For this reason, and considering our analyses are over large geographic areas, we use layers at 2.5 arc-minute (or roughly 5 square kilometre) resolution.

The extent of the environmental layers also plays a role in the resulting ENMs, especially in ENM algorithms such as Maxent and GARP that use pseudo-absence or 'background' datapoints. In practice, environmental extents are often limited to the region delimited by the inputted occurrence data. If other regions are of interest, both GARP and Maxent programs can transfer the niche model to a second extent. In the following analyses we delimit an extent, and in one case transfer the ENM to another continent to examine a species' invasiveness.

For various reasons (see Section 11.4) we only used Maxent (version 3.0.19; Phillips *et al.* 2006; Phillips and Dudik 2008) in the following case studies. In all examples we used the default parameters, including convergence threshold (10^{-5}) and maximum number of iterations (1,000) values. We let the program select both suitable regularization values and functions of environmental variables automatically, which it achieves based on considerations of sample size. We used the logistic output option, which results in a continuous probability output, ranging from 0 to 1.0, an indicator of relative suitability for the species, based on the principle of maximum entropy, as constrained by the input occurrence data.

After creating niche models under the above methods, a user could simply map the resulting continuous output, but often using a threshold for binning habitats as either 'suitable' or 'unsuitable' is desired or required. A number of different thresholds have been employed, some of which are arbitrary and some of which are based on inputted data (Pearson 2007). In ENM results that required thresholding, we used a least presence threshold (LPT). This commonly used threshold sets the threshold of suitability to be the highest possible value that still ensures no observed occurrences are considered unsuitable (i.e. omission rate is zero). Finally, statistical analyses can be calculated to examine the quality of the ENMs or to test hypotheses. A common measure of a model's performance is the AUC statistic, or the area under the receiver-operating curve (ROC). A model that is no better than random will have an AUC of 0.5, and a model that successfully predicts all inputted occurrences and no other areas would have an AUC of 1.0 (Pearson 2007). In addition, ENMs are usually evaluated by comparing the model to a separate set of occurrences. In such cases, occurrences are usually broken into training data and test data, for example using 70–75 per cent of the locations for training and 25–30 per cent of them for testing. However, if too few occurrences are inputted into the ENM algorithm (Pearson et al. 2007), this may result in the recovery of a poor model.

11.3   Examples of ecological niche models of parasites

11.3.1   Nested models: a tortoise–tick–haematozoan complex

The first case study that we examined was that of a tri-trophic pathogen–vector–host complex (Široký

et al. 2009). This complex includes the spur-thighed tortoise *Testudo graeca*, the tick *Hyalomma aegyptium*, and the apicomplexan *Hemolivia mauritanica*. *Hemolivia mauritanica* was originally described from *T. graeca* and its range is expected to overlap with that host (Landau and Paperna 1997; Široký et al. 2009). *H. aegyptium* predominantly parasitizes tortoises of the genus *Testudo*, and thus we may expect a strong correlation between the distributions of *H. mauritanica*, *H. aegyptium*, and *T. graeca*. Široký et al. (2009) conclude that there is in fact considerable overlap in distribution between *H. mauritanica* and *T. graeca*, and similarly found that all *T. graeca* in locations without *H. aegyptium* also lacked *H. mauritanica*, suggesting that no other vectors are used.

This parasite–host complex is an excellent candidate for ecological niche modelling, and we used the locality data in Široký et al. (2009) to generate novel models of the tortoise, the tick, and the apicomplexan parasite. We expected that all three would be similar, but because of the specialization in this system, would show a 'Russian doll'-type pattern, in which the distribution of the blood parasite, *H. mauritanica* should be equal to or smaller than that of the vector, *H. aegyptium*, which itself should be equal to or smaller than of the host, *T. graeca*. In this case study, we also conducted analyses that allowed for quantification of the similarities or differences between hosts, parasites, and pathogens.

To create tortoise, tick, and apicomplexan ENMs, we used the locality data from Table 1 of Široký et al. (2009), listed in degrees and minutes latitude and longitude. Niche models were created using Maxent (see Section 11.1 for more details), within the extent of 25–50 °N and 10 °W–55 °E. We used the software ENMTools (Warren et al. 2008) to compare the tortoise, tick, and apicomplexan niche models. Specifically, we measured niche overlap using the *I* statistic, conducting 100 pseudoreplicates to generate *p*-values (Warren et al. 2008). We also examined which of the 19 bioclimatic variables gave the largest contribution to the niche models via the 'Analysis of variable contributions' portion of Maxent outputs.

We found that the ecological niche models of *T. graeca*, *H. aegyptium* and *H. mauritanica* were very similar (Fig. 11.1). Testing model performance, the

AUC values for *T. graeca*, *H. aegyptium*, and *H. mauritanica* were 0.939, 0.964, and 0.979, respectively. Testing for niche overlap, we did not find a significant lack of overlap among predicted tortoise, tick, and apicomplexan distributions (Table 11.1). However, using a threshold that included all inputted locality points (least presence threshold, LPT), the tortoise niche model predicts a larger area than the tick niche model, which in turn predicts a larger area than the apicomplexan. Thus, the apicomplexan *H. mauritanica* has the smallest predicted distribution of the three species, consistent with the fact that *H. mauritanica* has the most complex life-cycle in the complex and is dependent on the distribution of its two hosts.

Široký *et al.* (2009) also suggest that both the tick and apicomplexan in this complex are more sensitive to aridity than tortoises. The ENM models corroborate this hypothesis, as the most important variable in all three taxa is precipitation-related. For *H. mauritanica* and *H. aegyptium*, the variable with the greatest contribution to the niche models was precipitation of the coldest quarter, while for *T. graeca*, it was precipitation of wettest month.

The ENMs of the tortoise–tick–apicomplexan complex show the usefulness of ENM in quantitatively examining host–parasite systems. Given locality data, such as in the study of Široký *et al.* (2009), niche modelling is a straightforward procedure that sheds more light on the co-distribution of parasites and their hosts. We particularly encourage all researchers to make available their specimen geo-references, as these data are critical for ENM analyses. Our ENM results broadly concur with the conclusions of Široký *et al.* (2009), but the use of ENM added additional insight to their findings. Because the tick and the apicomplexan in this complex appear to be host-specific, we might expect that their distributions would be constrained by their host and thus be similar but more limited than the host's distribution. However, because these niche models are estimating parasite distributions using climatic and not host data, it is possible to envision a scenario in which a parasite's estimated distribution is larger than its known host. This would indicate such a parasite has potential to invade new ranges and hosts.

The limited tick and apicomplexan distributions, relative to tortoise distributions, have a number of implications. For example, the northern coasts of Libya and Egypt are predicted to be suitable for *T. graeca* but only suitable in isolated pockets for *H. aegyptium* and not at all suitable for *H. mauritanica* (Fig. 11.1). This region may be an example where a host has escaped its parasite, and be a host source population under a paradigm of source-sink dynamics. In addition, based on these ENM results we would predict different levels of gene flow among these three species, and accordingly different phylogeographic patterns. Thus, our ENM results provide explicit spatial results that can be used to develop further hypotheses.

11.3.2 Modelling an invasive host and its potential biocontrol agents

The red imported fire ant *Solenopsis invicta* is a species native to southern South America that has invaded many regions around the world, with its spread across the southern United States exceptionally well-studied (Tschinkel 2006). Ecologists have been modelling its invasive potential for decades, both in the United States (Pimm and Bartell 1980; Stoker *et al.* 1994; Killion and Grant 1995; Morrison *et al.* 2004) and worldwide (Morrison *et al.* 2005). ENM methods have also been used to examine the spread of the red imported fire ant (Fitzpatrick *et al.* 2007, 2008; Peterson and Nakazawa 2008). We re-examined the distribution of this well-studied species, but also focused on two pathogens that infect *S. invicta*. The microsporidian species *Vairimorpha invictae* and *Thelohania solenopsae*, which infect *S. invicta* and other members of this species complex, have been identified in South America infecting fire ants in their native range. These obligate microsporidians are both candidates for biological control of imported fire ants, and after the discovery of *T. solenopsae* in the United States (Williams *et al.* 1998; Milks *et al.* 2008), this pathogen began

Table 11.1 Statistical comparison of tortoise, tick, and apicomplexan ENM predictions

	Tortoise vs. tick	Tortoise vs. apicomplexan	Tick vs. apicomplexan
I	0.883	0.751	0.823
p	0.990	0.140	0.760

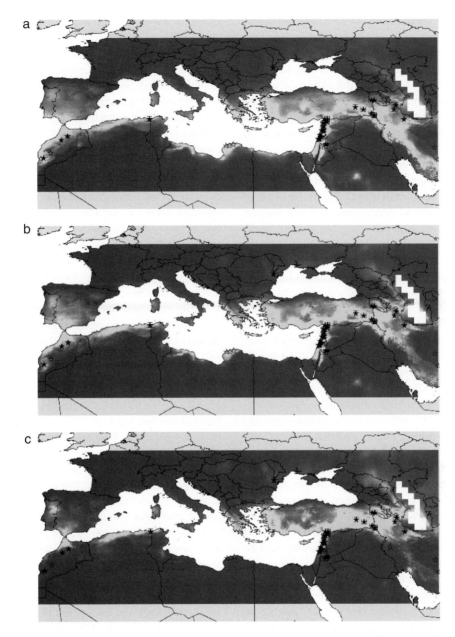

Figure 11.1   Maxent ENM predictions for (a) the tortoise *Testudo graeca*, (b) the tick *Hyalomma aegyptium*, and (c) the apicomplexan *Hemolivia mauritanica* (lighter shades indicate higher Maxent values). Hollow stars indicate locations with tortoises, solid stars locations with tortoise and ticks, and solid stars with surrounding squares locations with tortoise, tick, and apicomplexans.

being used by the USDA to control *S. invicta* (Briano *et al.* 2006). In a laboratory setting, fire ants infected with *V. invictae* also show high mortality rates (Briano and Williams 2002), making this pathogen another candidate for biological control.

We modelled the distribution of the ant *S. invicta* and the two fungal pathogens in their native ranges, and projected the ENMs to the south-eastern United States. We compared the ant ENMs to previously published results, and compared the distributions of the ant and the pathogens. If the predicted distribution of a pathogen is equal to or larger than that for the fire ant in the south-eastern United States, this would suggest that the pathogen may be a suitable control agent. In contrast, if a pathogen's ENM prediction is much smaller than the fire ant's, the pathogen may not be effective at controlling this species' spread and the ant may be capable of 'escaping' its parasite in its new range.

We created niche models for both fire ants and the pathogens using native locality data provided by researchers and an extent of South America of 10–40 °S and 45–75 °W. Models were then projected onto a larger area encompassing North and South America, including the south-eastern United States in the area 50 °N–40 °S and 45–125 °W. Projecting or transferring a niche model can lead to environmental parameters outside the range experienced in the original model. In such cases the ENM algorithm must cope with this issue; in Maxent these areas are termed 'clamping'. The predictive values in such areas are likely suspect (Phillips and Dudik 2008), and so we excluded all areas with clamping greater than 0.01 from maps of suitable habitat, a value that was shown to exclude the drastically different climatic regions in the Amazon Basin. Maps were constructed showing 'suitable' habitat where the Maxent predictive score was greater than the least presence threshold (LPT). Again, we also used the I statistic to measure niche overlap, in order to compare the projected fire ant and microsporidian pathogen ENM predictions in the south-eastern United States. P-values for this measure could not be calculated because ENMTools does not calculate pseudoreplicates of projected areas. We also examined which of the 19 bioclimatic variables gave the largest contribution to the niche models.

In their examination of niche models predicting the invasive potential of *S. invicta*, Peterson and Nakazawa (2008) argued that using a limited set of environmental layers results in less under-prediction. We, however, were interested in testing whether ENM predictions are feasible in parasite–host systems with more limited knowledge and in which *a priori* selections of particular environmental layers may not be appropriate. Thus, we used the standard 19 bioclimatic layers in this and other analyses. First, testing model performance, AUC values for *S.*

invicta, T. solenopsae, and *V. invictae* were 0.934, 0.941, and 0.978, respectively. Similar to the findings of Fitzpatrick *et al.* (2007, 2008), we found that our ENM of *S. invicta* when projected onto the south-eastern United States, under-predicts the known invaded area (Fig. 11.2a). When comparing the ENM predictions of the pathogens *T. solenopsae* and *V. invictae* to the fire ant ENM prediction, the *V. invictae* prediction was greatly reduced in area (Fig. 11.2). Quantitatively, the overlap between *T. solenopsae* and *S. invicta* predictions was also greater than the overlap between the *V. invictae* and *S. invicta* predictions ($I = 0.603$ vs. $I = 0.474$). The two variables with the greatest contribution to both the fire ant and *T. solenopsae* niche models were precipitation of the warmest quarter and temperature seasonality, while in *V. invictae* precipitation of warmest quarter and of the driest month contributed most to its ENM.

While extra care should be taken when transferring a niche model to another geographic area (see Section 11.4), this application gives ENM a powerful additional dimension for examining species distributions. In this case, we used it to compare the invasive potential of a host and pathogens that are candidates for its biological control. While locations of the fire ant in the invaded region could be used to model the host without need for transferring, projection is needed to predict a species' potential to invade an as-yet uninvaded area (Phillips 2008), making projection necessary to make ENM predictions of fire ant pathogens. We found that the fire ant and fire ant pathogen *T. solenopsae* have very similar ENM predictions. This suggests that this pathogen would likely spread to encompass all of the invaded fire ant populations in south-eastern United States and thus potentially be an effective biological control agent. Although we acknowledge that the fire ant is known to have invaded beyond our ENM prediction, and it is possible that it has undergone a niche shift in its newly invaded habitat (Fitzpatrick *et al.* 2007), because of the very similar predicted edge of distribution in the south-east United States and the same most important variables, the *T. solenopsae* pathogen is likely to be a more effective biocontrol agent and *V. invictae* perhaps should be de-emphasized by government agencies looking for additional biological control candidates.

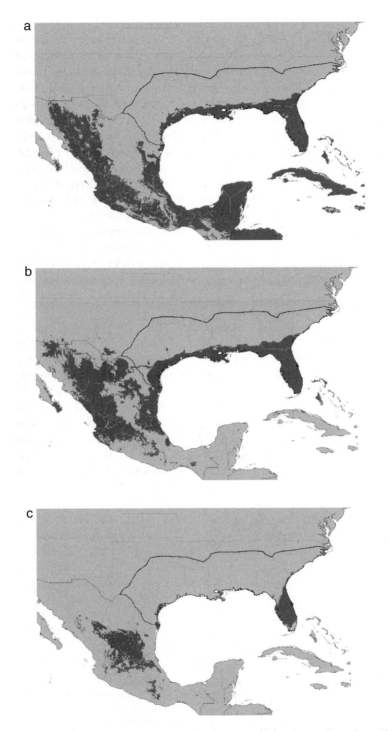

Figure 11.2   Maxent ENM predictions for (a) the red imported fire ant *Solenopsis invicta*, (b) the microsporidian pathogen *Thelohania solenopsae*, and (c) the microsporidian pathogen *Vairimorpha invictae*. The heavy line indicates the current known invaded range of *Solenopsis invicta*.

This study shows the utility of ENM in predicting invasive spread of pathogens. Other researchers have used ENM in a similar way, but examining human pathogens in a public health paradigm (Foley *et al.* 2008; Peterson *et al.* 2003). In this vein, if the ENM prediction of pathogen is larger than its current distribution or than its host species' range, then one could conclude that the pathogen has the potential to spread to native populations, or to new native host species. Furthermore, using transference in ENM prediction can similarly be used to predict the invasive potential of a pathogen, or any parasite or host species, under a specified climate shift.

11.3.3 Using niche modelling to predict the distribution of human malaria

ENM has not been widely used to characterize or predict the distributions of human diseases. Unlike the examples above that use wild animals as their definitive hosts, humans are practically ubiquitous on the planet and thus defy modelling. However, several studies have used ENM to look at the spatial distributions of human diseases that are transmitted by hematophagous insect vectors that might be limited by ecological or climatic factors, and diseases that primarily exist in non-human reservoir hosts (*c.f.* Peterson *et al.* 2002).

Human malaria is found in virtually all of the tropical regions of the world and is caused by five species of protozoan parasites in the genus *Plasmodium*. All mammalian malaria parasites use species of anopheline mosquitoes as their vectors (Martinsen *et al.* 2008). The study of the epidemiology of human malaria, predictions of its occurrence, and even efforts to control the disease have been hampered by the fact that most of the *Anopheles* species that transmit the parasites are part of species complexes that are taxonomically complicated and morphologically indistinguishable. Although progress has been made in identifying genetic markers that are useful for their identification and study (Norris 2002; Fontenille and Simard 2004), ENMs have also recently been implemented to test taxonomic hypotheses of malaria vectors. Foley *et al.* (2008) projected distributions for the Southeast Asian malaria vectors in the *Anopheles minimus*

complex, including that of *A. fluviatilis* species S. This species was originally designated as conspecific with *A. harrisoni* (Garros *et al.* 2005), but was later suggested to be distinct due to molecular differences (Singh *et al.* 2006). The ENMs for these two taxa showed significant overlap, thus the authors concluded that the hypothesis of conspecificity could not be rejected (Foley *et al.* 2008).

The study of the geographic distribution of malaria is nowhere as complicated as it is in Africa, where 29 named species of *Anopheles* have been implicated as vectors. Moffett *et al.* (2007) used ENMs to plot the predicted distributions of 10 of these species using Maxent and the 19 bioclimatic variables plus altitude and land cover layers. Although some degree of overlap between these species' distributions exists, there are several striking differences in the patterns generated by the models, which help to further illustrate the complexity of both the prediction of the risk of malaria under various climate change scenarios and the challenges to effective control. Moffett *et al.* (2007) also used ENM in combination with other data related to the malaria parasite's transmission to generate maps of the relative risk of the disease across the continent. Coefficients of 'human blood index', or the average proportion of the bloodmeals of a sample of mosquitoes that are found to be derived from humans and the density of the human population itself, were multiplied with each vector species' probability of occurrence; human density was found to be the most critical variable for predicting the risk of malaria in this model.

Despite the fact that modelling of human malaria should provide a simpler scenario due to the fact that the vertebrate host itself does not need to be modelled, the studies that have utilized ENM for predicting its distribution have, instead, highlighted the intricacy of these types of investigations. Not only are the niches of the mosquito vectors challenging to model due to the taxonomic complexities of these insects, but the interaction of the malaria parasite with its vector may also vary depending on ecological or climatalogical variables. The preference for human hosts by *Anopheles* species, for example, has been shown to decrease in the presence of irrigated rice cultivation (Muriu *et al.* 2008). Furthermore, although most studies have focused on the more virulent species of human malaria, *P. falciparum*, other species of malaria parasites may co-exist and compete within the insect vectors or human hosts and add additional complexities to attempts to make predictive models. Nonetheless, the fact that malaria is such an important and well-studied human disease may make it a likely candidate for additional development of new methodologies to more accurately model the distribution of parasites and pathogens. Combining the ecological niche of a relevant host of a parasite, generated through traditional ENM and based on occurrence data, with other data that are relevant to the transmission biology of a parasite, should enhance our abilities to more precisely predict the distribution of diseases.

11.4   Conclusions and prospects

Ecological niche modelling (sometimes known as species distribution modelling; Pearson 2007) has experienced a leap in its use since environmental data, species occurrence records, and modelling programs have become more refined and easier to use by biologists. While initial forays into ENM have been primarily conducted by ecologists and conservation biologists and implemented in studies that often model larger and comparatively well-studied organisms, their ease of use makes them easily transferred to all biological disciplines, including parasitology. We modelled a number of parasite systems to examine the applicability of ENM to parasite studies in general, and found that ecological niche modelling can indeed be applied to parasites. However, because of their unique ecological aspects, parasitologists should be especially mindful of the assumptions and caveats of ENM.

11.4.1   Caveats of ENM

Ecological niche modelling results in predictions of a species' suitable habitat based on inputted occurrence data and environmental parameters. As in all modelling, the creed of 'garbage in, garbage out' applies to ENM (Pearson 2007). We emphasize the need for researchers to double-check their occurrence points, as online databases may often contain spurious records due to clerical errors, taxonomic

revision, or imprecise initial recordkeeping. Additionally, the choice of inputted environmental parameters will affect the final prediction. While specific studies may call for specific environmental parameters (such as soil moisture, aspect, soil type), we have taken a generic approach and use a set of 19 bioclimatic variables based only temperature and precipitation. These are variables that have a direct influence on a species distribution, unlike elevation, for example (Pearson 2007). The extent and resolution of the area studied is also of great importance. The Worldclim variables we use are available globally at a 1 km resolution; finer-resolution data may lead to a more refined model, although the error of inputted occurrences may well be greater than the pixel size of the latest environmental layers. The extent used can be a major factor affecting predictions, and using overly small or large extents can result in poor predictions (VanDerWal *et al.* 2009).

While a large number of modelling programs exist, the two most widely used programs are GARP and Maxent (Elith *et al.* 2006; Pearson *et al.* 2007). We use Maxent because its algorithms are being actively improved upon (Phillips and Dudik 2008) and its outputs are highly amenable to post-modelling analysis and interpretation. For example, as standard output Maxent calculates a series of thresholds that have been used in previous biological studies and automatically calculates area under receiver operating curves (AUC), a commonly used statistic for evaluating model suitability. Lastly, Maxent outputs statistics and jack-knifing calculations showing which inputted environmental parameters are most relevant to the model. These outputs make Maxent a very powerful tool for evaluating and interpreting niche models.

11.4.2   ENM input variables

It is possible that, for many parasites, the climate variables we use in creating ENMs are of diminished importance relative to availability of hosts and other biotic factors. For example, drivers of species distributions may vary at different spatial scales. At broader scales environmental factors are thought to be more important, while biotic factors are more important at local scales (Heikkinen *et al.*

2007). Similarly, regarding parasite ecology, we thus expect that ENM will lead to more useful predictions in specialist parasites, particularly those with direct life-cycles. This is apparent in comparing our case studies. The directly transmitted microsporidian parasites of fire ants are likely the most sensitive to environmental factors, because part of their life-cycle is spent in the soil. The predictions of the specialist ectoparasites and apicomplexans found to infect *Testudo*, respectively, show high degrees of similarity, while generalist parasites that use multiple species of intermediate and final hosts might show a predicted distribution much larger than that of any one single host.

Modelling biotic interactions is also possible using ENM. For example, using a land cover layer as an input to the ENM can indirectly incorporate biotic interactions and can improve ENM predictions (Pearson *et al.* 2004; Luoto *et al.* 2007). To more directly incorporate biotic interactions, one could explicitly use a categorical layer indicating presence/absence of an organism thought to affect the taxon under study. This is still a nascent development in ENM, but one analysis using this technique is that of Heikkinen *et al.* (2007). In their study, they use land cover as well as woodpecker distributions as an input to ENMs of four owl species, due to owls' use of woodpecker holes for nest sites. These researchers found that both land cover and woodpecker distributions improved the owl ENMs at smaller spatial scales. Regarding modelling of parasite distributions, one could incorporate a host distribution in this manner. Host distributions are generally better understood relative to parasite distributions, and thus this information can be utilized in an ENM as described above. However, while parasite life-histories are generally more complex (making them amenable to incorporation of biotic factors in their ENMs), frequently their complete life-histories are often still poorly understood, and so incorporating unimportant biotic factors might result in less biologically relevant ENMs. Thus, while we believe that biotic factors are a potentially powerful addition to ENMs, we suggest initial examination of parasite ENMs should be made using bioclimatic or other climatological variables, and incorporation of biotic factors subsequent to these initial assays. In this vein, we limit our input variables in these initial forays into parasite and host ENMs to the 19 bioclimatic layers.

11.4.3   Future uses of ENM in modelling parasites

While the caveats of ecological niche modelling seem daunting, we believe that the benefits of having such spatially explicit predictions greatly outweigh any drawbacks, especially when the researcher recognizes and takes into account these caveats. Many uses of ENM involve making predictions without projecting the models onto secondary environmental layers. These include comparing host and parasite ENM predictions, as we showed in the first case study of the *Hemolivia/Hyalomma/Testudo* complex. Their similar predictions corroborate previous hypotheses that these species are tightly related. Comparing the distributions of a vector and those of several potential hosts for a disease can be used as compelling evidence for or against a given species' likelihood to function as a reservoir for the disease due to coupled ecological interactions (Peterson *et al.* 2002). In contrast, a lack of concordance between host and parasite ENM predictions could indicate that other biotic factors are involved. Another use of ENM is for survey purposes. ENMs can also be used to guide future surveys, which can lead to new records of known or potentially undocumented species (Raxworthy *et al.* 2003). This use is especially relevant in modelling understudied taxonomic groups, many of which are parasites.

Some ENM programs allow for a model made using an original set of environmental layers to be projected to a second set of layers (such as climate variables at a secondary location, or under future or past climate conditions), called transferability (Peterson *et al.* 2007b; Phillips 2008). In general, transferability leads to many more possible uses of environmental niche models. Note, however, that such projections require more assumptions. A critical assumption is that the variables affecting the species distribution at the time and location of the original model are the same as those affecting it in the projected area. Similarly, if different time scales are being examined, the projected prediction rests on the assumption of no niche evolution, which is a topic under much debate (Wiens and Graham 2005). ENM algorithms were first designed to perform optimally without projecting, and might have more errors after projecting (Phillips and Dudik 2008).

Perhaps the most exciting uses of transferred ENMs are their predictions of past or future species' distributions. In recent years climate modellers have made global GIS layers reconstructing paleoclimates(http://www.ncdc.noaa.gov/paleo/paleo.html) as well as climate predictions under various warming scenarios (http://www.ipcc-data.org). ENM projections to future climate scenarios allow biogeographers to predict the effects of climate change on individual species or entire communities (Thomas *et al.* 2004). Using paleoclimate data to reconstruct past distributions is also extremely powerful because it can be compared to ENM predictions based on fossils that do not need projection. Additionally, ENM paleoclimate projections can be used as an independent measure of historical biogeography. Many large taxonomic groups, as well as most parasites, often have very incomplete fossil records, and ENM gives a perspective that can be compared to genetic data (Waltari *et al.* 2007). After taking into account the myriad options available and its assumptions and caveats, ecological niche models can provide a fertile new dimension for exploration of parasite ecology, evolution, and biogeography. We thus see ENM becoming an important and increasingly used tool in the parasitologist's kit.

References

Araújo, M.B. and Williams, P.H. (2000). Selecting areas for species persistence using occurrence data. *Biological Conservation*, **96**, 331–45.

Briano, J.A. and Williams, D.F. (2002). Natural occurrence and laboratory studies of the fire ant pathogen *Vairimorpha invictae* (Microsporida: Burenellidae) in Argentina. *Environmental Entomology*, **31**, 887–94.

Briano, J.A., Calcaterra, L.A., Vander Meer, R., Valles, S.M., and Livore, J.P. (2006). New survey for the fire ant microsporidia *Vairimorpha invictae* and *Thelohania solenopsae* in southern South America, with observations on their field persistence and prevalence of dual infections. *Environmental Entomology*, **35**, 1358–65.

Cruz-Reyes, A. and Pickering-Lopez, J. (2006). Chagas disease in Mexico: an analysis of geographical distribution during the past 76 years - a review. *Memorias do Instituto Oswaldo Cruz*, **101**, 345–54.

Eisen, R.J. and Wright, N.M. (2001). Landscape features associated with infection by a malaria parasite (*Plasmodium mexicanum*) and the importance of multiple scale studies. *Parasitology*, **122**, 507–13.

Elith, J., Graham, C.H., Anderson, R.P. *et al.* (2006). Novel methods improve prediction of species' distributions from occurrence data. *Ecography*, **29**, 129–51.

Fitzpatrick, M.C., Dunn, R.R., and Sanders, N.J. (2008). Data sets matter, but so do evolution and biology. *Global Ecology and Biogeography*, **17**, 562–5.

Fitzpatrick, M.C., Weltzin, J.F., Sanders, N.J., and Dunn, R.R. (2007). The biogeography of prediction error: why does the introduced range of the fire ant over-predict its native range? *Global Ecology and Biogeography*, **16**, 24–33.

Foley, D.H., Rueda, L.M., Peterson, A.T., and Wilkerson, R.C. (2008). Potential distribution of two species in the medically important *Anopheles minimus* complex (Diptera: Culicidae). *Journal of Medical Entomology*, **45**, 853–60.

Fontenille, D. and Simard, F. (2004). Unraveling complexities in human malaria transmission dynamics in Africa through a comprehensive analysis of vector populations. *Comparative Immunology Microbiology and Infectious Diseases*, **27**, 357–75.

Garros, C., Harbach, R.E., and Manguin, S. (2005). Morphological assessment and molecular phylogenetics of the Funestus and Minimus Groups of *Anopheles* (*Cellia*). *Journal of Medical Entomology*, **42**, 522–36.

Guisan, A. and Thuiller, W. (2005). Predicting species distribution: offering more than simple habitat models. *Ecology Letters*, **8**, 993–1009.

Guralnick, R.P., Wieczorek, J., Beaman, R., and Hijmans, R.J. (2006). BioGeomancer: automated georeferencing to map the world's biodiversity data. *PLoS Biology*, **4**, 1908–09.

Heikkinen, R.K., Luoto, M., Virkkala, R., Pearson, R.G., and Körber, J.-H. (2007). Biotic interactions improve prediction of boreal bird distributions at macro-scales. *Global Ecology and Biogeography*, **16**, 754–63.

Hijmans, R.J., Cameron, S.E., Parra, J.L., Jones, P.G., and Jarvis, A. (2005). Very high resolution interpolated climate surfaces for global land areas. *International Journal of Climatology*, **25**, 1965–78.

Hijmans, R.J., Guarino, L., Bussink, C. *et al.* (2008). DIVA-GIS, version 5. A geographic information system for the analysis of biodiversity data. Available at http://www.diva-gis.org.

Killion, M.J. and Grant, W.E. (1995). A colony-growth model for the imported fire ant: potential geographic range of an invading species. *Ecological Modelling*, **77**, 73–84.

Kozak, K.H., Graham, C.H., and Wiens, J.J. (2008). Integrating GIS-based environmental data into evolutionary biology. *Trends in Ecology* and *Evolution*, **23**, 141–48.

Landau, I. and Paperna, I. (1997). The assignment of *Hepatozoon mauritanicum*, a tick-transmitted parasite of tortoise, to the genus *Hemolivia*. *Parasite*, **4**, 365–67.

Luoto, M., Virkkala, R., and Heikkinen, R.K. (2007). The role of land cover in bioclimatic models depends on spatial resolution. *Global Ecology and Biogeography*, **16**, 34–42.

Martinsen, E.S., Perkins, S.L., and Schall, J.J. (2008). A three-genome phylogeny of malaria parasites (*Plasmodium* and closely related genera): life history traits and host switches. *Molecular Phylogenetics and Evolution*, **47**, 261–73.

Milks, M.L., Fuxa, J.R., and Richter, A.R. (2008). Prevalence and impact of the microsporidium *Thelohania solenopsae* (Microsporidia) on wild populations of red imported fire ants, *Solenopsis invicta*, in Louisiana. *Journal of Invertebrate Pathology*, **97**, 91–102.

Moffett, A., Shackelford, N., and Sarkar, S. (2007). Malaria in Africa: vector species' niche models and relative risk maps. *PLoS ONE*, **9**, e824.

Morrison, L.W., Korzukhin, M.D., and Porter, S.D. (2005). Predicted range expansion of the invasive fire ant, *Solenopsis invicta*, in the eastern United States based on the VEMAP global warming scenario. *Diversity and Distributions*, **11**, 199–204.

Morrison, L.W., Porter, S.D., Daniels, E., and Korzukhin, M.D. (2004). Potential global range expansion of the invasive fire ant, *Solenopsis invicta*. *Biological Invasions*, **6**, 183–91.

Muriu, S.M., Muturi, E.J., Shililu, J.I. *et al.* (2008). Host choice and multiple blood feeding behaviour of malaria vectors and other anophelines in Mwea rice scheme, Kenya. *Malaria Journal*, **7**, 43.

Norris, D.E. (2002). Genetic markers for the study of the anopheline vectors of human malaria. *International Journal for Parasitology*, **32**, 1607–15.

Pearson, R.G. (2007). *Species' Distribution Modelling for Conservation Educators and Practitioners*. American Museum of Natural History, New York.

Pearson, R.G. and Dawson, T.P. (2003). Predicting the impacts of climate change on the distribution of species: are bioclimate envelope models useful? *Global Ecology and Biogeography*, **12**, 361–71.

Pearson, R.G., Dawson, T.P., and Liu, C. (2004). Modelling species distributions in Britain: a hierarchical integration of climate and land-cover data. *Ecography*, **27**, 285–98.

Pearson, R.G., Raxworthy, C.J., Nakamura, M., and Peterson, A.T. (2007). Predicting species distributions from small numbers of occurrence records: a test case using cryptic geckos in Madagascar. *Journal of Biogeography*, **34**, 102–17.

Peterson, A.T. (2006). Ecologic niche modelling and spatial patterns of disease transmission. *Emerging Infectious Diseases*, **12**, 1822–26.

Peterson, A.T. and Nakazawa, Y. (2008). Environmental data sets matter in ecological niche modelling: an example with *Solenopsis invicta* and *Solenopsis richteri*. *Global Ecology and Biogeography*, **17**, 135–44.

Peterson, A.T., Benz, B.W., and Papeş, M. (2007a). Highly pathogenic H5N1 avian influenza: entry pathways into North America via bird migration. *PLoS ONE*, **2**, e261.

Peterson, A.T., Papeş, M., and Eaton, M. (2007b). Transferability and model evaluation in ecological niche modelling: a comparison of GARP and Maxent. *Ecography*, **30**, 550–60.

Peterson, A.T., Sánchez-Cordero, V., Beard, C.B., and Ramsey, J.M. (2002). Ecologic niche modelling and potential reservoirs for Chagas disease, Mexico. *Emerging Infectious Diseases*, **8**, 662–7.

Peterson, A.T., Vieglais, D.A., and Andreasen, J.K. (2003). Migratory birds modeled as critical transport agents for West Nile virus in North America. *Vector-borne and Zoonotic Diseases*, **3**, 27–37.

Phillips, S.J. (2008). Transferability, sample selection bias and background data in presence-only modelling: a response to Peterson *et al.* (2007). *Ecography*, **31**, 272–78.

Phillips, S.J. and Dudik, M. (2008). Modelling of species distributions with Maxent: new extensions and a comprehensive evaluation. *Ecography*, **31**, 161–75.

Phillips, S.J., Anderson, R.P., and Schapire, R.E. (2006). Maximum entropy modelling of species geographic distributions. *Ecological Modelling*, **190**, 231–59.

Pimm, S.L. and Bartell, D.P. (1980). Statistical model for predicting range expansion of the red imported fire ant, *Solenopsis invicta*, in Texas. *Environmental Entomology*, **9**, 653–58.

Raxworthy, C.J., Martinez-Meyer, E., Horning, N. *et al.* (2003). Predicting distributions of known and unknown reptile species in Madagascar. *Nature*, **426**, 837–41.

Singh, O.P., Chandra, C., Nanda, N. *et al.* (2006). On the conspecificity of *Anopheles fluviatilis* species S with *Anopheles minimus* C. *Journal of Biosciences*, **31**, 671–77.

Široký, P., Mikulicek, P., Jandzik, D. *et al.* (2009). Co-distribution pattern of a haemogregarine *Hemolivia mauritanica* (Apicomplexa: Haemogregarinidae) and its vector *Hyalomma aegyptium* (Metastigmata: Ixodidae). *Journal of Parasitology*, **95**, 728–33.

Stockwell, D.R.B. and Peters, D.P. (1999). The GARP modelling system: problems and solutions to automated spatial prediction. *International Journal of Geographical Information Systems*, **13**, 143–58.

Stoker, R.L., Ferris, D.K., Grant, W.E., and Folse, L.J. (1994). Simulating colonization by exotic species: a model of the red imported fire ant (*Solenopsis invicta*) in North America. *Ecological Modelling*, **73**, 281–92.

Swenson, N.G. (2008). The past and future influence of geographic information systems on hybrid zone, phylogeographic and speciation research. *Journal of Evolutionary Biology*, **21**, 421–34.

Thomas, C.D., Cameron, A., Green, R.E. *et al.* (2004). Extinction risk from climate change. *Nature*, **427**, 145–48.

Tschinkel, W. R. (2006). *The Fire Ants*. Harvard University Press, Cambridge.

VanDerWal, J., Shoo, L.P., Graham, C.H., and Williams, S.E. (2009). Selecting pseudo-absence data for presence-only distribution modelling: how far should you stray from what you know? *Ecological Modelling*, **220**, 589–94.

Waltari, E. and Guralnick, R.P. (2009). Ecological niche modelling of montane mammals in the Great Basin, North America: examining past and present connectivity of species across basins and ranges. *Journal of Biogeography*, **36**, 148–61.

Waltari, E., Hijmans, R.J., Peterson, A.T., Nyári, A.S., Perkins, S.L., and Guralnick, R.P. (2007). Locating Pleistocene refugia: comparing phylogeographic and ecological niche model predictions. *PLoS ONE*, **2**, e563.

Warren, D.L., Glor, R.E., and Turelli, M. (2008). Environmental niche equivalency versus conservatism: quantitative approaches to niche evolution. *Evolution*, **62**, 2868–83.

Wiens, J.J. and Graham, C.H. (2005). Niche conservatism: integrating evolution, ecology, and conservation biology. *Annual Review of Ecology and Systematics*, **36**, 519–39.

Williams, D.F., Knue, G.J., and Becnel, J.J. (1998). Discovery of *Thelohania solenopsae* from the red imported fire ant, *Solenopsis invicta*, in the United States. *Journal of Invertebrate Pathology*, **71**, 175–76.

CHAPTER 12

The geography of defence

Serge Morand, Frédéric Bordes, Benoît Pisanu, Joëlle Goüy de Bellocq, and Boris R. Krasnov

12.1   Introduction

Parasites are not randomly distributed among and within host species and populations (see Poulin and Krasnov and Bordes *et al.* 2009 in this volume for examples) and, as parasites affect host fitness, one may expect that host defence may vary among hosts and according to host geographical distribution in association with differential parasitic pressures or parasite diversity. Hence, one may think that describing patterns of parasite distribution should be sufficient to predict the level of host defence. However, this prediction is not easily and immediately testable as it depends on the way of estimating parasitic pressures and host defences, the underlying hypotheses, and assumptions. To explore the geography of defence we have to first define host defence, how to quantify it, and how host defence is related to other biological or physiological functions. Second, we need to explore how host defence could be related to parasitic pressures. Only then we can make testable predictions on how host defence may vary among geographically distributed host species or populations. The main objective of this chapter is to review whether there are any strong geographical or latitudinal patterns of host defence ability both within and among vertebrate species.

12.2   What is defence and what to measure?

Hosts are subjected to various kinds of parasites and have evolved several means to control parasite infection and/or the detrimental effects of being parasitized. The first way is to avoid infection by recognizing and avoiding already infected hosts, infected food,

and/or contaminated places (Hart 1994; Loehle 1995). The second way, also linked to behaviour, is to develop specific activities aimed to remove ectoparasites actively, such as grooming, preening, or cleaning (Hart 1994). The efficiencies of such behavioural defences have been demonstrated in several host taxa such as mammals (Mooring *et al.* 2004), birds (Clayton *et al.* 2005), and fish (Arnal *et al.* 2001). The last way, and the last line of defence, involves the immune system—both innate and adaptive—which destroys and/or controls parasites by means of specialized cells and molecules. All these different types of defence, behavioural or immune, are supposed to be costly (Lochmiller and Deerenberg 2000). In this chapter, we will restrict our purposes to immune defences and more specifically how immune defences are investigated in natural conditions.

Immunoecology, or evolutionary immunology, has emerged as a new discipline that aims at studying immune defences in ecological context, where immunity is considered as a life-history trait, part of system of trade-offs, which should be optimized in order to maximize host fitness (Viney *et al.* 2005). Several techniques are used in immuno-ecological studies for the assessment of immuno-competence, where immuno-competence is defined as the general capacity of a host to mount an immune response against pathogens. These techniques can be divided into non-functional and functional measurements (Goüy de Bellocq *et al.* 2006).

12.2.1   Immune function, measuring immune investment and immuno-competence

The vertebrate immune system is composed of two components, namely innate and adaptive immunity. The effector cells of innate immunity, macrophages (i.e. monocytes in mammals) and granulocytes (i.e. neutrophils in mammals) constitute the first line of defence (Janeway *et al.* 1999). The adaptive immune system, in turn, is composed of cell-mediated and humoral components. The major effectors of cell-mediated immunity are T helper lymphocytes and cytotoxic T lymphocytes. B cells and T helper lymphocytes are the main effectors of humoral responses (Janeway *et al.* 1999). The costs of the induced immune responses are supposed to vary from low (humoral responses), variable (T-cell mediated responses), to high (systemic innate responses) depending on the level of systemic inflammation (see Lee and Klassing 2004). The efficiencies of these different types of defences seem not to be correlated (Lee and Klassing 2004) and vertebrate defence strategies will result from a combination of selective parasitic pressure exerted on different aspects of immune defence. From this perspective, white blood cell counts or leucocytes have often been used, notably in mammals, as a convenient measure of immune investment in numerous studies (Nunn *et al.* 2000, 2003; Preston *et al.* 2009). Despite the difficulty of inferring that high white blood cell levels imply higher immuno-competence or higher efficiency against pathogens, the authors assume that higher counts across species may be a reliable metric of a higher level of immune investment, at least in terms of immune system maintenance (Bradley and Jackson 2008).

Immune organs (thymus, spleen, bursa of Fabricius in birds) are also used for measuring immune response or immune investment. The spleen is one of the important organs of the immune systems. This organ is a major site for lymphocyte recirculation, presentation of antigens to T lymphocytes, and activation of immune response (John 1994).

There are many common assays used to measure immune activity (see review in Martin *et al.* 2008) but in the field of immuno-ecology, general immuno-competence is often assessed by spleen size or by the phytohaemagglutinin (PHA) test (Grasman 2002). Spleen size is used as a reliable metric of immuno-competence in birds (Saino *et al.* 1998; Morand and Poulin 2000) and fish (Kortet *et al.* 2003, Šimková *et al.* 2008), but more rarely in

mammals (Fernández-Llario *et al.* 2004). The PHA test consists of a subcutaneous injection of phytohaemagglutinin, which induces local cell stimulation and proliferation in the form of swelling (Stadecker and Leskowitz 1974), and measuring the amount of swelling after a period of 24 hours (Smits *et al.* 2001). The response involves macrophages, basophils, heterophils, and B lymphocytes, and it is orchestrated by cytokine secreted by T lymphocytes. Heterophils and macrophages infiltrate the injection site 1–2 hours after injection and the final stage of the response consists of a swelling at the injection site caused by the dense infiltration of macrophages, lymphocytes, basophils, and heterophils. The PHA response is then considered as an estimate of T cell-mediated immune response, and it has been shown to be a reliable tool for the assessment of immuno-competence and is widely used in field studies (Mendenhall *et al.* 1989; Webb *et al.* 2003; but see Kennedy and Nager 2008).

12.2.2 The importance of immuno-genetics

The capacity of hosts to control parasite infections is mainly dependent on particular immune genes. Among them, the major histocompatibility complex (MHC) is a multi-gene family controlling immunological self/nonself recognition in vertebrates. Some of the genes encode cell surface glycoproteins that present peptides of foreign and self-proteins to T lymphocytes, thereby controlling all specific immune responses, both cell and antibody mediated (Klein 1986). MHC genes are under selective constraints that contribute to maintaining remarkably high polymorphism at MHC loci. This MHC polymorphism is supposed to be maintained by disassortative mating preferences, maternalfoetal incompatibility, and pathogen-driven selection. Indeed, the main selective pressure affecting MHC diversity is generally assumed to arise from parasites, as MHC polymorphism is thought to contribute to differential resistance or susceptibility to infections in studies on humans and laboratory and natural populations of animals. The power of the immune responses against parasites is then strongly dependent on the immuno-genetics of hosts.

12.2.3   Costs of immune defence

Parasites affect their hosts by diverting nutrients, decreasing food intake, or simulating the activity of the immune system (Wakelin 1996). All these effects can potentially place stress on the energetic balance (Degen 2006). Several experimental studies have documented the energetic costs associated with parasitic infection (Khokhlova *et al.* 2002), and the existence of an energetic cost of mounting immune responses against parasites (Martin *et al.* 2003). Increase in parasite diversity seems also to be linked positively with basal metabolic rate (BMR), independently of body mass in mammals (Morand and Harvey 2000; but see Krasnov *et al.* 2004 for fleas).

As parasite impacts and immune activation may lead to a reduction of host fitness in terms of fecundity or survival (Hanssen *et al.* 2004), an optimal solution for the resulting energetic allocation trade-offs between immunity and other host physiological tasks is then expected (Ricklefs and Wikelski 2002; Bonneaud *et al.* 2003). Thus, facing great solicitation of the immune defences, which are constrained by allocation trade-offs, hosts that have been submitted to strong selective parasitic press ures have also been submitted to strong pressures to limit their immune-mediated energy expenditure.

12.3   The link between parasites and defences

12.3.1   Spleen size and cell-mediated responses

At the interspecific level, spleen size and PHA response were found to be both positively related to parasitism in birds (Morand and Poulin 2000; Møller and Rozsa 2005) and fish (Šimková *et al.* 2008). This relationship indicates that hosts invest in immune response according to parasite pressure, estimated either by parasite prevalence or by parasite species richness. At the intra-specific level, both spleen size and PHA response were also found to be related to parasite load (Shutler *et al.* 1999; Christe *et al.* 2000; Navarro *et al.* 2003). In other words, individuals with a large spleen or a high PHA response harboured a lower parasite load than individuals with relatively small spleen or low PHA response.

However, the relationship between spleen size and parasite load was questioned. Some studies suggest that a larger spleen size could be indicative of a high capacity of immunological response (Fernández-Llario *et al.* 2004; Goüy de Bellocq *et al.* 2007), others suggest higher spleen size is indicative of a physiological response to current infection (sometimes leading to splenomegaly) (Brown and Brown 2002; Smith and Hunt 2004). Larger spleens could result from phenotypic response to current infection rather than a product of evolutionary history.

The link between spleen size and parasitism is even more complex in mammals as spleen, in addition to having immune function (i.e. in the lymphoid tissue in the white pulp part of the organ), is also a reservoir of red blood cells (i.e. in the red pulp of the organ) (Corbin *et al.* 2008). There is only one study to date (Corbin *et al.* 2008 for red deer) which has clearly established a positive correlation between red pulp and white pulp mass, allowing the inference that spleen mass as a whole is a positive function of these two parts. These authors then suggest that spleen mass is a reliable metric of lymphoid tissue composition and can be used as an indicator of immuno-competence. Despite this important contribution, there is still not a clear link between spleen size and parasite loads in mammals. For example, Malo *et al.* (2008) have established a negative correlation between parasitic loads (*Elaphostrongylus cervi* larvae excretion) and spleen size in red deer, when others have found a positive relationship (Garside *et al.* 1989 for *Babesia microti*). The results of Malo *et al.* (2008) suggest that individuals with large spleens are more capable of maintaining lower parasite levels, which is just the opposite result to the precedent hypothesis.

Studying PHA response at the intra-specific level in a shrew *Crocidura russula*, Goüy de Bellocq *et al.* (2007) highlighted the importance of the composition of the helminth community, which seemed an important component of the PHA response. The PHA response was not related to total helminth intensity of infection, but either negatively or positively correlated with abundance of one or other taxonomic group of parasites (cestodes or nematodes). The authors suggested that this observation could be explained by a differential modulating effect on immunity by the two types of helminths (suppressing effect versus simulating effect on the PHA response). Whatever the mechanisms, this may suggest that host immunity is a key element in the structure of the helminth community by mediating the coexistence and/or competition between distantly phylogenetically related parasite taxa.

12.3.2 Diversity at immune genes

Several studies in fish, birds, and mammals have shown that the variability of some immune genes, such as the MHC, is strongly and positively linked to parasite diversity (Prugnolle *et al.* 2006; Šimková *et al.* 2006; Goüy de Bellocq *et al.* 2008). In intra-specific comparisons, recent studies have shown associated diversities of MHC alleles and parasites. In natural populations of three-spined sticklebacks (*Gasterosteus aculeatus*), parasite diversity was found to be positively correlated with MHC class IIB diversity (Wegner *et al.* 2003). A more recent study on wild Atlantic salmon (*Salmo salar*) showed a positive correlation between genetic diversity at MHC and bacterial diversity in the local environment (Dionne *et al.* 2007, but see below). A positive relationship between HLA class I genetic diversity and pathogen richness in human populations was reported by Prugnolle *et al.* (2006). All these studies suggest that host populations or species facing high parasite diversity are likely to experience relatively high selection pressures for maintaining polymorphism at MHC genes, when compared with those facing low parasite diversity.

However, there are very few studies that have shown a positive relationship between allelic diversity at MHC genes and parasite diversity in interspecific comparisons. The studies of Goüy de Bellocq *et al.* (2008) for rodents and Šimková *et al.* (2006) for cyprinid fish were among the first to directly establish a positive correlation between genetic diversity at MHC and parasite diversity among host species.

12.4 Predictions in a geographical context: a review of hypotheses and tests

12.4.1 Latitudinal gradients in parasitic loads: tropical versus temperate zones

Theoretically, parasite load may vary geographically for various reasons related to both host or parasite

traits. We may expect clinal variation in parasitic loads as a result of a greater exposure to parasites in the tropics. For example, host geographical variability in reproductive effort may affect parasite transmission and parasite load as an increase in parasite transmission is often observed when the reproductive effort is high (Pelletier *et al.* 2005). From this perspective, it was also predicted that in areas with a high host birth rate, or favourable environments for host metabolism, as is supposed to be the case in the tropics, high levels of parasitism may evolve because the detrimental effects of parasitism are reduced (Hochberg and Van Baalen 1998). There is also a possibility that parasite mortality resulting from harsh conditions in winter far from equator may be a strong constraint for parasite transmission (i.e. many infective stages live off their definitive hosts) at northern latitudes. The hypothesis that parasite diversity is higher in the tropics is, however, mainly supported by verbal arguments (Guernier *et al.* 2004; Salked *et al.* 2008; see Bordes *et al.* in this book for review). The reality of our still poor knowledge of clinal variation in parasite richness or load is clearly illustrated by the recent study of Salked *et al.* (2008), establishing that the intensity of a blood parasite *Hepatozoon hinulinae* in its lizard host (*Eulamprus quoyii*) is higher in the tropics and claiming: 'As far as we are aware, this is the first study that compares parasite load in both tropical and temperate zones in a single host-parasite system'.

In terms of immune investment, it has been supposed that a close link with parasitic pressures or parasite risks should be higher in tropics. This hypothesis is called 'the adjustment to parasitic pressures' with higher investment in immune function expected in species inhabiting parasite-rich areas (Hasselquist 2007). For example, animals undergoing extensive migration, or living in the tropics, show greater investments in immunity than their counterparts that do not migrate, or live in temperate regions (Møller 1998; Møller and Erritzøe 1998; Martin *et al.* 2004) as, among various ecological reasons, animals living in the tropics are thought to face greater pressures from parasites than animals living in temperate regions. Indeed, measures of immune function in bird species from tropical and temperate zones revealed that tropical birds had higher levels of circulating leucocytes and larger spleen sizes for a given body size than their temperate counterparts (Møller 1998). Moreover, a clear pattern of enhanced humoral immunity at lower latitudes is observed, but it is not the case for cell-mediated immune responses (Hasselquist 2007). In birds, latitudinal patterns of immune responses are also supposed to be linked with life-history strategies (Lee 2006). Broadly speaking, slow-pace-of-life species (i.e. tropical species), having longer life expectancy and reduced investment per breeding attempt, are predicted to invest more in immune defence than fast-pace-of-life species (i.e. most temperate species). Here again, species with slow life-history strategies seem to sustain higher humoral immunity (Hasselquist 2007). The proximate factors (i.e, parasites or life-history) for enhanced immunity near the tropics are then hard to distinguish in birds.

In terms of polymorphism of immune genes, Dionne *et al.* (2007) tested the hypothesis that MHC genetic diversity increases with temperature along a latitudinal gradient in response to pathogen selective pressure in wild Atlantic salmon. Their results showed that diversity at the alleles of MHC class II increased with temperature, which conforms to latitudinal gradient. Moreover, MHC diversity was also found to increase significantly with local bacterial diversity, suggesting that genetic diversity at MHC class II copes with pathogen diversity in rivers associated with different thermal regimes (Dionne *et al.* 2007).

Relationships between climate and parasite abundance may be due to the indirect effects of temperature or photoperiod on the host's immune system (Raffel *et al.* 2006). Animals use photoperiod to adapt to annual changes in their environment. Adjustments in reproduction and in immunity may permit individuals to survive severe seasonal changes in high-latitude environments. Pyter *et al.* (2005) investigated reproductive and immune adjustments in meadow voles (*Microtus pennsylvanicus*) by manipulating their environmental conditions in the laboratory, and predicted that short days would result in elevated immune and reproductive responses in animals from areas situated at high latitudes compared with animals from areas situated at low latitudes. Whereas results on reproduction supported the prediction, this was not the case for immune responses. Photoperiod affected

reproductive and immune systems differently, suggesting that immune responses may reflect other environmental factors. Similarly, by doing transplanting experiments, Martin *et al.* (2004) showed that immune activity varied in latitude in passerines, and that this variation was more than just acclimations of bird populations to different environments, as the relative differences in the PHA response between populations from different latitudinal origins were maintained in captivity. More recently, Owen-Ashley *et al.* (2008) presented evidence for genetic control of immune responses across latitude in the white-crowned sparrow (*Zonotrichia leucophrys*), but their study did not provide support for environmental (photoperiodic) regulation.

In conclusion, all these studies suggest the existence of latitudinal gradients in immuno-competence. Moreover, these variations across latitudinal gradients seem to have a strong selection origin, probably due to parasite pressures rather than abiotic factors (temperature or photoperiod). Unfortunately, most studies to date are strongly biased toward hosts (mainly fish and birds). Moreover, and more importantly, the reality of higher parasitic pressures at tropical latitudes has still to be demonstrated as most of these studies ignore parasites *per se* (i.e. direct parasites pressures were not tested). Following Owen and Clayton (2007) we can then also ask: 'Where are the parasites in the PHA response?' and claim to investigate the existence of such general geographical patterns of immuno-competence within and among species to bring new insights in this issue.

12.4.2   Lessons from insular populations

Parasite species diversity of native host species has been shown to be poorest in insular compared to continental populations. This pattern has led to the hypothesis of lesser parasitic pressures on islands, with hosts expressing lowered immune defences as an evolutionary response to the energetic costs of mounting or maintaining immune functions (Lindström *et al.* 2004). Additionally, the low genetic diversity of these small and isolated host populations have been linked to an increase in parasite infection rates (Meagher 1999; Whiteman *et al.* 2006).

The immuno-ecology of insular populations has, however, been poorly studied, and the studies have produced contradictory results. For example, Matson (2006) studied the immune functions of 15 phylogenetically related bird species between the North American continent and oceanic islands (Hawaii, Bermuda, and Galápagos) and found no reduced immunity in these inbred host populations characterized by depauperate parasite communities. Thus, lessons from insular population are not straightforward, and still need to be investigated. More specifically, Matson (2006) highlighted that immune functions of insular hosts could be reorganized, or maladapted, favouring innate and inducible and not acquired humoral defences and immunological memory.

12.5 New investigations in mammals

As reviewed above, mammals are poorly investigated. In this last part of this chapter, we investigate first, at the interspecific level, the existence of a latitude gradient in mammal immune defence and second, at the intra-specific level, the pattern of mammal immuno-competence. For this, we use our unpublished data.

12.5.1 Latitudinal gradient, leucocytes, and metabolism in mammal species

We investigated the existence of latitudinal gradient in mammal defences taking into account energetic constraints (i.e. BMR). We obtained data on circulating levels of three classes of white blood cells (WBC) from the ISIS database (International Species Information System Physiological Reference Values, CD-ROM, Minnesota Zoological Garden, Apple Valley, MN, 1999). This database has previously been used to investigate correlations between immune defences and ecological traits in primates (see Nunn et al. 2000 for more details). WBC data were given as number of cells per 10^{-9} l of blood. Data on host longevity are extracted from Carey and Judge (2000) and data on BMR from McNab (1988). We obtained data on range latitude (i.e. mammal distribution expressed in latitude) and mid-latitude (i.e. mean latitudinal distance from the equator) from online GIS databases. We tested our predictions using a comparative method based on independent contrasts, with phylogeny of mammals coming from the phylogenetic supertree of Binida-Emonds et al. (2007). We obtained 72 independent contrasts.

We performed multiple regression analysis on leucocyte counts as a function of host body mass, BMR, longevity, range latitude, and mid-latitude. We found that only BMR and body mass correlated with leucocyte counts (Table 12.1). Importantly, there was no latitudinal gradient in immune cell-mediated defences. A negative relationship between leucocyte counts and BMR, corrected for host body mass (Fig. 12.1a), suggested that cell-mediated defence does not increase at the expense of metabolism. Finally, mammal BMR, corrected for host

Table 12.1 Multiple regression analyses on potential determinants of leukocyte counts in mammals (host longevity, host body mass, basal metabolic rate, latitudinal range, and latitudinal mid-range) using independent contrasts. The first analysis included all potential independent variables. The second is based on a selection procedure of the most informative variables (using Bonferroni-corrected significance level = 0.05/number of variables included in the model). Note that these analyses were conducted on only 72 independent contrasts.

Analysis	Independent variables	*b* value (slope)	*p*	*df;F, df, r (p)*
All variables	Host body mass	4.78	0.005	
	BMR	−4.08	0.04	
	Host longevity	−0.36	0.85	
	Latitudinal range	0.01	0.52	
	Latitudinal mid-range	−0.02	0.40	
				5,67; 3.69, 0.46 (0.005)
Selection procedure	Host body mass	4.31	0.004	
	BMR	−4.39	<0.001	
				2,70; 8.90, 0.45 (0.0004)

body mass, was positively correlated with mid latitude of mammals (Fig. 12.1b), a latitudinal pattern already observed in mammals.

These results suggest that predictions dealing with a higher investment in immune defence at tropical latitudes seem not to hold for mammals, in contrast to birds. It is not so surprising given that there is no strong support for latitudinal pattern in parasite species richness in this clade (see Bordes *et al.* in this volume). These results, however, stressed a potential intriguing link between investment in adaptive immune defences and reduced energy requirements in mammals. Clearly, if there is any relationship between saving energy and

adaptative immunity in mammals, it should encourage future research in this field.

12.5.2 Geographic variation of defences and parasite diversity among populations of a rodent species

Goüy de Bellocq *et al.* (2003) investigated the distribution of parasites among populations of the woodmouse *Apodemus sylvaticus* and found a nested pattern of parasite species assemblages. In other words, parasite species in depauperate assemblages constituted subset samples occurring in richer parasitic assemblages. Differential colonization and

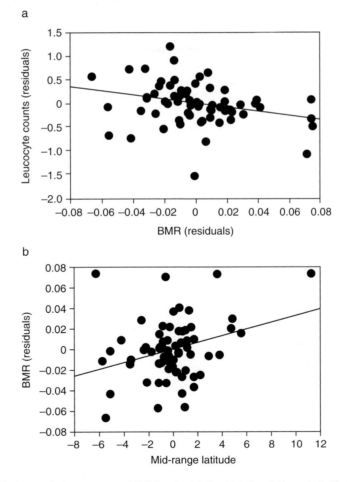

Figure 12.1   (a) Relationships between leucocyte counts and BMR (basal metabolic rate), both variables controlled for host body mass, in mammals (using independent contrasts) ($F_{1,71}$=6.35, r=0.29, p=0.014). (b) Relationships between BMR (basal metabolic rate), controlled for host body mass, and latitudinal mid-range distribution in mammals (using independent contrasts) ($F_{1,71}$=7.049, r=0.30, p=0.009).

extinction probabilities seem to be the most likely causes of nestedness in parasite species assemblages, and Goüy de Bellocq *et al.* (2003) hypothesized this nested pattern as related to both the life-cycle of the parasites and the geographic distribution of the rodent. Because this geographic distribution of this species is not independent of its history (Michaux *et al.* 2003), the nestedness of parasite assemblages of the woodmouse seem also to reflect the past history of this host species. The woodmouse is then a model of choice to investigate the geographical pattern of host defences at intraspecies level (Goüy de Bellocq *et al.* 2005).

We hypothesized that, in the wild, selection on immune investment can change spatially as a function of the structure of local parasite communities. We investigated the effect of parasite diversity on immuno-competence, appreciated by spleen size and PHA skin test, and on genetic polymorphism at MHC class II genes. Data were obtained on 222 woodmice trapped in 8 different localities, including island and mainland populations. A total of 28 parasite species were recorded in the 8 rodent populations. Allelic diversity at MHC genes was obtained from Goüy de Bellocq *et al.* (2005).

At the level of immune genes, a positive trend was found between the number of MHC alleles and parasite species richness. We performed a discriminant analysis on MHC alleles harboured by individual rodents and found that the woodmouse populations were statistically discriminated by their MHC allelic composition ($F_{78,662} = 2.089$; $p < 0.0001$). Subsequently, a positive relationship was found between the Mahalanobis distances based on the composition of MHC haplotypes and the differences in parasite species richness between host populations (Fig. 12.2), suggesting that local parasite diversity is linked to local diversity at MHC genes. Using spleen size, we found that mean spleen mass was positively related to total parasite species richness among woodmouse populations ($p = 0.03$).

At the level of cell-mediated immunity, the PHA response was not correlated to total parasite richness among woodmouse populations ($p > 0.05$), suggesting that parasite species accumulation is not the key factor that enhanced immuno-competence. In order to account for the structure of the parasite assemblage (presence/absence of each parasite species, and parasite intensity), we performed a Principal Component Analysis (PCA) on the matrix of the number of each parasite species by each individual host. The first component axis of this PCA may then represent a measure of the parasite diversity that includes not only the presence/absence of each parasite species (the parasite species richness) but also the individual number of each parasite species. Similarly a PCA was performed on

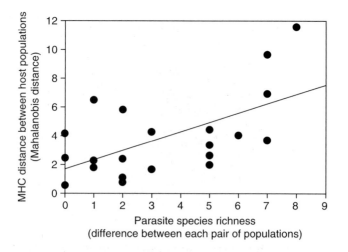

Figure 12.2 Relationship between Mahalanobis distances of the woodmouse populations based on their MHC alleles and their differences in parasite species richness (permutation test, $p = 0.01$).

the PHA response of each individual host, taking into account its body mass. The first component axis of this second PCA may then represent a measure of immuno-competence. We then regressed these two variables and found that the structure of the helminth community is significantly related to the immuno-competence ($p=0.001$; Fig. 12.3). This suggests that the structure of the parasite community (i.e. richness and intensity) and not the species richness *per se* is related to level of immune investment.

Our results supported the idea that at the intra-specific level (and contrary to the inter-specific level) some patterns in immuno-competence seem to emerge in mammals whatever the measure of immune activity adopted. Importantly, parasite diversity (i.e. parasite species richness) seems to be a good predictor of immune investment measured with spleen size or MCH diversity. On the contrary, cell mediated immunity seems rather to be influenced by parasitic loads *per se*, notably estimated by combining parasitic diversity and parasitic intensities.

12.5.3   Islands: the case of recent man-made introduced rodent species

An extreme case of insular populations of rodent is the house mouse *Mus musculus.* introduced on the sub-Antarctic islands almost two centuries ago (Chapuis *et al.* 1994). On the Kerguelen archipelago, the domestic mouse *M. m. domesticus* lives on dense plant communities over the entire area of the main island and on the seven smallest islands of the archipelago (Pisanu *et al.* 2001). Mice populations have been monitored on two islands of the archipelago, following rabbit eradication on Cochons Island (165 ha) in 1997, and both rabbit eradication and domestic cat (*Felis silvestris*) control on Guillou Island (155 ha) in 1994. The mouse is now the single mammal species living on both islands, and individuals are free of any helminth infections on Cochons Island, whereas on Guillou Island they are infected by a single species of intestinal nematode, the oxyurid *Syphacia obvelata* (Pisanu *et al.* 2001; Pisanu and Chapuis 2003). The population dynamics of *S. obvelata* mainly depend on the age structure of its host population and particularly on the male reproductive status during the summer breeding season (Pisanu *et al.* 2002, 2003).

A comparison of the immune response between uninfected and infected mice on both islands has been conducted during the austral summer months between January and March in 2003. In order to avoid ontogenetic effects on measurements of immunity and on *S. obvelata* burden (Pisanu *et al.* 2002, 2003), only adult mice (i.e. above 20 g) were

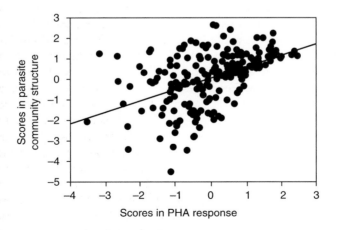

Figure 12.3   Relationship between parasite community structure and the PHA response in the woodmice. Both variables were obtained by using the scores of the first axis of two Principal Component Analyses (PCA). The first PCA was performed on the matrix of the number of each parasite species by each individual host. The second PCA was performed on the PHA response of each individual host, taking into account its body mass (permutation test, $p=0.01$).

selected in the following analysis. Cellular immune response was measured by skin response to the PHA injection in hind feet. Mice were sampled during the same periods in similar habitats (closed communities of *Acaena magellanica*, see Chapuis *et al.* 2004) on both islands (see Pisanu *et al.* 2002; Chapuis *et al.* 2001 for details on mouse population monitoring). We expected that adult mice, that is, large body sized individuals, free of helminths on Cochons Island would have a lowered immune response compared to infected mice on Guillou Island.

A total of 23 mice above 20 g were trapped on Cochons Island, all helminth free, and 13 on Guillou Island, of which 10 were infected by *S. obvelata* (on average, 11±4 worms per infected mouse). Body mass and body length of rodents were similar between the two islands, as well as skin reaction to the PHA (Table 12.2). Contrary to our initial expectation, uninfected mice on Cochons Island did not show depressed cellular immune response. Thus, mice on Cochons Island were able to develop an innate and inducible cellular immunity in the absence of any helminth, and also of any ectoparasite, which have never been recovered from mice on the Kerguelen archipelago. The other known and unique infectious agent present in mice on Kerguelen is the murine cytomegalovirus (MCMV, a herpes-like virus infecting mice), that has been found in mice living in the scientific station of Port-aux-Fran's (Booth *et al.* 1993), 30 km east and far from our studied islands. Immune functions expressed by mice on Cochons Island could therefore have been induced by the MCMV infections. In conclusion, mice introduced on Guillou and Cochons Islands maintained cellular immune responses, even under a very low parasitic diversity.

Table 12.2 Body size and immunological measures from house mice captured in summer 2003 on Cochons and Guillou islands, Kerguelen, sub-Antarctic.

Island	Cochons	Guillou
Number of mice	23	13
Body mass (g)	23±1	23±1
Body length (mm)	91±1	90±1
PHA (mm)	0.30±0.04	0.42±0.05
PBS (PHA control)	0.03±0.02	0.11±0.03

12.5.4 What are geographical gradients in mammalian anti-parasitic defence?

Our results have tried to link more directly three pieces of a fascinating puzzle: parasitic pressures *per se* (estimated by parasite species richness and parasite intensities), immuno-competence, and host biogeography. Taken together with all previous studies (notably Nunn 2002; Nunn *et al.* 2003; Goüy de Bellocq *et al.* 2007, 2008; Malo *et al.* 2008; Preston *et al.* 2009) we are rather confident in inferring three emerging patterns in the geography of defence in mammals. First, the MCH diversity seems to be highly correlated with parasite diversity in this clade within and among species. Second, leucocytes counts may represent a valuable parameter to infer parasite diversity pressures and parasitic loads across species when spleen size seems to be potentially related both to current fights against infestations (with large spleens reflecting stronger possibilities limit parasite infestations) and also to parasite diversity pressure. Third, the PHA response seems to be rather strongly influenced by the structure of parasite community (i.e. richness, but also intensity and taxonomic groups of the parasites) and not only by the species richness as a whole.

12.6 Conclusion

Despite considerable studies related to immuno-competence, general patterns seem rather hard to find, with important discrepancies between vertebrate clades. It may suggest that vertebrate clades do not deal with parasitic pressures in a stereotyped way. There is also a strong possibility that all the measures of immuno-competence or immune investment do not capture the same reality. Clearly, immune investment is probably highly dependent on time scale adaptations.

If immunity is a life trait subjected to inherent trade-offs, it has probably been under strong selective pressures across evolutionary time. Among them, parasite diversity and constraints to save energy may have played a major role. Diversity at the MCH genes, but also leucocyte counts, could be good predictors of such constraints, notably among species.

Importantly, however, parasites are always, at any time and any place, a strong constraint in the host environment, which probably leads to the present ecological constraints. The PHA response and spleen size may then be alternative and interesting measures of immune investment in particular ecological situations, notably dealing with spatial heterogeneity in parasitic loads (prevalence or intensities). These hypotheses, however, need to be validated in future empirical studies.

Finally, a still fascinating and intriguing question remains. Does immuno-competence structure parasite communities or does parasite diversity enhance immune investment? As an example, host defence-mediated competition can influence the structure of parasite communities and may play a part in the evolution of parasite diversity (Bush and Malenke 2008). Here again we need field studies clearly dealing with this problematic and controlling for all variables.

References

Arnal, C., Côté, I.M. and Morand, S. (2001). Why clean and be cleaned? The importance of client ectoparasites and mucus in Caribbean cleaning symbioses. *Behavioural Ecology and Sociobiology*, **51**, 1–7.

Binida-Emonds, O.R.P., Cardillo, M., Jones, K.E. *et al.* (2007). The delayed rise of present-day mammals. *Nature*, **446**, 507–12.

Bonneaud, C.J., Mazuc, G., Gonzalez, C., Haussy Chastel, O., Faivre, B., and Sorci, G. (2003). Assessing the cost of mounting an immune response. *American Naturalist*, **161**, 367–79.

Booth, T.W.M., Scalzo, A.A., Carrello, C. *et al.* (1993). Molecular and biological characterization of new strains of murine cytomegalovirus isolated from wild mice. *Archives of Virology*, **132**, 209–20.

Bradley, J.E. and Jackson, J.A. (2008). Measuring immune system variation to help understand host-pathogen community dynamics. *Parasitology*, **135**, 807–23.

Brown, C.R. and Brown, M.B. (2002). Spleen volume varies with colony size and parasite load in a colonial bird. *Proceedings of the Royal Society of London B*, **269**, 1367–73.

Bush, S.E. and Malenke, J.R. (2008). Host defence mediates interspecific competition in ectoparasites. *Journal of Animal Ecology*, **77**, 558–64.

Carey, J.R. and Judge, D.S. (2000). *Longevity Records: Life Spans of Mammals, Birds, Amphibians, Reptiles, and Fish. Monographs on Population Aging, Vol. 8*. University Press, Odense.

Chapuis, J.-L., Boussès, P., and Barnaud, G. (1994). Alien mammals, impact and management in the French sub-Antarctic islands. *Biological Conservation*, **67**, 97–104.

Chapuis, J.-L., Frenot, Y., and Lebouvier, M. (2004). Recovery of native plant communities after eradication of rabbits from the sub-Antarctic Kerguelen Islands, and influence of climate change. *Biological Conservation*, 117, 167–79.

Clayton, D.H., Moyer, B.R., Bush, S.E. *et al.* (2005). Adaptive significance of avian beak morphology for ectoparasite control. *Proceedings of the Royal Society of London B*, **272**, 811–17.

Corbin, E., Vicente, J., Martin-Hernando, M.P., Acevedo, P., Perez-Rodriguez, L., and Gortazar, C. (2008). Spleen mass as a measure of immune strength in mammals. *Mammal Review*, **38**, 108–15.

Christe, P., Arlettaz, R., and Vogel, P. (2000). Variation in intensity of a parasitic mite (*Spinturnix myoti*) in relation to the reproductive cycle and immunocompetence of its bat host (*Myotis myotis*). *Ecology Letters*, **3**, 207–12.

Degen, A.A. (2006). Effects of macroparasites on the energy budget of small mammals. In S. Morand, B.R. Krasnov, and R. Poulin, eds. *Micromammals and Macroparasites: From Evolutionary Ecology to Management*, pp. 371–99. Springer Verlag, Tokyo.

Dionne, M., Miller, K.M., Dodson, J.J., Caron, F., and Bernatchez, L. (2007). Clinal variation in MHC diversity with temperature: evidence for the role of host-pathogen interaction on local adaptation in Atlantic salmon. *Evolution*, **61**, 2155–64.

Fernandez-Llario, P., Parra, A., Cerrato, R., and de Mendoza, J.H. (2004). Spleen size variations and reproduction in a Mediterranean population of wild boar (*Sus scrofa*). *European Journal of Wildlife Research*, **50**, 13–17.

Garside, P., Benhke, J.M., and Rose, R.A. (1989). The immune response of male DSN hamster to a primary infection with *Ankylostoma ceylanicum*. *Journal of Helminthology*, **63**, 251–60.

Grasman, K.A. (2002). Assessing immunological function in toxicological studies of avian wildlife. *Integrative Comparative Biology*, **42**, 34–42.

Goüy de Bellocq, J., Sara, M., Casanova, J. C., Feliu, C., and Morand, S. (2003). A comparison of the structure of helminth communities in the woodmouse, *Apodemus sylvaticus*, on islands of the Western Mediterranean and continental Europe. *Parasitology Research*, **90**, 64–70.

Goüy de Bellocq, J., Delarbre, C., Gachelin, G., and Morand, S. (2005). Allelic diversity at the Mhc-DQA locus of woodmouse populations (*Apodemus sylvaticus*) present in the islands and mainland of the north Mediterranean. *Global Ecology and Biogeography*, **14**, 115–22.

Goüy de Bellocq, J., Krasnov, B.R., Khokhlova, I.S., Ghazaryan, L., and Pinshow, B. (2006). Immunocompetence and flea parasitism of a desert rodent. *Functional Ecology*, **20**, 637–46.

Goüy de Bellocq, J., Ribas, A., Casanova, J.-C., and Morand, S. (2007). Immunocompetence and helminth community of the white-toothed shrew, *Crocidura russula*, from the Montseny Natural Park, Spain. *European Journal of Wildlife Research*, **53**, 315–20.

Goüy de Bellocq, J., Charbonnel, N., and Morand, S. (2008). Coevolutionary relationship between helminth diversity and MHC class II polymorphism in rodents. *Journal of Evolutionary Biology*, **21**, 1144–50.

Guernier, V., Hochberg, M.E., and Guégan, J.F. (2004). Ecology drives the worldwide distribution of human infectious diseases. *PloS Biology*, **2**, 740–46.

Hanssen, S.A., Hasselquist, D., Folstad, I., and Erickstad, K.E. (2004). Costs of immunity: immune responsiveness reduces survival in a vertebrate. *Proceedings of the Royal Society of London B*, **271**, 925–30.

Hart, B.L. (1994). Behavioral defense against parasites: Interaction with parasite invasiveness, *Parasitology*, **109**, S139–51.

Hasselquist, D. (2007). Comparative immunoecology in birds: hypotheses and tests. *Journal of Ornithology*, **148**, 571–82.

Hochberg, M.E. and van Baalen, M. (1998). Antagonistic coevolution over productivity gradients. *American Naturalist*, **152**, 620–34.

Janeway, C.A, Travers, P., Walport, M., and Capra, J.D. (1999). *Immunobiology: The Immune System in Health and Disease*, 4th Edn. Current Biology Publications, New York.

John, J.L. (1994). The avian spleen: a neglected organ. *Quarterly Review of Biology*, **69**, 327–51.

Kennedy, M.W. and Nager, R.G. (2006). The perils and prospects of using phytohaemagglutinin in evolutionary ecology. *Trends in Ecology and Evolution*, **21**, 653–55.

Khokhlova, I.S, Krasnov, B.R, Kam, M., Burdelova, N.I., and Degen, A.A. (2002). Energy cost of ectoparasitism: the flea *Xenopsylla ramesis* on the desert gerbil *Gerbillus dasyurus*. *Journal of Zoology*, **258**, 349–54.

Klein, J. (1986). *Natural History of the Major Histocompatibility Complex*. John Wiley & Sons, New York.

Kortet, R., Taskinen, J., Sinisalo, T., and Jokinen, I. (2003). Breeding-related seasonal changes in immunocompetence, health state and condition of the cyprinid fish, *Rutilus rutilus*, L. *Biological Journal of the Linnean Society*, **78**, 117–27.

Krasnov, B.R., Shenbrot, G.I., Khokhlova, I.S., and Degen, A.A. (2004). Flea species richness and parameters of host body, host geography and host "milieu". *Journal of Animal Ecology*, **73**, 1121–28.

Lee, K.A. (2006). Linking immune defence and life history at the level of the individual and the species. *Integrative Comparative Biology*, **46**, 1000–15.

Lee, K.A. and Klasing, K.C. (2004) A role for immunology in invasion biology. *Trends in Ecology and Evolution*, **19**, 523–29.

Lindström, K.M., Foufopoulos, J., Pärn, H., and Wikelski, M. (2004). Immunological investments reflects parasite abundance in island populations of Darwin's finches. *Proceedings of the Royal Society of London B*, **271**, 1513–19.

Lochmiller, R. and Deerenberg, C. (2000). Trade-offs in evolutionary immunology: just what is the cost of immunity? *Oikos*, **88**, 87–98.

Loehle, C. (1995). Social barriers to pathogen transmission in wild animal populations. *Ecology*, **76**, 326–35.

Malo, A.F., Roldan, E.R.S., Garde, J.J. *et al.* (2008). What does testosterone do for red deer males? *Proceedings of the Royal Society of London B*, **276**, 971–80.

Martin, L.B., Scheuerlein, A., and Wikelski, M. (2003). Immune activity elevates energy expenditure of house sparrows: a link between direct and indirect costs? *Proceedings of the Royal Society London B*, **270**, 153–58.

Martin, L.B., Pless, M., Svoboda, J., and Wikelski, M. (2004). Immune activity in temperate and tropical house sparrows: a common-garden experiment. *Ecology*, **85**, 2323–31.

Martin, L.B., Weil, Z.M., and Nelson, R. (2008). Seasonal change in vertebrates immune activity: mediation by physiological trade-offs. *Philosophical Transactions of the Royal Society of London B*, **363**, 321–39.

Matson, K.D. (2006). Are there differences in immune function between continental and insular birds? *Proceedings of the Royal Society of London B*, **273**, 2267–74.

McNab, B.K. (1988). Complications inherent in scaling the basal rate of metabolism in mammals. *Quarterly Review of Biology*, **63**, 25–54.

Meagher, S. (1999). Genetic diversity and *Capillaria hepatica* (Nematoda) prevalence in Michigan deer mouse populations. *Evolution*, **53**, 1318–24.

Mendenhall, C.L, Grossman, C.J., Roselle, G.A. *et al.* (1989) Phytohemagglutinin skin test responses to evaluate in vivo cellular immune function in rats. *Proceedings of the Society for Experimental Biology and Medicine*, **190**, 117–20.

Michaux, J.R., Magnanou, E., Paradis, E., Nieberding, C., and Libois, R. (2003). Mitochondrial phylogeography of the woodmouse (*Apodemus sylvaticus*) in the western Palearctic region. *Molecular Ecology*, **12**, 685–97.

Møller, A.P. (1998). Evidence of larger impact of parasites on hosts in the tropics: investment in immune function within and outside the tropics. *Oikos*, **82**, 265–70.

Møller, A.P. and Erritzøe, J. (1998). Host immune defense and migration in birds. *Evolutionary Ecology*, **12**, 945–53.

Møller, A.P. and Rozsa, L. (2005). Parasite biodiversity and host defenses: chewing lice and immune response of their avian hosts. *Oecologia*, **142**, 169–76.

Mooring, M.S., Blumstein, D.T., and Stoner C.J. (2004). The evolution of parasite-defence grooming in ungulates. *Biological Journal of the Linnean Society*, **81**, 17–37.

Morand, S. and Harvey, P.H. (2000). Mammalian metabolism, longevity and parasite species richness. *Proceedings of the Royal Society of London B*, **267**, 1999–2003.

Morand, S. and Poulin, R. (2000). Nematode parasite species richness and the evolution of spleen size in birds. *Canadian Journal of Zoology*, **78**, 1356–60.

Navarro, C., Marzal, A., de Lope, F. and Møller, A.P. (2003). Dynamics of an immune response in house sparrows *Passer domesticus* in relation to time of day, body condition and blood parasite infection. *Oikos*, **101**, 291–98.

Nunn, C.L. (2002). A comparative study of leukocyte counts and disease risk in primates. *Evolution*, **56**, 177–90.

Nunn, C.L., Gittleman, J.L., and Antonovics, J. (2000). Promiscuity and the primate immune system. *Science*, **290**, 1168–69.

Nunn, C.L., Gittleman, J.L., and Antonovics, J. (2003). A comparative study of white blood cell counts and disease risk in carnivores. *Proceedings of the Royal Society of London B*, **270**, 347–56.

Owen, J.P. and Clayton, D.H. (2007). Where are the parasites in the PHA response? *Trends in Ecology and Evolution*, **22**, 228–29.

Owen-Ashley, N.T, Hasselquist, D., Raberg, L., and Wingfield, J.C. (2008). Latitudinal variation of immune defense and sickness behavior in the white-crowned sparrow (*Zonotrichia leucophrys*). *Brain Behavior and Immunity*, **22**, 614–25.

Pelletier, F., Page, K.A., Ostiguy, T., and Festa-Bianchet, M. (2005). Fecal counts of lungworm larvae and reproductive effort in Bighorn Sheep, *Ovis canadensis*. *Oikos*, **110**, 473–80.

Pisanu, B. and Chapuis J.-L. (2003). Helminths from introduced mammals on sub-Antarctic islands. Antarctic biology in a global context, *Proceedings 8th SCAR International Biology Symposium*, Amsterdam, 240–43.

Pisanu, B., Chapuis, J.-L., and Durette-Desset, M.-C. (2001). Helminths from introduced small mammals on Kerguelen, Crozet, and Amsterdam Islands (Southern Indian ocean). *Journal of Parasitology*, **87**, 1205–08.

Pisanu, B., Chapuis, J.-L., Durette-Desset, M.-C., and Morand, S. (2002). Epizootiology of *Syphacia obvelata* from a domestic mouse population on sub-antarctic Kerguelen archipelago. *Journal of Parasitology*, **88**, 645–49.

Pisanu, B., Chapuis, J.-L., and Perin, R. (2003). *Syphacia obvelata* infections and reproduction of male domestic mice *Mus musculus domesticus* on a sub-Antarctic Island. *Journal of Helminthology*, **77**, 247–53.

Preston, B.T., Capellini, I., McNamara, P., Barton, R.A., and Nunn, C.L. (2009). Parasite resistance and the adaptive significance of sleep. *BMC Evolutionary Biology*, **9**, 7.

Prugnolle, F., Manica, A., Charpentier, M., Guégan, J.-F., Guernier, V., and Balloux, F. (2006). Pathogen-driven selection and worldwide HLA class I diversity. *Current Biology*, **15**, 1022–27.

Pyter, L.M., Weil, Z.M., and Nelson R.J. (2005). Latitude affects photoperiod-induced changes in immune response in meadow voles (*Microtus pennsylvanicus*). *Canadian Journal of Zoology*, **83**, 1271–78.

Raffel, T.R., Rohr, J.R., Kiesecker, J.M., and Hudson, P.J. (2006). Negative effects of changing temperature on amphibian immunity under field conditions. *Functional Ecology*, **20**, 819–28.

Ricklefs, R.E and Wikelski, M. (2002). The physiology/life history nexus. *Trends in Ecology and Evolution*, **17**, 462–68.

Saino, N., Calza, S., and Møller, A.P. (1998). Effects of a dipteran ectoparasite on immune response and growth trade-offs in barn swallow, *Hirundo rustica*, nestlings. *Oikos*, **81**, 217–28.

Salked, D.J., Trivedi, M., and Schwarzkopf, L. (2008). Parasite loads are higher in the tropics: temperate to tropical variation in a single host-parasite system. *Ecography*, **31**, 538–44.

Shutler, D., Alisauskas R.T., and McLaughlin, J.D. (1999). Mass dynamics of the spleen and other organs in geese: measures of immune relationships to helminths? *Canadian Journal of Zoology*, **77**, 351–59.

Šimková, A., Ottová, E., and Morand, S. (2006). MHC variability, life-traits and parasite diversity of European cyprinid fish. *Evolutionary Ecology*, **20**, 465–77.

Šimková, A., Lafond, T., Ondrackova, M., Jurajda, P., Ottová, E., and Morand, S. (2008). Parasitism, life history traits and immune defence in cyprinid fish. *BMC Evolutionary Biology*, **8**, 29.

Smith, K.G. and Hunt, J.L. (2004). On the use of spleen mass as a measure of avian immune system strength. *Oecologia*, **138**, 28–31.

Smits, J.E., Bortolotti, G.R., and Tella, J.L. (2001). Measurement repeatability and the use of controls in PHA assays: a reply to Siva-Jothy and Ryder. *Functional Ecology*, **15**, 814–17.

Stadecker, M.J. and Leskowitz, S. (1974). The cutaneous basophil response to mitogens. *Journal of Immunology*, **113**, 496–500.

Viney, M.E., Riley, E.M., and Buchanan, K.L. (2005). Optimal immune responses: immunocompetence revisited. *Trends in Ecology and Evolution*, **20**, 665–69.

Wakelin, K. (1996). *Immunity to Parasites*. Cambridge University Press, Cambridge.

Webb, R.E., Leslie, D.M., Lochmiller, R.L., and Masters, R.E. (2003). Immune function and hematology of male cotton rats (*Sigmodon hispidus*) in response to food supplementation and methionine. *Comparative Biochemistry and Physiology A*, **136**, 577–89.

Wegner, K.M., Reusch, T.B.H., and Kalbe, M. (2003). Multiple parasites are driving major histocompatibility complex polymorphism in the wild. *Journal of Evolutionary Biology*, **16**, 224–32.

Whiteman, N.K., Matson, K.D., Bollmer, J.L., and Parker, P.G. (2006). Disease ecology in the Galápagos Hawk (*Buteo galapagoensis*): host genetic diversity, parasite load and natural antibodies. *Proceedings of the Royal Society of London B*, **273**, 797–804.

CHAPTER 13

Evolutionary landscape epidemiology

Julie Deter, Nathalie Charbonnel, and Jean-François Cosson

13.1 Introduction: what is evolutionary landscape epidemiology?

Infectious diseases have long been studied by scientists and public managers from clinical points of view, including causes of diseases, prevention of illness, or means to facilitate recovery. These last decades, our understanding of disease occurrence and dynamics has been improved by the recognition of an unequal distribution of parasites in space and of relationships between environmental factors and disease cases. This has led to the development of 'landscape or spatial epidemiology', a field of research which can be traced back to John Snow's work on cholera in London (Snow 1855). By projecting case reports on maps, Snow highlighted that all cases were clustered around a street pump and deduced that cholera was a water-borne pathogen. Later, considering the influence of biotic and abiotic factors on ecological processes, Pavlovsky (1966) stated that landscape strongly affects the spatio-temporal distribution, abundance, and dispersal of hosts and parasites. Landscape epidemiology was then proposed as an integrative approach that aimed to understand the spatial spread of disease agents by analysing both spatial patterns and environmental risk factors (Pavlovsky 1966). The 'BAM diagram' introduced for host–parasite interactions by Soberon and Peterson (2005) provides a useful conceptual framework for understanding the geography of diseases at the scale of the landscape (Fig. 13.1a). It predicts that the distribution of a species (i.e. a parasite) is the overlap between favourable biotic conditions (i.e. presence of competent hosts and other interacting parasites), abiotic factors (i.e. temperature) controlling the survival of free-living stages, and mobility capacities allowing the presence of the species in the appropriate areas. These last years, landscape epidemiology and the modelling of disease dynamics have largely benefited from two scientific breakthroughs. New computing technologies such as geographic information systems or remote sensing have allowed important advances in infectious disease epidemiology. They are described in Part V of this book. Besides this, the combination of metapopulation theory (proposed by Levins 1969) and epidemiological models has greatly improved our understanding of disease dynamics. In these models, a metapopulation of parasites is described as a set of populations distributed over distinct patches represented by either host individuals or host populations, and connected to varying degrees by dispersal (Hess 2002; Fig. 13.1b). This framework has provided fundamental predictions about the probability of disease diffusion or persistence in different situations (e.g. Ostfeld *et al.* 2005 for a review).

A detailed knowledge of the genetics of both host and parasite is essential to predict the outcomes of host–parasite interactions in natural populations, either in terms of population dynamics or evolutionary interactions (see Charbonnel *et al.* 2006 for a review on macroparasites and micromammals). In particular, studying the variability of outcomes generated by genotype x genotype interactions might help disentangle the effects of genetic and environmental factors on the spatial risks of disease emergence. These considerations have led to the 'evolutionary landscape epidemiology' discipline. Its main challenge is the understanding of how both ecology and evolution of host–parasite interactions shape disease distribution, dynamics, and severity over complex geographic landscapes. This connection between epidemiological processes and

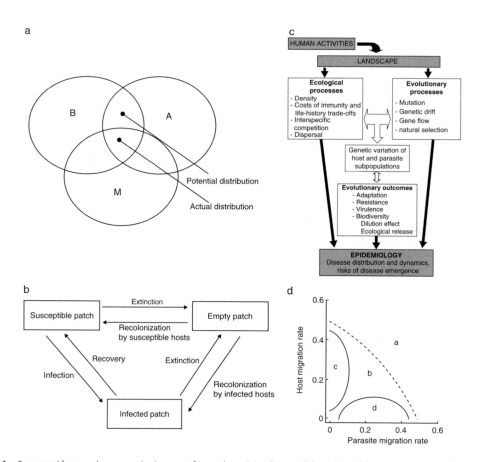

Figure 13.1 Conceptual frameworks representing important factors determining the spatial distribution of diseases at the scale of the landscape. (a) A simple heuristic 'BAM' diagram. It focuses on the interactions among: *A*, which summarizes abiotic variables that circumscribe the potential geographic distribution of a parasite including temperature, moisture, and soil chemistry; *B*, which adds biotic considerations concerning other species such as hosts or other parasite communities; and *M*, which indicates limitations due to restricted dispersal abilities. Modified after Peterson (2008). A virtual actual distribution is represented, as well as the maximal potential one. (b) A metapopulation model. It incorporates disease dynamics by considering a subdivision of habitats into three types: empty patches (no hosts and no parasites), susceptible patches (hosts but no parasites), and infected patches (hosts and parasites). Arrows between patches indicate the most important ecological processes involved in the disease dynamics. Modified after Hess *et al.* (2002). (c) A schematic framework for evolutionary landscape epidemiology. It illustrates the main determinants of disease distribution and severity in heterogeneous landscapes. The left part represents ecological processes influencing parasite abundance and transmission. The right one summarizes microevolutionary processes influencing the outcome of host–parasite interactions. (d) A theoretical framework illustrating selection mosaics and coevolutionary hotspots across landscapes. This graph represents the results of a simulation study. It reports the combinations of host and parasite migration rates resulting in host or parasite local adaptation. There is a weak or an absence of local adaptation in zones a and b. Zones c and d respectively represent areas where parasite and host local adaptation may potentially occur. The dashed line between zones a and b shows the threshold values of host and parasite migration rates above which no fluctuations of gene frequencies are observed. Modified after Gandon and Michalakis (2002).

host–parasite evolution is central to comprehending many issues, such as the evolution of virulence, the emergence of new diseases, and the design of vaccines (Grenfell 2004). Four evolutionary processes, namely mutation, genetic drift, gene flow, and natural selection, act within and among populations to shape genetic variation of hosts and

parasites (Figure 13.1c). Landscape structure is likely to influence these processes. Obviously, it may affect population density, which is related to genetic drift, or movement patterns of hosts or vectors, which are related to gene flow (landscape genetics; see Manel *et al.* 2003). Local environmental differences may also mediate variations in mutation rates (e.g. Weaver *et al.* 1994). Finally, the geographic mosaic theory of coevolution (Thompson 1999) shows that natural selection acting on host–parasite interactions varies among populations. It is partly explained by genotype x genotype x environment interactions in fitness of interacting species. These spatial variations in natural selection and gene flow result in local coevolutionary dynamics (Figure 13.1d) and generate complex distributions of coldspots/hotspots and adaptation/maladaptation patterns across landscapes (see Fig. 13.2, Thrall *et al.* 2002).

It is obvious that all ecological and evolutionary processes act together to shape disease epidemiology. To gain clarity, we first describe how these processes individually affect the spatial distribution and dynamics of diseases across heterogeneous landscapes. We focus on the local scale as the larger scales are investigated in Parts I and II of this book. We then illustrate how the combination of these processes may drive the evolution of disease through host resistance, parasite virulence, or

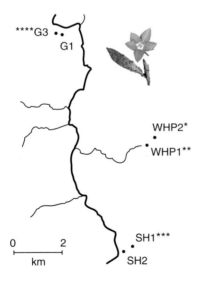

Figure 13.2 Location of *Linum marginale* and *Melampsora lini* populations studied for local adaptation in Australia. Each fungus population exhibited significant variation in its ability to infect the six host populations. Contrast tests for differences in the ability of pathogens to attack their own versus other host populations were significant for four of the six pathogen populations (* – $p < 0.05$, ** – $p < 0.01$, **** – $p < 0.0001$). For pathogen populations SH1, WHP1, and WHP2, mean virulence was greater on sympatric than allopatric host populations, contrary to G3 for which the reverse was observed. Modified after Thrall *et al.* (2002).

impact of biodiversity. Finally, we use this ecological and evolutionary framework to investigate the particular impact of global changes on the risks of disease emergence.

A distinction between infectious diseases, which are caused by pathogens referred to as microparasites (generally viruses, bacteria, fungi, and protozoa), and parasitic diseases, caused by larger size parasites referred to as macroparasites (mainly worms) is sometimes found in the literature. However, recent works tend to consider both kinds of transmissible diseases within a unified epidemiologic framework. In this line, we will use 'parasite' as a generic terminology for both micro- and macroparasites.

13.2   Ecological and evolutionary processes

13.2.1   Favourable habitats, patch size, and genetic drift

Vector-borne and zoonotic diseases often display clear spatial patterns simply because favourable habitats of vectors or hosts are linked to environmental factors that vary in space. It has been evidenced for the distribution of the hantavirus *Puumala*. Linard *et al.* (2007) showed that the spatial distribution of Puumala virus infections were positively associated with forested areas in Belgium. This result was confirmed by Deter *et al.* (2007) at a finer scale in France. This habitat-specific distribution could, however, not be mediated by preferences of the host only, but also by those of *Puumala* virus. Epidemiological models and experiments have shown that this hantavirus might survive outside its host for extended periods (Kallio *et al.* 2006; Sauvage *et al.* 2003). Moreover, *Puumala* virus survival is expected to be higher in forests where humidity, pH, and protection against sunlight are highly favourable. Such habitat effect on the probability of hantavirus transmission was further illustrated considering *Tula* virus and its host, the common vole *Microtus arvalis*. Indeed, infected voles were mainly found in forest edges although *M. arvalis* preferentially lives in grasslands (Deter *et al.* 2008).

In turn, the heterogeneity of favourable habitats, that is, landscape structure, affects population size and density of hosts and parasites, which are major drivers of disease epidemics. Epidemiological models are interested in determining the threshold below which a parasite cannot invade a population of susceptible individuals (McCallum et al. 2001). In this context, the basic reproduction number R_0 is defined as the number of secondary infections produced after the introduction of an infected host (or a mature parasite for macroparasites) into a susceptible population. R_0 allows the definition of the minimum threshold (host population size or density) necessary for the persistence of a parasite in the host population ($R_0=1$). Generally, the eradication of an infectious disease becomes possible when model parameters are able to reduce R_0 below 1. R_0 can be used to determine the Critical Community Size (CCS). Bartlett (1960) defined the CCS as 'the size of the community for which measles is as likely as not to fade out after a major epidemic'. Below the CCS, infection often becomes extinct between outbreaks and must be reintroduced from an external source. Studies based on this concept have highlighted cities that harbour enough individuals to allow the persistence of particular diseases (e.g. Bartlett 1960 for measles). They have also emphasized the role of large urban centres in diffusing epidemics to small towns and to rural districts ('cities and villages' paradigm; see Grenfell and Bolker 1998 for measles).

These examples illustrate how landscape heterogeneity may act directly on disease spatial variation and persistence by shaping the geographic distribution of hosts, vectors, and parasites. Indirect effects also exist as landscape heterogeneity can affect the genetic variability of hosts and parasites. This impact of landscape on the global genetic diversity of hosts and parasites is essential in epidemiology as it is a key component of their evolutionary potential. From a neutral point of view, low levels of genetic variability might be observed in small, isolated patches, and can be associated with inbreeding depression, which sometimes results in a reduction of the ability to mount an immune response (O'Brien and Evermann 1988). This phenomenon, when concerning hosts, will increase the probability that parasites spread in genetically depauperate populations (de Castro and Bolker 2005). Landscape heterogeneity can also lead to an increase of the global genetic diversity. It may concern

parasites with habitat-specific requirements differing among strains (e.g. swine fever virus, Weesendorpa et al. 2008). Abiotic characteristics may also influence the local genetic diversity of parasites through their impact on mutation rates, as evidenced for temperature and the equine encephalite virus (e.g. Weaver et al. 1994). This local genetic diversity is of main importance in epidemiology as it may result in greater competition among strains of parasites, and thus in an increase of parasite virulence (see Galvani 2003 for a review). Finally, experimental and theoretical works suggest that high levels of host genetic diversity might induce variability in host susceptibility and consequently, may decrease the impact of some of their parasites (O'Brien and Evermann 1988).

13.2.2 Dispersal and gene flow

Host and parasite dispersal is one of the key processes influencing spatial disease dynamics. Based on the metapopulation theory, epidemiological models now investigate the probabilities of disease persistence/extinction and diffusion in heterogeneous landscapes (Hess et al. 2002; see also Fig. 13.2).

Habitat connectivity, that is, the degree to which a landscape facilitates or prevents movements among patches depending on the presence of corridors and barriers, is likely to influence spatial patterns of disease spread and persistence (Thrall and Burdon 1997). Briefly, the probability of parasite extinction is high in a given patch if landscape connectivity is low, and for the whole metapopulation if landscape connectivity is high because the risk of host extinction then becomes elevated. Only intermediate levels of connectivity seem to result in long-term persistence of parasites through the rescue effect (e.g. Hagenaars et al. 2004). These conclusions have been largely explored in the *Silene alba* and *Ustilago violacea* studies (review in Carlsson-Granér and Thrall 2002). Identifying dispersers may also help in understanding disease epidemiology. Milne et al. (2008) demonstrated how feral pig population movements between water sources influence the transmission of the classical swine fever. During the dry season, the concentration of herds around water sources reduces the overall disease transmission. Using a simulation model, the authors showed that outbreaks could spread through adult male herds, which travel over great distances and may 'connect' otherwise isolated family herds. All these studies show that disease eradication strategies can be improved by focusing on particular populations, social groupings, or individuals mostly involved in parasite transmission. In this context, landscape genetics (Manel *et al.* 2003) might help in identifying both high-risk populations based on their genetic connection to infected populations, and landscape characteristics reducing host gene flow and consequently disease spread (e.g. Blanchong *et al.* 2008).

From the parasite perspectives, genetics might be the only approach providing insights into the geographic origin of outbreaks. It also gives the opportunity to depict retrospectively the epidemiological history of parasites and their evolutionary changes during disease spread (Archie *et al.* 2009). Studies of rabies viruses illustrate these different contributions of landscape genetics. Genetic and epidemiological models based on a 30-year survey of raccoon rabies virus (RRV) epizootics in North America revealed the pattern of disease dynamics, with the number of infections increasing exponentially during certain periods, then stabilizing (Biek *et al.* 2007). Besides this, the molecular epidemiology analysis of RRV revealed different RRV genetic lineages, each associated with a specific path of spread (Biek *et al.* 2007). This result indicates that the persistence of RRV in its enzootic stage did not depend on regular immigration of infected individuals, and that landscape characteristics, such as continuous mountain ranges and waterways, had a pronounced negative effect on RRV spread.

Dispersal underlies gene flow, the movement of genes between populations. In consequence, it acts as a force homogenizing genetic variation among populations and counteracting the stochasticity mediated by genetic drift within population. Therefore it is an important microevolutionary process affecting coevolutionary dynamics. This point is described in detail in the next paragraph.

13.2.3 Selection and adaptation

As parasites reduce the fitness of the hosts they infect, and hosts develop defence strategies to prevent these infections or limit their effects, selection is an important evolutionary force driving disease distribution and dynamics. Host or parasite fitness may indeed be interpreted in terms of epidemiological parameters such as reproductive rate or mortality. Parasites are expected to become adapted to their local hosts, because they often exhibit shorter generation times and higher abundance than their hosts (Hamilton *et al.* 1990). Local adaptation is defined here as a higher mean fitness of a population on its own habitat than on a remote one. The study of spatial patterns of adaptation (i.e. local adaptation) has led to more complex views of host–parasite coevolutionary processes in heterogeneous landscapes. Many field studies fail to demonstrate parasite local adaptation and even detect local parasite maladaptation (see Kaltz and Shykoff 1998 for review). These geographic patterns of selection mosaics, highly divergent even among narrow geographic places, are the main predictions of the geographic mosaic theory of coevolution (Thompson 1999). Since Gandon *et al.* (1996), numerous theoretical works based on metapopulation modelling have been developed to predict the conditions of genetic drift, gene flow, and selection promoting local adaptation. These works have considered two prominent models to describe host–parasite interaction: the matching allele (a host is resistant unless the parasite matches all of its interaction alleles) and the gene-for-gene (the host is resistant if it recognizes at least one protein of the parasite) models. Population sizes, generation times, mutation, and migration rates of both hosts and parasites may shape local adaptation through their effects on the evolutionary potential of each species (Gandon and Michalakis 2002). Moreover, migration rates are particularly important as the ratio of host to parasite migration rates strongly shapes the pattern of local adaptation when selection intensities are similar for hosts and parasites. Only relatively high levels of gene flow will promote local adaptation, by introducing genetic variability upon which selection can act. Too high migration rates will prevent independent evolution in local populations, and thus local adaptation will not emerge (Gandon and Michalakis 2002). These predictions have been confirmed by empirical studies (meta-analysis studies such as Hoeksema and Forde 2008). Most of these models did not consider spatial heterogeneity in abiotic environments. This variability is nevertheless of main importance in host–parasite coevolutionary processes as the quality or amount of resource available for the host may determine energetically trade-offs between immunity and other life-history traits (Sheldon and Verhulst 1996). The outcomes of genotype x genotype interactions have also been shown to vary with environment, in particular with temperature (see Laine and Tellier 2008). Spatial heterogeneity in abiotic factors might then influence the emergence of local adaptation in host–parasite metapopulation (e.g. Gandon and Nuismer 2009).

Besides these effects of local adaptation, selection can also affect epidemiological processes by influencing genetic diversity of hosts or parasites. Selective effects may homogenize parasite strains over large areas. For example, Biek *et al.* (2007) found a clear decrease with time in the diversity of epizootic hemorrhagic disease virus strains in deer. Five genetic groups were identified throughout eastern USA between 1978 and 1997, then four between 1994 and 1997, and one between 1998 and 2001. This reduction of diversity might be accompanied by changes of parasite virulence or transmission, which will in turn modify the probability of persistence or diffusion of the disease (Galvani 2003).

13.3 Contributions of evolutionary landscape epidemiology

13.3.1 Insights into the evolution of host resistance

Resistance is defined as the detrimental effects of host defence on parasites (Dieckmann *et al.* 2002) and is an important epidemiological parameter.

Landscape heterogeneity can induce spatial variation in host resistance through physiological or genetic processes. First, abiotic factors may influence the availability and quality of food resources, which, in turn, affect body condition and immune defences (Sheldon and Verhulst 1996). For example, in cyclic populations of montane water voles *Arvicola scherman*, high levels of population

abundance induce high stress levels, which result in detrimental body condition and immune function, and participate in further population decline (Charbonnel *et al.* 2008). Second, landscape heterogeneity could influence the evolution of host resistance by its effects on parasite community. Insights are coming from co-infection studies. As an example, Corby-Harris and Promislow (2008) showed that survival of naturally isolated populations of *Drosophila melanogaster* experimentally inoculated with two bacteria (*Lactococcus lactis* and *Pseudomonas aeruginosa*) was positively linked to their previous exposition to species-rich bacterial communities. In humans, interactions between parasites could explain the spatial distribution of many diseases. The absence of smallpox disease cases in dairymaids was explained by high rates of cowpox virus infections, which were transmitted by cattle and conferred a cross-protected immunity (Riedel 2005). Inversely, the distribution of certain diseases can be positively linked to the presence of other diseases. The most obvious examples concern opportunistic diseases, including tuberculosis re-emerging with AIDS epidemics (Vall Mayans *et al.* 1997). Finally, landscape heterogeneity may shape variation in host resistance through different outcomes of genotype–environment interactions (see Schulenburg *et al.* 2009 for a review)

In line with the geographic mosaic theory of coevolution, theoretical models have shown that resistance in host populations may either lead to parasite extinction or explain disease persistence, depending on the spatial heterogeneity of host population structure (Haagenars *et al.* 2004). In particular, resistant populations can act as 'reservoirs' and be at the origin of epidemics when they enter into contact with more susceptible populations. The presence of plague resistant genotypes in black rat populations is thus likely to explain both epidemics of urban plague in ports of Madagascar in the 1990s after 70 years of the disease absence and the persistence of the disease on Malagasy Highlands (Duplantier *et al.* 2003). Modelling studies have confirmed this hypothesis, showing that persistence could occur if 50 per cent of resistant rats were present within populations (Keeling and Gilligan 2000).

13.3.2 Insights into the evolution of parasite virulence

Virulence characterizes the detrimental effect of parasitic exploitation on the host. The different definitions generally found in the literature correspond to the processes through which parasites exploit their host to promote their own multiplication and transmission (Dieckmann *et al.* 2002). (Table 13.1) As virulence affects host life-history traits such as mortality or reproduction, it strongly influences disease epidemiology and host evolution (review in Galvani 2003). Reciprocally, virulence may vary in space and time as the result of complex interactions among evolutionary, ecological, and epidemiological processes.

Optimal virulence, which confers an optimal fitness to parasites, is not zero but results from a trade-off between 'how fast' and 'how long' the parasite transmits: intermediate levels of virulence might be beneficial to parasites if the costs associated with host death are balanced by an increase in the instantaneous rate of transmission (McKinnon and Read 2004b). This concept is supported by the virulence of myxoma virus, which decreased over time after its introduction in Australia (Fenner *et al.* 1956).

Landscape heterogeneity will impose strong selection on virulence through its influence on spatial structuring (and consequently the probability of parasite transmission), and on within-host dynamics of parasite competition. The assumption of a positive relationship between virulence and trans-missibility is inherent to the trade-off described above. Indeed, a parasite that could not disperse to another host patch before it goes extinct would not persist. Experimental studies confirmed that a rapid increase in connectivity may select for parasite strains with higher virulence (Boots and Mealor 2007). Some empirical studies also corroborate this assumption (review in McKinnon and Read 2004a). Parasites with long-ranging dispersal, such as water-borne or air-borne parasites, often sustain greater virulence than those directly transmitted (e.g. Boots and Sasaki 1999). Besides, parasite migration and transmissibility influence parasite genetic variability within host. High levels of migration may lead to competition of differentiated strains infecting the same host. This usually results in the selection of parasite strains exploiting host resources the most rapidly before they become depleted, and thus in higher levels of virulence (Read and Taylor 2001). Inversely, low parasite migration or transmissibility may increase the relatedness of parasite strains within hosts. This is expected to reduce competition and consequently virulence. However, virulence is not so easy to figure out, and more complex outcomes can occur, depending on the nature of competition and its effects on parasite strain fitness (Read and Taylor 2001).

These insights of evolutionary landscape epidemiology on the evolution of virulence are particularly important for the design of public health policies (vaccination campaign, antibiotics etc.) with regard to virulence management.

Table 13.1 The different definitions for virulence are linked to the mechanisms used by parasites to damage the host (Dieckmann *et al.* 2002).

A virulent parasite may...	Involved mechanism	Definition
Enter into the host	*Gaining entrance*	Damages strongly depend on the capacity of parasites to enter the host (especially in plants, battle between resistance genes in the host and genes in the parasite to overcome that resistance). Often little variation is found in the damage inflicted on hosts by different parasite strains once they have gained entrance into the host.
Spread within the host population	*Local spreading*	The harm depends on the transmission within the local population—which, in turn, depends both on the local transmission rate and on the damage inflicted on individual hosts.
Kill the host	*Killing the host*	The exploitation of hosts results in their death
Impact host fitness	*Impairing other life-history characteristics*	Detrimental impacts of the parasites affect host fitness, through a decrease in host fecundity, or a change in its mobility or well-being, but more rarely involve death.

13.3.3 Impact of host, vector, parasite, and predator diversity on disease epidemiology

Host–parasite interactions are rarely reduced to a one-to-one species interaction. Hosts share their environment with other species that may belong to the host spectrum of their parasites. The same holds for parasites, which may coexist with other parasites within their host population (including the co-occurrence of different strains of the same pathogen species). Spatial variation in the biodiversity of these host and parasite communities may affect the outcomes of epidemiology in different ways. A large body of theory concerns the influence of vector or host species diversity on disease epidemiology. This biodiversity might first determine the genetic variability of parasites, which we previously described as an important factor for epidemiology and evolution of diseases. As an example, West Nile virus (WNV) genetic diversity was greater in natural green areas than in residential sites in Chicago. This result was partly explained by the higher biodiversity of avian species in natural sites and of different replicative properties of WNV in these avian species (Bertolotti *et al.* 2008). Therefore, even at the small spatial scale of a city, heterogeneity in landscape shapes variation in parasite genetic diversity through host biodiversity. Host heterogeneity might also select for lower virulence if parasites are more adapted to one host genotype than to the others (see Galvani 2003). This is particularly important when designing agricultural or farming practices to prevent outbreaks of virulent diseases. Host or vector biodiversity might also influence epidemiological parameters such as parasite transmission through the 'dilution effect' (Ostfeld and Keesing 2000; Schmidt and Ostfeld 2001). Under four conditions (vector species are generalist with regard to definitive hosts, transmission is horizontal, host susceptibility is species specific, and the most susceptible host species is the more abundant), host biodiversity limits infection rates of vectors (Schmidt and Ostfeld 2001). Indeed transmission to and from the most competent host reservoir is diluted due to the presence of less suitable hosts. The dilution effect has been demonstrated in several vector-borne diseases, among which the most famous are the Lyme disease due to

Borrelia burgdorferi (LoGiudice *et al.* 2003) and the WNV fever (see Ezenwa *et al.* 2006). The 'dilution effect' also highlights the potential outcomes of host species introduction on disease epidemiology. If introduced species are less competent than native ones, they will induce a dilution effect and then reduce the epidemiological risk of disease. Inversely, if introduced species are more competent reservoirs than native ones, the epidemiological risk might be amplified (Keesing *et al.* 2006).

The role of vector diversity in disease epidemiology is less investigated. Both intra-species genetic diversity and species diversity may play a role in disease persistence and spread as they induce variability in parasite transmission within and between host species (Power and Flecker 2007). However, how this vector diversity might affect disease dynamics remains to be explored mathematically.

Reciprocally, landscape heterogeneity, through variation in abiotic conditions, may shape parasite species biodiversity, which might in turn influence host biodiversity and community composition. Generalist parasites exhibiting varying levels of virulence according to host species will mediate apparent competition among hosts and, consequently, the abundance of less resistant host species may decrease (e.g. Tompkins *et al.* 2000). On the other hand, specialist parasites are expected to become adapted to the most common host species, which will in turn decrease in abundance through the effects of virulence. This will reduce parasite mediated competition among this host community, and maintain high levels of host biodiversity (Hudson *et al.* 2006).

Finally, spatial variability in host antagonist communities including predators will induce changes in epidemiological risks. For example, Lyme disease has emerged in re-forested areas in North America since the 1990s. The white-footed mouse *Peromyscus leucopus* is by far the most competent host for the bacteria transmission via the tick vector *Ixodes scapularis*. Within large forest fragments, the high diversity of competitors and predators regulated the population dynamics of mice, and the transmission risk was therefore low. Alternatively, the mice display high densities within small reforested fragments, which are depleted in competitors and predators. Such high densities induce high transmission levels of the bacteria and therefore high transmission risks to humans (Allan *et al.* 2003; Davis *et al.* 2005).

13.4 Applications of evolutionary landscape epidemiology: global changes and preventive actions

13.4.1 Human activities and landscape disturbance

Human activities such as deforestation, agriculture, farming, hunting, and fishing may contribute to landscape disturbance and impact on the epidemiological risks of infectious disease emergence. These changes in landscape structure usually affect all the ecological and evolutionary processes described above with complex consequences. Malaria is a relevant disease to illustrate this link between changes in landscape and epidemiology. In the last century, large use of quinine, drying of marshes, and improvement of housing conditions induced a drastic decrease of malaria cases in southern France where the disease was endemic. Nowadays, only imported cases are reported in France, although a very competent vector is still present in the Camargue (Ponçon *et al.* 2007). However, landscape changes could reverse the situation. In Kenya for example, deforestation could enhance malaria risk by increasing mosquito density, biting frequency, and by enhancing the vectorial capacities of *Anopheles gambiae* mosquitoes (Afrane *et al.* 2008).

Landscape disturbance may change the diversity of species present in an ecosystem, and consequently strongly impact the epidemiology of infectious diseases. Fragmentation of natural habitats and agriculture development are the principal human activities contributing to such changes (Rosenzweig 1995). One common effect is the local extinction of predators, resulting in the loss of top-down natural control in ecological communities. This, in turn, usually results in an increased abundance of species (Terborgh *et al.* 2001) potentially involved in disease agent transmission (rodents, insects, ticks, weeds). This phenomenon, called 'ecological release' can explain the occurrence of many emerging zoonotic diseases in areas exhibit-

ing abrupt or episodic social and ecological changes (Wilcox and Gubler 2005). The example of the Nile perch *Lates niloticus* introduction in Lake Victoria (East Africa) illustrates this point, and more specifically how the introduction of new species for livestock farming or fishing may unbalance food webs and increase epidemiological risks. This introduction was initially committed to improve the local economic development, but it surprisingly led to epidemics of schistosomiasis. Community studies have shown that the introduction of this species caused a reduction in the diversity of cichlid fishes, which, in turn, induced an important increase of their prey, such as snails, among which are the intermediate hosts of *Schistosoma sp.* (Constantin De Magny *et al.* 2008).

Besides, the desire for increasing resource production is often at the origin of a homogenization of the landscape, leading to ecosystems of large patch size and high connectivity, where disease outbreaks can easily persist and spread out (Gunderson and Holling 2002).

Landscape disturbance may induce closer contacts between reservoirs of parasites and humans, or between domestic and feral animals that carry disease agents. Such changes might affect epidemiological cycles and selective pressure acting on parasites through 'spillover' events, that is, the possibility for a parasite to infect a new host. A landscape genetics study suggested that forest fragmentation in Uganda increases cross-species transmission rates of bacteria by enhancing the ecological overlap among host species (Goldberg *et al.* 2008). Indeed both humans and livestock harboured *Escherichia coli* bacteria that were more genetically similar to bacteria found in non-human primates living in remnant forests within habitat mosaics of human settlements, than to those sampled on non-human primates living in undisturbed forests. Close contact between non-human and human primates following environmental changes in Central West Africa were also the reason behind cross-species transmission of lentiviruses and the AIDS pandemic (Heeney *et al.* 2006). Changes in epidemiological cycle and selected viral adaptations have subsequently facilitated human-to-human transmission. Similarly, the development of intensive poultry and pig breeding in countryside provides a favourable context for host switching from wild reservoirs to animals living in close contact with humans. A lot of evidence support pigs as intermediate hosts of inter-species transmission or as mixing vessels for avian and human influenza viruses and the emergence of new strains with human pandemic potential (Ito *et al.* 1998).

Finally, it is also of concern that worldwide pandemics may establish into local epidemics in areas where spillover, parasite evolution, and changes in epidemiological cycles are favoured (emerging disease 'hotspots'). As mentioned above, landscape structure is central to this question, as heterogeneity and connectivity may favour contact between original and new potential reservoirs and humans. The changing ecology of murine typhus in southern California and Texas over the past 30 years is a good example of made-man landscape changes (here, urban and suburban expansion toward rural areas) affecting infectious disease outbreaks. In these areas, the classic rat–flea–rat cycle of *Rickettsia typhi* has been replaced by a peri-domestic animal cycle involving free-ranging cats, dogs, and opossums and their fleas (Azad *et al.* 1997).

All these examples highlight the links between human activities, biodiversity, and spatial proximity between species on the one hand, and circulation and cross-species transmission of parasites on the other hand. From the public health and conservation points of view, further extensive surveillance of animals is important to better understand the animal reservoir in wildlife and the inter-species transmission events that can lead to outbreaks. Identified risky behaviours could then be targeted to reduce disease emergence and transmission.

13.4.2 Climate changes

Climate changes have considerable effects on ecosystems, communities, and populations (Walther *et al.* 2002). In particular, there is increasing evidence that frequency and severity of disease emergence are influenced by global warming and extreme meteorological events (Daszak 2000; Jones *et al.* 2008). The numerous correlations observed between warmer and wetter conditions associated with El Nino events and outbreaks of emerging infectious diseases, including malaria or cholera, illustrate these points (Kovats *et al.* 2003).

Climate changes can lead to disease emergence by different processes. First, warming has recently caused changes in species distribution and abundance, in particular those of hosts, vectors, or parasites (Epstein 1999; Parmesan 1996). Warming might thus favour transmission by allowing the presence, in sufficient density, of reservoir hosts, competent vectors, and parasites in the same area at the same time. Emerging diseases can then concern pre-existing parasites in areas where hosts or vectors were not present or were at too low densities (e.g. leishmaniasis in the Mediterranean basin with the geographic expansion of sandflies), or introduce parasites that would find the favourable temperature conditions to emerge (e.g. malaria or dengue). Second, warming induces important modifications of parasite epidemiological parameters, including accelerated vector and parasite life-cycles, increased transmission and virulence (e.g. for plant pathogens Harvell *et al.* 2002). Third, climate changes are expected to modify host susceptibility to infection. Warming might induce stress on hosts through the impoverishment of water and resources. This could lead to energy trade-offs that might occur at the expense of the ability to mount immune responses. Amphibians provide an important illustration of these impacts of climatic changes on host susceptibility, in particular because global warming has been suggested as one of the main factors for amphibian decline due to infectious diseases (Pounds *et al.* 2006). Seasonal surveys show that environmental temperature strongly influences the amphibian immune system and susceptibility to parasites (see Raffel *et al.* 2006). Climate changes, including both warming and changing climatic variability, can accentuate the emergence of infectious diseases in amphibians. Indeed, warming has induced an increase of common toad metabolic rate during hibernation in the United Kingdom. Consequently, due to this physiological stress, body condition and immuno-competence are weakened during amphibian spring emergence and increase the risk of disease emergence (Reading 2006). In Costa Rica, dry weather forces amphibians to gather near water sources, thus increasing their probability of being attacked by parasitic flies or fungi (Pounds *et al.* 1999). Besides this, increased variability in climatic conditions might lead to longer or more frequent periods of immune suppression in amphibians, which could exacerbate amphibian declines (Raffel *et al.* 2006).

Simultaneously, climate changes will also reduce the emergence risk of certain infectious diseases. Looking at amphibians again, global warming will limit the spread of chytrid epizootics, which are strongly involved in amphibian population declines in Central America and Australia. Indeed, this parasite requires cool temperatures and humidity, so its geographic distribution is expected to shrink (Harvell *et al.* 2002). Epidemiological risks associated with fungal entomopathogens might also decline as hot and dry conditions hamper their growth (Harvell *et al.* 2002).

13.4.3 Preventive actions

Studying ecological and evolutionary processes governing spatial epidemiology of diseases should help managers to develop preventive actions. Such measures may consist of managing landscapes in order to influence host communities, reservoir populations, and the dispersal of infectious hosts, in a manner designed to reduce epidemiological risks. Understanding 'who' and 'when' disperses with the parasite is important for control strategies, which would be more successful if focused on the right target (for example adult male herds of feral pig populations during the dry season to reduce the risk of swine fever—see above). Adverse effects may occur when the epidemiological system is not well understood. In the UK, Eurasian badgers *Meles meles* are principal wildlife reservoirs for the agent of bovine tuberculosis (TB), a serious disease of cattle and other mammals including humans. Therefore an intuitive solution to reduce TB transmission rates has always been to cull badgers and reduce their densities. However, a recent synthesis revealed that culling made the problem worse by upsetting badger social structure and increasing rates of movement and consequently disease transmission (McDonald *et al.* 2008).

In parallel to landscape changes, continuous improvements in hygiene and education conditions, added to the use of pesticides (against vectors), antibiotics, and vaccines, have drawn back numerous parasitic and infectious diseases. These measures impact the dispersal of parasites and their abilities to invade a host population. Unfortunately, such preventive actions are not uniformly developed and depend on the richness of the countries/socio-economic categories involved. This still strongly explains the spatial distribution of certain parasitic and infectious diseases (Jones *et al.* 2008). Vaccination is able to totally change the spatio-temporal dynamics of diseases by reducing the number of susceptible hosts. It may even lead to the eradication of diseases (e.g. smallpox in the 1980s). However, vaccination strategy, if it is not well thought out, can produce perverse effects like unexpectedly increasing the number of infectious people (Choisy *et al.* 2006), enhanced virulence (e.g. McKinnon and Read 2004a), the occurrence of vaccine-favoured variants that may be implicated in the re-emergence of disease (review in Gandon and Day 2008), or the emergence of resistant strains Gandon (Iwami *et al.* 2009). Depending on landscapes and epidemiological cycles, optimal vaccination strategies probably exist and need to be explored mathematically before being set *in natura* (Gandon and Day 2009).

13.5 Perspectives

Emerging diseases may bring three kinds of emergence together:

(1) established infectious diseases undergoing increased incidence;
(2) newly discovered infections; and
(3) newly evolving (newly occurring) infections (McMichael 2004).

While wars and conquests were the most important causative factors of disease emergence 200–500 years ago, environmental changes have been identified as some of the main drivers of these increased epidemiological risks these last decades (Fig. 13.3).

In this chapter, we have shown how landscape may influence disease epidemiology through ecological and evolutionary processes acting at relatively fine scales (see Fig. 13.1c). We have highlighted the importance of the evolutionary approach, which has brought insights into the biological processes

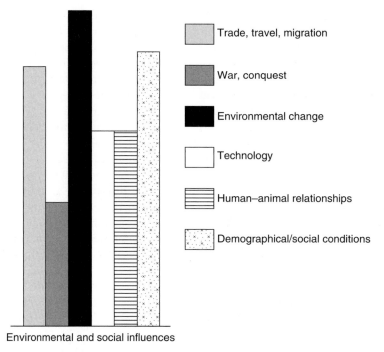

Figure 13.3 Indicative relative importance of various major environmental and social influences on infectious disease emergence at present time. The top 12 environmental changes, in descending order, were as follows: (1) agricultural development; (2) urbanization; (3) deforestation; (4) population movement; (5) introduced species/pathogens; (6) biodiversity loss; (7) habitat fragmentation; (8) water and air pollution; (9) road building; (10) impacts of HIV/AIDS; (11) climatic changes; and (12) hydrological changes, including dams. Modified after McMichael (2004).

involved in disease emergence and has provided a better understanding of some patterns observed. Investigating both processes and patterns are an essential prerequisite for the development of modelling that aims to predict the changes of epidemiological risks associated with landscape disturbance, global changes, or management policies. The importance of a systemic view including different spatial and temporal scales, and the associated processes favouring zoonotic and infectious diseases, is highlighted by a growing number of authors (e.g. Constantin De Magny *et al.* 2008). Jones *et al.* (2008) recently analysed the factors underlying the origins of 335 emerging infectious diseases, and demonstrated non-random global patterns, confirming the role of particular socio-economic, environmental, and ecological factors. These studies therefore lead to important predictions in the areas or categories of diseases with high risks of emergence. Another important factor influencing human (but not only

human) disease distribution is travelling. More and more mathematical models take human movements into account in the study of epidemics, and aim at predicting the effects of particular containment measures, including travel restriction (Colizza and Vespignani 2008).

There are many challenges remaining to be addressed, and we will detail here two ways to better take advantage of the evolutionary information provided by molecular epidemiology. First, a better assessment of epidemiological risks could be gained from the molecular *a priori* characterization of natural wildlife reservoir susceptibility/vector competence. This information might be achieved from immuno-genomics, the analysis of genomic polymorphism in specific recognition and immune regulation. It focuses on immune defence genes and on the variability of outcomes generated by genome–genome interactions between and within host and parasite species. One of the leading goals of immuno-genomics has been to understand the genetic basis of susceptibility/resistance to infectious diseases. Combined with spatio-temporal surveys of natural populations, that is, molecular epidemiology, knowledge of immuno-genomic polymorphism could provide a key insight into the factors that determine the appearance, spread, and distribution of resistance/immuno-modulating alleles within populations and across geographic areas, and consequently into geographical epidemiological risks at the intra-specific level. On another hand, combined with phylogenies, immuno-genomics might help in identifying the host genetic characteristics associated with the possibility of parasites, including viruses, to replicate or persist within their hosts (e.g. for a review on hantaviruses Schöenrich *et al.* 2008, and on arenaviruses Radoshitzky *et al.* 2008). Such an approach could provide immuno-genomic key parameters to identify species that would be likely to be reservoirs for a given parasite. This information could help in designing preventive management policies to control spill-over events and therefore epidemics in ecosystems.

Second, the spatial analyses of epidemiological risks based on geographic information systems and remote sensing still mainly focus on socio-economic or environmental factors and ecological processes. Molecular epidemiological information including both the distribution of genetic variation associated with host susceptibility, vector competence or parasite virulence, and the outcomes of genetic x environment interactions, should be incorporated in these approaches for a better consideration of evolutionary processes, and consequently, a better evaluation of spatial epidemiological risks.

References

Afrane, Y.A., Little, T.J., Lawson, B.W., Githeko, A.K., and Yan, G. (2008). Deforestation and vectorial capacity of *Anopheles gambiae* giles mosquitoes in malaria transmission, Kenya. *Emerging Infectious Diseases*, **14**, 1533–38.

Allan, B.F., Keesing, F. and Ostfeld, R.S. (2003). The effect of habitat fragmentation on Lyme disease risk. *Conservation Biology*, **17**, 267–72.

Archie, E.A., Luikart, G. and Ezenwa, V.O. (2009). Infecting epidemiology with genetics: a new frontier in disease ecology. *Trends in Ecology and Evolution*, **24**, 21–30.

Azad, A.F., Radulovic, S., Higgins, J.A., Noden, B.H., and Troyer, J.M. (1997). Flea-borne rickettsioses: ecologic considerations. *Emerging Infectious Diseases*, **3**, 319–27.

Bartlett, M.S. (1960). The critical community size for measles in the United States. *Journal of the Royal Statistical Society A*, **123**, 37–44.

Bertolotti, L., Kitron, U.D., Walker, E.D. *et al.* (2008) Fine-scale genetic variation and evolution of West Nile Virus in a transmission "hot spot" in suburban Chicago, USA. *Virology*, **374**, 381–9.

Biek, R., Henderson, J.C., Waller, L.A., Rupprecht, C.E., and Real, L.A. (2007). A high-resolution genetic signature of demographic and spatial expansion in epizootic rabies virus. *Proceedings of the National Academy of Sciences of the USA*, **104**, 7993–38.

Blanchong, J.A., Samuel, M.D., Scribner, K.T., Weckworth, B.V., Langenberg, J.A., and Filcek, K.B. (2008). Landscape genetics and the spatial distribution of chronic wasting disease. *Biology Letters*, **4**, 130–33.

Boots, M. and Mealor, M. (2007). Local interactions select for lower pathogen infectivity. *Science*, **315**, 1284–86.

Boots, M. and Sasaki, A. (1999). 'Small worlds' and the evolution of virulence: infection occurs locally and at a distance. *Proceedings of the Royal Society of London B*, **266**, 1933–38.

Carlsson-Granér, U. and Thrall, P.H. (2002). The spatial distribution of plant populations disease dynamics and evolution of resistance. *Oikos*, **97**, 97–110.

Charbonnel, N., Göuy de Bellocq, J., and Morand, S. (2006). Immunogenetics of micromammal-macroparasite interactions. In: S. Morand, B.R. Krasnov and R. Poulin, eds. *Micromammals and Macroparasites: From Evolutionary Ecology to Management*, pp. 401–42. Springer Verlag, Tokyo.

Charbonnel, N., Chaval, Y., Berthier, K. *et al.* (2008). Stress and demographic decline: a potential effect mediated by impairment of reproduction and immune function in cyclic vole populations. *Physiological and Biochemical Zoology*, **81**, 63–73.

Choisy, M., Guégan, J.-F., and Rohani, P. (2006). Dynamics of infectious diseases and pulse vaccination: teasing apart the embedded resonance effect. *Physica D*, **223**, 26–35.

Colizza, V. and Vespignani, A. (2008). Epidemic modelling in metapopulation systems with heterogeneous coupling pattern: Theory and simulations. *Journal of Theoretical Biology*, **251**, 450–67.

Constantin De Magny, G., Durand, P., Renaud, F., and Guégan, J.-F. (2008). Health ecology: a new tool, the Macroscope. In F. Thomas, F. Renaud, and J-F. Guégan, eds. *Ecology and Evolution of Parasitism*. Oxford University Press, Oxford, pp. 129–48.

Corby-Harris, V. and Promislow, D.E.L. (2008). Host ecology shapes geographical variation for resistance to bacterial infection in *Drosophila melanogaster*. *Journal of Animal Ecology*, **77**, 768–76.

Daszak, P., Cunningham, A.A., and Hyatt, A.D. (2000). Emerging infectious diseases of wildlife - threats to biodiversity and human health. *Science*, **287**, 443–49.

Davis, S., Calvet, E., and Leirs, H. (2005). Fluctuating rodent populations and risk to humans from rodent-borne zoonoses. *Vector-Borne and Zoonotic Diseases*, **5**, 305–14.

De Castro, F. and Bolker, B. (2005). Mechanisms of disease-induced extinction. *Ecology Letters*, **8**, 117–26.

Deter, J. (2007). *Ecologie de la Transmission de Parasites (Virus, Nématodes) au sein d'une Communauté de Rongeurs à Populations Cycliques. Conséquences sur la Santé Humaine* PhD thesis, Université de Montpellier II, Montpellier.

Deter, J., Chaval, Y., Galan, M. *et al.* (2008). Kinship, dispersal and hantavirus transmission in bank and common voles. *Archives of Virology*, **153**, 435–44.

Dieckmann, U., Metz, J.A.J., Sabelis, M.W., and Sigmund, K. (2002). *Adaptive Dynamics of Infectious Diseases: In Pursuit of Virulence Management*. Cambridge University Press, Cambridge.

Duplantier, J.-M., Catalan, J., Orth, A., Grolleau, B., and Britton-Davidian, J. (2003). Systematics of the black rat in Madagascar: consequences for the transmission and distribution of plague. *Biological Journal of the Linnean Society*, **78**, 335–41.

Epstein, P.R. (1999). Climate and health. *Science*, **285**, 347–8.

Ezenwa, V.O., Godsey, M.S., King, R.S.J., and Guptill, S.C. (2006). Avian diversity and West Nile virus: testing associations between biodiversity and infectious disease risk. *Proceedings of the Royal Society of London B*, **273**, 109–17.

Fenner, F., Day, M.F., and Woodroofe, G.M. (1956). The epidemiological consequences of the mechanical transmission of myxomatosis by mosquitoes. *Journal of Hygiene*, **54**, 284–303.

Galvani, A.P. (2003). Epidemiology meets evolutionary ecology. *Trends in Ecology and Evolution*, **18**, 132–39.

Gandon, S. and Nuismer, S.L. (2009). Interactions between genetic drift, gene flow, and selection mosaics drive parasite local adaptation. *American Naturalist*, **173**, 212–24.

Gandon, S. and Day, T. (2008). Evidences of parasite evolution after vaccination. *Vaccine*, **26**, 4–7.

Gandon, S. and Day, T. (2009). The evolutionary epidemiology of vaccination. *Journal of the Royal Society Interface*, **4**, 803–19.

Gandon, S., Capowiez, Y., Dubois, Y., Michalakis, Y., and Olivieri, I. (1996). Local adaptation and gene-for-gene coevolution in a metapopulation model. *Proceedings of the Royal Society of London B*, **263**:1003–09.

Gandon, S. and Michalakis, Y. (2002). Local adaptation, evolutionary potential and host-parasite coevolution: interactions between migration, mutation, population size and generation time. *Journal of Evolutionary Biology*, **15**, 451–62.

Goldberg, T.L., Gillespie, T.R., Rwego, I.B., Estoff, E.L., and Chapman, C.A. (2008). Forest fragmentation as cause of bacterial transmission among nonhuman primates, humans, and livestock, Uganda. *Emerging Infectious Diseases*, **14**, 1375–82.

Grenfell, B.T. (2004). Unifying the epidemiological and evolutionary dynamics of pathogens. *Science*, **303**, 327–32.

Grenfell, B.T. and Bolker, B.M. (1998). Cities and villages: infection hierarchies in a measles metapopulation. *Ecology Letters*, **1**, 63–70.

Gunderson, L.H. and Holling, C.S. (2002). *Panarchy: Understanding Transformations in Systems of Humans and Nature*. Island Press, Washington.

Hagenaars, T.J., Donnelly, C.A., and Ferguson, N.M. (2004). Spatial heterogeneity and the persistence of infectious diseases. *Journal of Theoretical Biology*, **229**, 349–59.

Hamilton, W. D., Axelrod, R., and Tanese, R. (1990). Sexual reproduction as an adaptation to resist parasites (a review). *Proceedings of the National Academy of Sciences of the USA*, **87**, 3566–73.

Harvell, C.D., Mitchell, C.E., Ward, J.R. *et al.* (2002) Climate warming and disease risks for terrestrial and marine biota. *Science*, **296**, 2158–62.

Heeney, J.L., Dalgleish, A.G., and Weiss, R.A. (2006). Origins of HIV and the evolution of resistance to AIDS. *Science*, **313**, 462–66.

Hess, G.R., Randolph, S.E., Arneberg, P. *et al.* (2002). Spatial aspects of disease dynamics. In P.J. Hudson, A. Rizzoli, B.T. Grenfell *et al.*, eds. *The Ecology of Wildlife Diseases*, pp. 102–18. Oxford University Press, Oxford.

Hoeksema, J. D. and Forde, S. E. (2008). A meta-analysis of factors affecting local adaptation between interacting species. *American Naturalist*, **171**, 275–90.

Hudson, P.J., Dobson, A.P., and Lafferty, K.D. (2006). Is a healthy ecosystem one that is rich in parasites? *Trends in Ecology and Evolution*, **21**, 381–85.

Ito, T., Couceiro, J.N., Kelm, S. *et al.* (1998). Molecular basis for the generation in pigs of influenza A viruses with pandemic potential. *Journal of Virology*, **72**, 7367–73.

Iwami, S., Suzuki, T., and Takeuchi, Y. (2009). Paradox of vaccination: is vaccination really effective against avian flu epidemics? *PLoS One*, **4**, e4915.

Jones, K.E., Patel, N.G., Levy, M.A. *et al.* (2008). Global trends in emerging infectious diseases. *Nature*, **451**, 990–94.

Kallio, E.R., Klingström, J., Gustafsson, E. *et al.* (2006). Prolonged survival of Puumala hantavirus outside the host: evidence for indirect transmission via the environment. *Journal of General Virology*, **87**, 2127–34.

Kaltz, O. and Shykoff, J.A. (1998). Local adaptation in host-parasite systems. *Heredity*, **81**, 361–70.

Keeling, M.J. and Gilligan, C.A. (2000). Bubonic plague: a metapopulation model of a zoonosis. *Proceedings of the Royal Society of London B*, **267**, 2219–30.

Keesing, F., Holt, R.D., and Ostfeld, R.S. (2006). Effects of species diversity on disease risk. *Ecology Letters*, **9**, 485–98.

Kovats, R.S., Bouma, M.J., Hajat, S., Worrall, E., and Haines, A. (2003). El Niño and health. *Lancet*, **362**, 1481–89.

Laine, A.L. and Tellier, A. (2008). Heterogeneous selection promotes maintenance of polymorphism in host-parasite interactions. *Oikos*, **117**, 1281–88.

Levins, R. (1969). Some demographic and genetic consequences of environmental heterogeneity for biological control. *Bulletin of the Entomological Society of America*, **15**, 237–40.

Linard, C., Lamarque, P., Heyman, P. *et al.* (2007). Determinants of the geographic distribution of Puumala virus and Lyme borreliosis infections in Belgium. *International Journal of Health Geography*, **2**, 15.

LoGiudice, K., Ostfeld, R.S., Schmidt, K.A., and Keesing, F. (2003). The ecology of infectious diseases: effects of host diversity and community composition on Lyme disease risk. *Proceedings of the National Academy of Sciences of the USA*, **100**, 567–71.

Manel, S., Schwartz, M.K., Luikart, G., and Taberlet, P. (2003). Landscape genetics: combining landscape ecology and population genetics. *Trends in Ecology and Evolution*, **18**, 189–97.

McCallum, H., Barlow, N., and Hone, J. (2001). How should pathogen transmission be modelled? *Trends in Ecology and Evolution*, **16**, 295–300.

McKinnon, M.J. and Read, A.F. (2004a). Immunity promotes virulence evolution in a malaria model, *PLoS Biology*, **2**, e230.

McKinnon, M.J. and Read, A.F. (2004b). Virulence in malaria: an evolutionary viewpoint. *Philosophical Transactions of the Royal Society of London B*, **359**, 965–86.

McMichael, A.J. (2004). Environmental and social influences on emerging infectious diseases: past, present and future. *Philosophical Transactions of the Royal Society of London B*, **359**, 1049–58.

Milne, G., Fermanis, C., and Johnston, P. (2008). A mobility model for classical swine fever in feral pig populations. *Veterinary Research*, **39**, 53.

O'Brien, S.J. and Evermann, J.F. (1988). Interactive influence of infectious-disease and genetic diversity in natural populations. *Trends in Ecology and Evolution*, **3**, 254–59.

Ostfeld, R.S. and Keesing, F. (2000). Biodiversity and disease risk: the case of Lyme disease. *Conservation Biology*, **14**, 722–28.

Ostfeld, R.S., Glass, G.E., and Keesing, F. (2005). Spatial epidemiology: an emerging (or re-emerging) discipline. *Trends in Ecology and Evolution*, **20**, 328–35.

Parmesan, C. (1996). Climate change and species' range. *Nature*, **382**, 765–66.

Pavlovsky, E.N. (1966). *The Natural Nidality of Transmissible Disease*. University of Illinois Press, Urbana.

Ponçon, N., Toty, C., L'Ambert, G. *et al.* (2007). Biology and dynamics of potential malaria vectors in Southern France. *Malaria Journal*, **6**, 18.

Pounds, J.A., Fogden, M.P.L., and Campbell, J.H. (1999). Biological response to climate change on a tropical mountain. *Nature*, **398**, 611–15.

Pounds, J.A., Bustamante, M.R., Coloma, L.A. *et al.* (2006). Widespread amphibian extinctions from epidemic disease driven by global warming. *Nature*, **439**, 161–67.

Power, A.G. and Flecker, A.S. (2007). The role of vector diversity in disease dynamics. In: R.S. Ostfeld, F. Keesing, V.T. Eviner, eds. *Infectious Disease Ecology*, pp. 30–47. Princeton University Press, Princeton.

Radoshitzky, S.H., Kuhn, J.H., Spiropoulou, C.F. *et al.* (2008). Receptor determinants of zoonotic transmission of New World hemorrhagic fever arenaviruses. *Proceedings of the National Academy of Sciences of the USA*, **105**, 2664–69.

Raffel, T.R., Rohr, J.R., Kiesecker, J.M., and Hudson, P.J. (2006). Negative effects of changing temperature on amphibian immunity under field conditions. *Functional Ecology*, **20**, 819–28.

Read, A.F. and Taylor, L.H. (2001). The ecology of genetically diverse infections. *Science*, **292**, 1099–102.

Reading, C.J. (2006). Linking global warming to amphibian declines through its effects on female body condition and survivorship. *Oecologia*, **151**, 125–31.

Riedel, S. (2005). Edward Jenner and the history of smallpox and vaccination. *Baylor University Medical Center Proceedings*, **18**, 21–5.

Rosenzweig, M.L. (1995). *Species Diversity in Space and Time*. Cambridge University Press, Cambridge.

Sauvage, F., Langlais, M., Yoccoz, N.G., and Pontier, D. (2003). Modelling hantavirus in fluctuating populations of bank voles: the role of indirect transmission on virus persistence. *Journal of Animal ecology*, **72**, 1–13.

Schmidt, K.A. and Ostfeld, R.S. (2001). Biodiversity and the dilution effect in disease ecology. *Ecology*, **82**, 609–19.

Schoenrich, G., Rang, A., Lütteke, N., Raftery, M. J., Charbonnel, N., and Ulrich, R. G. (2008). Hantavirus-induced immunity in rodent reservoirs and humans. *Immunological reviews*, **225**, 163–89.

Schulenburg, H., Kurtz, J., Moret, Y., and Siva-Jothy, M.T. (2009). Introduction. Ecological immunology. *Philosophical Transactions of the Royal Society of London B*, **364**, 3–14.

Sheldon, B.C. and Verhulst, S. (1996). Ecological immunology: costly parasite defences and trade-offs in evolutionary ecology. *Trends in Ecology and Evolution*, **11**, 317–21.

Snow, J. (1855). *On the Mode of Communication of Cholera*. John Churchill, London.

Soberón, J. and Peterson, A.T. (2005). Interpretation of models of fundamental ecological niches and species' distributional areas. *Biodiversity Informatics*, **2**, 1–10.

Terborgh, J., Lopez, L., Nuñez, P. *et al.* (2001). Ecological meltdown in predator-free forest fragments. *Science*, **294**, 1923–26.

Thompson, J.N. (1999). Specific hypotheses on the geographic mosaic of coevolution. *American Naturalist*, **153**, S1–S14.

Thrall, P.H. and Burdon, J.J. (1997). Host-pathogen dynamics in a metapopulation context: the ecological and evolutionary consequences of being spatial. *Journal of Ecology*, **85**, 74–53.

Thrall, P.H., Burdon, J.J., and Bever, J.D. (2002). Local adaptation in the *Linum marginale-Melampsora lini* host-pathogen interaction. *Evolution*, **56**, 1340–51.

Tompkins, D.M., Greenman, J., Robertson, P., and Hudson, P. (2000). The role of shared parasites in the exclusion of wildlife hosts: *Heterakis gallinarum* in the ring-necked pheasant and the grey partridge. *Journal of Animal Ecology*, **69**, 829–41.

Vall Mayans, M., Maguire, A., Miret, M., Alcaide, J., Parrón, I., and Casabona, J. (1997). The spread of AIDS and the re-emergence of tuberculosis in Catalonia, Spain. *AIDS*, **11**, 499–505.

Walther, G.R., Post, E., Convey, P. *et al.* (2002). Ecological responses to recent climate change. *Nature*, **416**, 389–95.

Weaver, S.C., Hagenbaugh, A., Bellew, L.A. *et al.* (1994). Evolution of alphaviruses in the eastern equine encephalomyelitis complex. *Journal of Virology*, **68**, 158–69.

Weesendorpa, E., Stegemanb, A., and Loeffena, W.L.A. (2008). Survival of classical swine fever virus at various temperatures in faeces and urine derived from experimentally infected pigs. *Veterinary Microbiology*, **132**, 249–59.

Wilcox, B.A. and Gubler, D.J. (2005). Disease ecology and the global emergence of zoonotic pathogens. *Environmental Health and Preventive Medicine*, **10**, 263–72.

PART IV

Invasion, Insularity, and Interactions

CHAPTER 14

The geography of host and parasite invasions

Kevin D. Lafferty, Mark E. Torchin, and Armand M. Kuris

14.1 Introduction

This volume demonstrates that there can be strong geographic patterns for some parasite communities. For instance, one general prediction is that the similarity of parasite communities should decrease as a function of distance (Soininen *et al.* 2007; see also Chapter 9 of this volume). Species invasions are fundamental biogeographical processes that have occurred through geological time via long distance dispersal and through historical biotic exchange (Vermeij 2005). Now, the globalization of the world's economies is dramatically increasing the rate of invasions. Parasites can be lost, transferred, and gained when their hosts invade a new location. To what extent do these invasions affect parasite biogeography?

The 'Enemy Release Hypothesis' predicts that colonizing populations can benefit from a lack of natural enemies compared to populations within their original range (Elton 1958). Studies of contemporary species invasions indicate that most of the parasite species a colonist might bring with it are either left behind during the colonization process, lost shortly thereafter, or cannot survive in the new and different habitat (Dobson and May 1986; Torchin *et al.* 2003). The invasion process can 'filter out' parasites and pathogens that occur in an invading host's native range through several mechanisms (Keane and Crawley 2002).

Herein, we start with the general premise that the extent of enemy release in a host population should increase in distant, novel habitats where colonizing hosts face the greatest obstacles to establishment (Blossey and Notzhold 1995; Keane and Crawley 2002; Torchin and Mitchell 2004). Host-specific parasites will not likely be awaiting a colonizer because they need to be brought in with the colonizing species. Escape from parasites should increase with the isolation of a new habitat because few infected colonists will reach isolated areas (or colonizing groups will arrive less frequently) and isolated areas will be more likely to differ environmentally from the colonist's native range. Perhaps, as a result of this, endemic species on islands often lack infectious diseases (Van Riper III *et al.* 1986; Fallon *et al.* 2005).

We first consider patterns of escape from natural enemies in the context of historical 'natural' invasions, asking whether the distribution of parasites of native mice on islands was affected by biogeography. We then use a model to illustrate how the biogeography of escape from natural enemies might increase host speciation rates. Finally, we consider escape from natural enemies in contemporary human-mediated invasions, asking whether biogeography influences the extent to which introduced species escape from natural enemies, namely, parasites.

14.2 Does island size or distance from mainland affect parasitism of island mice?

Smith and Carpenter's (2006) study, evaluating the transfer of helminth parasites from introduced black rats (*Rattus rattus*) to native deer mice (*Peromyscus maniculatus*) on the California Channel Islands, provides a good starting point for understanding how geography can affect enemy release. Here, parasitological surveys of native mice were replicated across a single island archipelago. Each island has its own described subspecies of mouse that colonized the islands naturally, and with assistance from Native Americans over 1,000 years ago. The authors note that of the 40 genera of helminths known to affect deer mice in North America, only 5 genera occur in the Channel Islands. Introduced rats appear to have introduced one of these (*Trichuris muris*). Thus, we were able to ask whether the authors' measures of parasite richness and summed prevalence correspond to the geographic variables of island size and distance from mainland. We predicted that parasite richness and summed prevalence in mice should increase with island size and decrease with distance from the mainland.

Consistent with our prediction, helminth richness in mice declined with the distance from the mainland (Fig. 14.1, Table 14.1). In particular, the putatively introduced nematode was absent from the four most distant islands. However, there was no significant effect of island size on helminth richness. Summed prevalence also declined with distance from the mainland (Table 14.2). However, surprisingly, summed prevalence declined with island size, par-

ticularly on distant islands. These patterns were driven mostly by two species, *Hymenolepis* sp. and *Pterygodermatite peromysci*. Why did island size fail to have the predicted effect on parasite richness and prevalence? Larger islands could have more predators or competitors and this could reduce host density, making transmission less efficient on larger islands. Information on mouse density (not presently available) would be necessary to evaluate this hypothesis. In addition, large islands have more habitat diversity, increasing the likelihood of sampling locations with few parasites, assuming, as is usually the case, that parasites are spatially aggregated.

We also analysed data on the prevalence of hantavirus in mice from the same islands (Jay and Ascher, 1997). Unlike for helminths, viral prevalence was not affected by the distance to the mainland, but increased significantly with island size ($F_{1,6}$=9.5, r^2=0.62, Estimate=0.22±0.07, t-ratio=3.1, p=0.022; Fig. 14.2). This pattern could result if life-time immunity strongly influences the dynamics of viral infections. On small islands viral infections, unlike helminth infections, are more likely to run out of susceptible hosts and go extinct. This is partly what leads to a strong association between viral diversity and landmass size in human populations (Constantin De Magny *et al.* 2009).

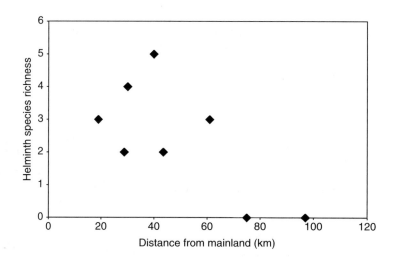

Figure 14.1 A decrease in helminth species richness in native Channel Island deer mice with distance from mainland California using parasite data from Smith and Carpenter (2006). See Table 14.1 for statistics.

Table 14.1   Species richness of mouse helminths in relation to island size (km²) and island distance from mainland (km) (islands: Anacapa [3 km², 19 km], San Miguel [38 km², 40 km], Santa Barbara [3 km², 61 km], Santa Rosa [215 km², 44 km], Santa Cruz [250 km², 29 km], Santa Catalina [194 km², 30 km], San Clemente [147 km², 75 km], San Nicolas [60 km², 97 km]) (r^2=0.71, $F_{3,4}$=3.2).

Term	Estimate	Std Error	t Ratio	Prob>|t|
Intercept	6.55	1.490	4.39	0.0117
Area	−0.01	0.006	−1.69	0.1672
Distance	−0.06	0.021	−2.97	0.0413
Area×Distance	−0.00	0.000	−1.04	0.3553

Table 14.2   Summed prevalence of mouse helminths in relation to island distance and island size (r^2=0.87, $F_{3,4}$=9.1).

Term	Estimate	Std Error	t Ratio	Prob>|t|
Intercept	184.39	24.806	7.43	0.0017
Area	−0.31	0.101	−3.01	0.0395
Distance	−1.86	0.360	−5.17	0.0067
Area×Distance	−0.01	0.005	−2.82	0.0480

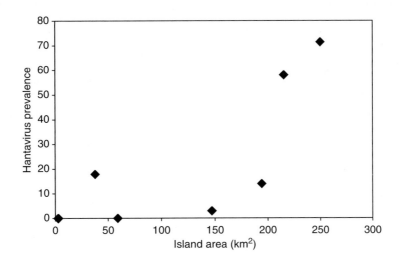

Figure 14.2   Increase in Hantavirus prevalence in native Channel Island deer mice with island size, using data from Jay and Ascher (1997). See text for statistics.

14.3   The biogeography of enemy release and host speciation

Allopatric speciation, such as for the suite of island endemic mice species (or subspecies), described above, requires isolation from gene flow. This can be promoted by dispersal to remote locations and establishment of populations at those places. An obvious impediment to allopatric speciation is the rarity of colonists. If initial colonists are few,

demographic stochasticity and Allee effects can prevent establishment (Lande *et al.* 2003; Williamson 1996). Lack of adaptation to a new location (Holt *et al.* 2005; Peterson 2003) and/or low genetic diversity (Drake 2006; Briskie and Mackintosh 2004) make the establishment of colonists even less likely, particularly if they face competition from resident species already adapted to local conditions (Price 2008). However, an increased population growth rate resulting from enemy release might help compensate for, and can theoretically exceed, the cost of demographic stochasticity resulting from a small initial population size (Drake 2003).

Our models (see Appendix) indicated that enemy release could increase the probability of speciation several fold because species that dispersed to an isolated habitat were more likely to establish if they left some of their natural enemies behind (Fig. 14.3). A higher probability of establishment in isolated locations increased the probability of the persistence of isolated populations, which was a prerequisite for allopatric speciation in the model. To a certain extent, these results derive from simple logic. Any factor that aids the establishment of arriving colonists in isolated areas should increase the potential for speciation. Escape from natural enemies may be transient on evolutionary time scales, but it may buy invaders needed time to colonize and adapt to novel environments.

The magnitude of the effect of natural enemies in speciation, expressed by the differences in the curves in Fig. 14.3, depends chiefly on the extent that parasites negatively affect demographic performance. Although ecologists historically viewed parasites as benign (Lack 1954), recent models (Anderson and May 1978; May and Anderson 1978), field studies (Canter and Lund 1948; Fenner and Ratcliffe 1965; Lemly and Esch 1984; Lafferty 2004) and experiments in the laboratory (Park 1948; Keymer 1981; Scott and Anderson 1984) and field

(Dobson and Hudson 1992; Lafferty 1993; Fitze *et al.* 2004) indicate that some parasites can greatly affect host density (Tompkins and Begon 1999). For instance, the nematode *Heligmosomoides polygyrus* increases host mortality and can reduce lab mouse densities 20 fold (Scott 1987) and native populations of the European green crab infected with a castrating parasitic barnacle (*Sacculina carcini*) have, on average, one third the crab biomass of uninfected populations (Torchin *et al.* 2001).

In addition to facilitating colonization and establishment, a lack of parasites and pathogens presents a shift in selective forces that shape evolution. Faced with fewer enemies initially and different suites of enemies over time, rapid diversification of founding populations can also influence rates of speciation (Ricklefs and Bermingham 2008). While a lack of natural enemies might foster genetic differentiation and eventual speciation, losing the legacy of past enemies also puts a species at risk if parasites eventually catch up with their hosts. In particular, if selection for defences is relaxed (Wolfe *et al.* 2004), and if genetic variation is low in isolated populations (Lyles and Dobson 1993), then susceptibility to both new and former pathogens may increase. The current susceptibility of native island endemics to mainland pathogens (Warner 1969) is a good example of a process that, in geological time, has been called the taxon cycle (Ricklefs and Cox 1972). In addition, local natural enemies might eventually

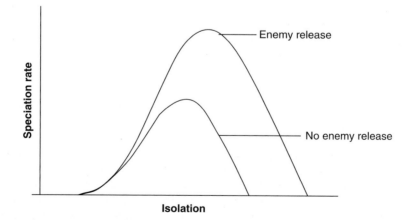

Figure 14.3 Relationship between speciation and isolation for no enemy release (*b*=0) and enemy release (*b*=1) in the full model. All model parameters set to 1 except *p* and *q*, which were set to 0.5.

evolve to be able to use new hosts, particularly if those hosts become abundant (Tabashnik 1983; Zietara and Lumme 2002). Therefore, while our model uses the rate of species creation, it does not consider the extent to which new species will persist over time as conditions change. However, under natural conditions, invasions of competitors or natural enemies that could lead to extinction of island endemics seem to be sufficiently infrequent at isolated locations, promoting a relatively long persistence for island endemics (Price 2008).

Often, allopatric speciation events probably do not involve dispersal and establishment processes. Rather, the genetic isolating mechanisms are geological, including stream capture, orogeny, rifting, and the formation of new barriers such as the Isthmus of Panama (a barrier to marine species). Populations separated by these processes likely retain a substantial part of their original parasitofauna and enemy release is probably less important in influencing diversification.

14.4   The biogeography of enemy release and human-mediated biological invasions

Globalization of the world's economies is dramatically increasing the rate of biological invasions and homogenizing the earth's biota on a global scale. Escape from the effects of parasites is a common explanation given for the success of introduced species. Although invaders can also accumulate natural enemies from the communities they invade, accumulation does not generally make up for escape, and invaders often have fewer parasites than where they are native (Torchin and Mitchell 2004). To what extent does biogeography affect patterns of enemy release? First, it is conceivable that some or all of what appears to be enemy release is explainable through biogeographic processes. For instance, if species tend to invade from the tropics to temperate regions or from mainlands to islands, their parasite fauna could be reduced in accordance with general biogeographic patterns. If this were the case, differences in parasite communities among host populations would depend more on latitude, distance between populations, and landmass size than on whether a species was historically present

in a particular location. Clearly, enemy release and biogeography are not mutually exclusive hypothesis—each may contribute to the differences seen among parasite communities.

To evaluate the extent to which biogeography explains parasite release, we used the data from Torchin *et al.* (2003) providing information on the parasites of 26 diverse animal taxa in native and introduced populations from around the globe. Here, since we were interested specifically in the phenomenon of enemy release (as opposed to community similarities), we consider two measures of the parasite communities; relative species richness and summed prevalence. We use a standard measure of parasite species richness since we compared measures across a diverse range of host taxa (which varied in their parasite richness). For this, we calculated richness as a proportion relative to the total number of parasite species found in all studies in the native range of that host species as per Torchin *et al.* (2003). Similarly, to provide an indication of the unweighted cumulative extent of parasitism (or potential impact of parasitism on a host population) that each host experiences, we used summed prevalence (sum of the prevalence of all parasite species for each host species; see Torchin *et al.* 2003). We took several approaches to investigating this question. We first examined whether species origin (native or introduced) and geographical factors (latitude and land area) explained the parasite load in a particular host population. We then evaluated whether there were differences in enemy release between terrestrial and aquatic invaders. We also considered how distance and landmass area influenced differences in parasitism between pairs of native and introduced populations.

14.4.1   How do latitude, landmass area, and population origin (native or introduced) affect parasitism?

We evaluated how origin (native or introduced), location (longitude, latitude), and land area explained relative parasite species richness and summed parasite prevalence. We expected that parasitism would be higher near the equator (due to latitudinal diversity gradients), higher on large landmasses (due to negative effects of isolation on diversity), and higher for native populations than for introduced populations (due to enemy release). The factors we considered in a least squared general linear model were:

(1) species;
(2) taxonomic group (molluscs, crustaceans, fishes, amphibians and reptiles (=herps), birds, mammals);
(3) latitude (absolute);
(4) longitude;
(5) landmass type (island or mainland); and
(6) origin (native vs. introduced).

We nested species within taxonomic group. We also included the first order interactions among taxa, longitude, latitude, and landmass type. Assumptions of the general linear model were met after using the square root transformation for the sum of prevalence and the angular transformation for relative species richness.

The only hypothesis consistently supported was origin. Introduced populations had lower parasite species richness (back transformed LSqM=23.8±0.2%) than did native populations (LSqM=45.5±2%). Taxon and species were also significant main effects (note that due the standardization among species, the main effects for taxon and species indicate differences in the variance of species richness, which was not a hypothesis we were considering; Table 14.3). There were several significant interactions among the main effects. Unlike the other taxonomic groups, parasite richness in herps and mammals was not affected by origin. Parasite richness decreased with latitude for birds, mammals, and molluscs, but increased with latitude for herps. Parasite richness tended to increase with land area for herps and mammals, but declined with land area for birds and freshwater fishes. Parasite richness declined, on average, with latitude for native populations (as theory predicted), but increased with latitude for introduced populations. The interaction between area and latitude indicated that the effect of land area on parasite richness was negative near the equator and positive at the poles (or alternatively that richness declined with latitude on islands but increased with latitude on continents).

Similarly, introduced populations had a lower summed prevalence of parasites (back transformed LSqM=50.8±0.8%) than did native populations (LSqM=118.4±0.4%). This form of enemy release varied among taxa, with herps, mammals, and molluscs showing weak effects in comparison to crustaceans, birds, and fishes (Table 14.4). Latitude and land area did not have a consistent significant effect on summed prevalence. Summed prevalence increased slightly with landmass area for crustaceans, fish, mammals, and herps, but decreased with landmass area for birds and molluscs. Just as for parasite richness, summed prevalence tended to decline with landmass area near the equator and increase with landmass area near the poles. The summed prevalence of parasites varied among taxa

Table 14.3 Variation in relative species richness among populations (r^2=0.57, $F_{45,241}$=7.02, p<0.0001).

Source	df	Sum of Squares	F	Prob>F
Origin	1	0.81	23.1	<0.0001
Abs Lat	1	0.00	0.1	0.8020
Area	1	0.03	0.9	0.3319
Taxon	5	0.43	2.5	0.0317
Species[Taxon]	19	2.55	3.9	<0.0001
Taxon×Origin	5	1.06	6.2	<0.0001
Taxon×Abs Lat	5	0.52	3.1	0.0109
Taxon×Area	5	0.48	2.8	0.0177
Origin×Abs Lat	1	0.08	2.3	0.1304
Origin×Area	1	0.15	4.5	0.0352
Abs Lat×Area	1	0.24	7.0	0.0085

Table 14.4 Variation in summed prevalence among populations (r^2 = 0.70, $F_{45,232}$ =11.8; p <0.0001).

Source	df	Sum of Squares	F	Prob>F
Origin	1	212.7	19.8	<0.0001
Abs Lat	1	4.2	0.4	0.5312
Area	1	2.5	0.2	0.6271
Taxon	5	1388.6	25.8	<0.0001
Species[Taxon]	19	1022.8	5.0	<0.0001
Taxon×Origin	5	306.6	5.7	<0.0001
Taxon×Abs Lat	5	113.0	2.1	0.0664
Taxon×Area	5	171.4	3.2	0.0085
Origin×Abs Lat	1	19.1	1.8	0.1847
Origin×Area	1	4.9	0.5	0.5001
Abs Lat×Area	1	103.6	9.6	0.0022

with mammals; herps, birds, and fish having higher summed prevalence than crustaceans and molluscs. Whether a population was native or introduced was the best predictor for both measures of parasitism and while geographical effects were present, they were inconsistent among taxonomic groups.

14.4.2   Aquatic versus terrestrial invasions

Perhaps the most important aspect of biogeography is the interplay of land and water. Aquatic and terrestrial invasions might have different patterns of enemy release. Our analysis of data from Torchin *et al.* (2003) indicated that while there was no difference in accumulation of novel parasites by introduced species across habitats, a significantly higher extent of escape from natural enemies occurred in aquatic relative to terrestrial systems (Fig. 14.4). These results are consistent with Soininen *et al.* (2007) who demonstrate a greater similarity of ecological communities (i.e. initial similarity) in freshwater and marine communities relative to terrestrial systems. Another factor contributing to the aquatic-terrestrial differences may be a greater likelihood of multiple introductions for the terrestrial compared

to the aquatic species. Species such as rats and starlings were introduced multiple times with a much greater likelihood of successful transport and establishment of their native parasites (Torchin *et al.* 2003). While aquatic invaders are also introduced multiple times, they are often introduced as larval forms free from parasites (Torchin *et al.* 2001; Torchin and Lafferty 2008).

14.4.3   How do distance and difference in landmass area affect enemy release in invasive species?

As predicted by island biogeography theory, dispersal of organisms to remote locations, especially islands, is infrequent (MacArthur and Wilson 1967). Similarly, parasite release should vary with distance from the source of invasion, where distant colonists to novel locations escape a greater proportion of natural enemies relative to those invading close to home. To further explore how biogeography might affect enemy release for introduced species, we evaluated how distances between populations and differences in landmass areas among populations correlated with differences in parasitism. For each

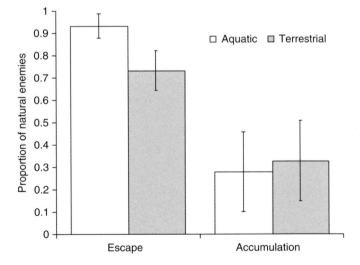

Figure 14.4   Parasite release in aquatic and terrestrial habitats. In an analysis of 16 aquatic (white bars) and ten terrestrial (grey bars) animal taxa, aquatic animals escaped a higher proportion of their natural enemies from their native range (93 per cent), compared with terrestrial animals (73 per cent) (*t*=−4.02, *p*=0.0005). Error bars are 95 per cent confidence intervals. Aquatic and terrestrial animals did not differ in the proportion of natural enemies accumulated from their new range (expressed as a fraction of the parasites that they had in their home range). Enemy release is escape minus accumulation. Data from Torchin *et al.* (2003).

of the 26 species in Torchin *et al.* (2003), we constructed a matrix to pair each introduced site with each native site. Note that these pairings did not imply the source and target of an invasion, but were merely a comparison of introduced and native populations. For each pair of sites, we quantified two measures of 'release', first, the difference in relative species richness between the native location and the introduced location and second, the difference in the summed prevalence of parasitism between the native location and the introduced location. Our measures of isolation of introduced populations were the log distance between the sites in kilome-

tres, and the proportional change in log landmass area (a value only meaningful for species that invaded new landmasses). We expected that enemy release would increase with the distance between native and introduced sites and with a shift from large landmasses to small landmasses.

Full models of all species were dominated by strong interactions between taxon (or species) and biogeography, indicating no general effect of biogeography (distance and area) on enemy release. The large number of contrasting patterns in this analysis led us to analyse each species separately (Table 14.5). Limited numbers of replicate popula-

Table 14.5 The contribution of distance and area and their interaction to the species richness and prevalence of parasites in introduced populations (Mc—mollusc, H—reptile/amphibian, C—crustacean, F—fish, Mm—mammal, B—bird).

Group	Species	Release	N	r^2	Distance	Area	Interaction
Mc	Batillaria	Richness	60	0.19	NS		NS
	Batillaria	Prevalence	60	0.29	NS		NS
H	Bufo	Richness	12	0.78	NS	NS	*(+)
	Bufo	Prevalence	12	0.37	NS	NS	NS
Mc	Bythinia	Richness	12	0.95	NS	*(−)	NS
	Bythinia	Prevalence	12	0.75	NS	NS	NS
C	Carcinus	Richness	204	0.02	NS	NS	NS
	Carcinus	Prevalence	204	0.05	NS	NS	NS
Mc	Dreissena	Richness	96	0	NS		NS
	Dreissena	Prevalence	96	0.03	NS		NS
F	Gambusia	Richness	14	0.5	NS	NS	*(−)
	Gambusia	Prevalence	14	0.4	NS	NS	NS
Mc	Ilyanassa	Richness	11	0.04	NS		
	Ilyanassa	Prevalence	11	0.27	NS		
H	Lepidodactylus	Richness	20	0.12	NS	NS	NS
	Lepidodactylus	Prevalence	20	0.35	NS	NS	NS
Mc	Littorina lit	Richness	32	0.2	NS		*(−)
	Littorina lit	Prevalence	32	0.03	NS		NS
Mm	Oryctolagus	Richness	108	0.17	*(+)	NS	NS
	Oryctolagus	Prevalence	108	0.2	NS	NS	NS
F	Poecilia latipinna	Richness	16	0.14	NS	NS	NS
	Poecilia latipinna	Prevalence	16	0.27	NS	NS	NS
Mc	Potamopyrgus	Richness	30	0.22	*(+)	*(+)	NS
	Potamopyrgus	Prevalence	30	0	NS	NS	NS
Mm	Rattus	Richness	36	0.14	NS	NS	NS
	Rattus	Prevalence	36	0.06	NS	NS	NS
B	Sturnus vulgaris	Richness	18	0.24	NS	NS	NS
	Sturnus vulgaris	Prevalence	18	0.32	NS	NS	NS
Mm	Trichosurus	Richness	14	0.02	NS		NS
	Trichosurus	Prevalence	14	0.93	***(+)		
Mm	Vulpes	Richness	24	0.28	NS	NS	NS
	Vulpes	Prevalence	24	0.57	NS	NS	NS

tions of some species (*Anas platyrhynchos, Cancer novaezelandiae, Hemigrapsus sanguineus, Melanoides tuberculata, Passer domesticus, Perca fluviatilis,* and *Rana catesbeiana*) and insufficient variation in biogeography of another (*Onchorynchus mykiss*) prevented separate analyses for some species. Again, the species-level results did not provide strong evidence for an effect of biogeography on enemy release (Table 14.5). Neither did a similar taxon-level analysis. Of the 16 species analysed, 2 (*Oryctolagus cuniculus* and *Potamopyrgus antipodarum*) showed positive associations between release (in terms of richness) and distance. One species (*Trichosurus vulpecula*) showed a positive association between release (in terms of summed prevalence) and distance. And one species (*Potamopyrgus antipodarum*) showed a positive association between release (in terms of richness) and reduction in landmass. Counter to expectations, *Bythinia tentaculata* showed a negative association between release (in terms of richness) and reduction in landmass. Given the large number of independent tests in our analysis, it is questionable whether any of these effects are biologically meaningful. These results indicate that there are no general geographical patterns for enemy release in our global dataset. Thus, for contemporary biological invasions, simply moving far away from the native range, or from a mainland to an island, may not lead to release. Instead, enemy release probably results from local host-specific interactions, which can be easily broken simply by transport to a location just outside the native range.

14.5   Conclusion

Geography is likely an important factor explaining the current distribution of natural parasite communities. Distance between populations might be directly related to propagule pressure in some instances. In particular, our analysis of parasites in mice across a small archipelago is consistent with island biogeography because we expect that, here, distance from the mainland relates to declining propagule pressure of hosts and parasites. Landmass size, on the other hand, may be more important for infectious agents like viruses that provoke life-time immunity because island populations may be able to escape such parasites if susceptible hosts quickly become limited. Enemy release, in addition to shaping variation in parasitism, particularly in remote locations, may have played an important part in generating biodiversity by increasing the chance that hosts could become established in remote locations. We may see similar evolutionary responses in contemporary human-mediated invasions (Huey *et al.* 2005).

The main factor explaining enemy release in our invasions database was whether a species was native or introduced. Geographical factors explained little of the variation in parasite release for a particular host species in our study. We emphasize that our comparison controls for the effect of species and taxon, so it is not necessarily at odds with other studies that find parasite communities among species to vary with geography. For instance, it is conceivable that parasites of herps as a group could decline with latitude while the parasites of a single frog species might not strongly vary with latitude. Our results strongly indicate that any study of parasite biogeography needs to control for the origin of the host populations. Whether host populations are native or introduced may override other geographical factors commonly used to evaluate parasite community similarity.

For introduced species, distance and landmass area were generally not strong indicators of enemy release from parasites. Unlike natural dispersal, where isolation is a direct function of distance, human mediated invasions break normal dispersal barriers and provide corridors for rapid and often frequent invasion. For instance, ships regularly and frequently transport species long distances between different biotic provinces (Carlton and Geller 1993; Cohen and Carlton 1998). As a result, propagule pressure or the number of individuals and number of times a species is introduced to a novel location by humans (Williamson 1996; Lonsdale 1999), may be less dependent on distance and other geographical factors. Placing parasite communities in a biogeographical context has deepened our understanding of the ecology and evolution of host–parasite interactions. Now, as humans continue to homogenize the earth's biota, insights generated from species invasions will be key to further this understanding.

Appendix

We used mainland–island models (one simple and one more complex, hereafter called the full model) where allopatric speciation results from long-distance dispersal, establishment, and subsequent isolation. In the models, a source population exists on a mainland with associated islands that vary in distance from the mainland. Assuming no island-to-island dispersal (an assumption often violated in nature), the probability of dispersal from the mainland declines with the isolation of the island from the mainland. Some islands may be so isolated that dispersing species never reach them. For closer islands, a dispersal event from the mainland has four potential outcomes over geological time: failure to establish, persistence without speciation, allopatric speciation, and extinction. As a result, whether an island supports the mainland species, a derived species, or no species, should correspond to the degree of isolation. A hump-shaped relationship between isolation and speciation will result if individuals rarely or never colonize the most isolated islands and gene flow prevents divergence on islands near the mainland.

The models track the probability, in an arbitrary unit of time, that an island will experience a colonization of individuals destined to become a new species. For simplicity, we refer to this as the 'probability of speciation'. In the models, the probability of speciation is a product of the probabilities of dispersal, D, establishment (including persistence), E, and the probability, B, that barriers to gene flow will allow speciation to occur, or $S=DEB$.

One possible way to describe a probability of dispersal to a new location is e^{-x} (MacArthur and Wilson 1967), where x is a measure of isolation from the mainland (which itself is a function of distance, and dispersal ability). The models assume that rates of speciation are not constrained by an initially small population size.

The initial number of colonizers, n, and the initial instantaneous growth rate in the new location, r, will determine the success of a dispersal event, such that a simple model for the probability of establishment is $E=1-e^{-m}$. One way to isolate the effects of enemy release or establishment is to describe population growth rate as $r=m-p$, where p is the extent that natural enemies depress population growth, and m is the residual growth rate of the species without natural enemies. Note that $m>p$ is a condition necessary for speciation.

Even if a species colonizes a new location, it needs time in isolation from gene flow for drift, natural selection, and/or random mutations to lead to enough genetic differentiation that reproductive barriers to the parent species evolve (i.e. reproductive isolation). We express these complex and little-known functions in a single term, z, which represents the time needed for a successful colonist to speciate. The probability of speciation is a function of speciation rates in isolation and the probability of isolation, or $B=(1-e^{-x})^z$.

A simple model for the probability of speciation is, therefore, $S=DEB$, or

$$e^{-x}\left(1-e^{-x}\right)^z \left(1-e^{n(p-m)}\right) \qquad \text{(Eq. 14.1)}$$

To find the level of isolation, x^*, that results in the maximum amount of speciation, we set $\delta S/\delta x$ to zero and solved for x. Substituting x^* for x and then solving for S indicated the height of the peak in the relationship between speciation and isolation. We then investigated the sign of $\delta S/\delta p$ to determine how speciation was affected by natural enemies, p, under the expectation that natural enemies would reduce S by reducing establishment/persistence.

Several parameters in this simple model might co-vary with isolation and lead to interactions that are more complex. For this reason, we created an expanded model to model enemy release explicitly.

Because species may be less physiologically adapted to areas far from where they are native, the residual growth rate in the new habitat without natural enemies, m, might decline with isolation, x, where a is a measure of the association between isolation and habitat suitability, such that residual growth rate is me^{-ax}. Such an effect of distance on reduced suitability could be ameliorated by a lack of competitors (or biotic resistance) at remote locations. A reduction in biotic resistance with isolation (Darwin 1872), while not explicitly treated in this chapter, might have an effect on speciation similar to an increase in enemy release with isolation (Mack 1996).

Because few individuals are likely to simultaneously colonize remote areas, the initial population size, n, might decline with isolation (due to the rarity of colonization), x, where N is the size of the source population and c is the strength of the association between N and x such that initial population size is Ne^{-cx}.

Enemy release as a function of isolation: The sampling effect (fewer colonists bring a smaller proportion of the available natural enemies) (Drake 2003), was represented as the probability of an infectious disease accompanying a set of colonists $= (1-q^{e^{-cx}N})^b$, where q is the proportion of uninfected individuals in the native population and b is a measure of enemy release (e.g. for $b=0$, natural enemies always accompanied dispersers, for $b=1$, the probability of a natural enemy dispersing depended directly on the sampling effect). There are two other mechanisms for enemy release with isolation. Distant environments may be unsuitable for natural enemies; for instance, they might lack necessary intermediate hosts for parasites. In addition, few colonizers may be below the host threshold density needed for parasite transmission. Because, like the sampling effect, unsuitability and the initial number of colonizers should tend to decrease with x, these distinct mechanisms could have an additive effect on the extent to which isolation interacts with escape from natural enemies to promote speciation.

Because distant (especially different) habitats may foster more rapid genetic change through adaptation, time to speciation in isolation, z, might decline with isolation, x, where f is a measure of the extent that time to speciation decreases with isolation from the source population, such that time to speciation is ze^{-fx}. A full model of the probability of speciation, S, is, therefore:

$$e^{-x}\left(1-e^{e^{-cx}N\left(-e^{-ax}m+p\left(1-q^{e^{-cx}N}\right)^b\right)}\right)\left(1-e^{-x}\right)e^{-fxz} \qquad \text{(Eq. 14.2)}$$

Calculating partial derivatives indicated the conditions for which speciation increased or decreased with the different variables (evaluated for conditions where $r>0$). We then used second-order mixed derivatives to determine how other variables (a, c, f) influenced the effect of enemy release on speciation. For simplification, we only report non-trivial results. We also used numerical simulations to graphically

explore how release affected the relationship between speciation and isolation.

In the simple model, the relationship between speciation, S, and isolation, x, was hump-shaped with a peak (for non-trivial values) at $x=\text{Log}(1+z)$. The peak resulted from gene flow preventing speciation in areas that were not isolated because isolated areas were only rarely colonized. Because p was missing from the solution, the location of the speciation peak was unaffected by natural enemies (so long as $p<m$). However, the height of the speciation peak was $(1-e^{-m+p})nz^z(1+z)^{1+z}$ and, therefore, decreased with the effects of natural enemies, p. In the simple model, $\delta S/\delta p$ was always negative for $x>0$, indicating that natural enemies always decreased the probability of speciation. In other words, natural enemies in the new location decreased the probability of speciation (by decreasing the chance of a successful colonization) but would not shift the distance at which the probability of speciation was maximal. Consequently, these results indirectly suggested that enemy release would increase the height of the speciation curve.

In the full model, just as for the simple model, $\delta S/\delta p$ was negative, indicating that enemies decreased speciation rates. Simulations of the full model exhibited a strong hump-shaped relationship between isolation and speciation that was consistent with the analytical results of the simple model. However, in the full model, an increase in the impact of natural enemies shifted the peak to the left along the isolation axis, indicating that natural enemies inhibited speciation in more isolated locations. These results suggested that enemy release could affect both the height (increasing it) and the location of the speciation curve (shifting it to the right).

In the full model, $\delta S/\delta b$ was always positive for $x>0$, showing that release from natural enemies always increased the probability of speciation. In addition, enemy release shifted the peak of the speciation curve to the right, indicating that enemy release differentially facilitated speciation in isolated locations (Fig. 14.3). The shift resulted from the compensatory effect of enemy release decreasing the probability of host establishment in distant areas that were otherwise ideal for speciation. Not surprisingly, $\delta S/\delta b \delta p$ was always positive. Hence, the effect of enemy release on speciation increased with the impact of natural enemies.

The partial derivatives of the full model showed how other variables affected the probability of speciation, while the second order mixed derivatives indicated how these other variables affected the relationship between enemy release and speciation. $\delta S/\delta a$ was negative, demonstrating that if isolated habitat were less suitable, the probability of speciation decreased (because unsuitable habitat reduced establishment). $\delta S/(\delta b \delta a)$ was positive, indicating that the strength of the negative association between habitat unsuitability and speciation increased the extent to which enemy release increased speciation. $\delta S/\delta f$ was positive, suggesting that a decrease in time to speciation with isolation (due to natural selection in novel environments) increased the probability of speciation (rapid rates of evolution allowed reproductive isolation to outpace gene flow). $\delta S/(\delta b \delta f)$ was also positive, indicating that faster speciation in more distant habitats increased the extent that enemy release increased speciation (because enemy release increased establishment in isolated areas). $\delta S/\delta c$ was positive for larger values of p and negative for smaller values of p. This partial derivative represented the trade-off between the value of large population sizes for establishment (if parasites had low pathology) and the value of small population sizes for leaving parasites behind (if parasites were highly pathogenic). $\delta S/(\delta b \delta c)$ was positive because a reduction in the number of colonists with isolation increased the contribution of enemy release to speciation.

References

Anderson, R.M. and May, R.M. (1978). Regulation and stability of host-parasite population interactions. I. Regulatory processes. *Journal of Animal Ecology*, **47**, 219–47.

Blossey, B. and Notzhold, R. (1995). Evolution of increased competitive ability in invasive nonindigenous plants: a hypothesis. *Journal of Ecology*, **83**, 887–89.

Briskie, J. and Mackintosh, M. (2004). Hatching failure increases with severity of population bottle-necks in birds. *Proceedings of the National Academy of Sciences of the USA*, **101**, 558–61.

Canter, H.M. and Lund, J.W.G. (1948). Studies on plankton parasites I. Fluctuations in numbers of *Asterionella formosa* Hass in relation to fungal epidemics. *New Phytologist*, **47**, 238–61.

Constantin De Magny, G.C., Renaud, F., Durand, P., and Guégan, J.-F. (2009). Health ecology: a new tool, the macroscope. In F. Thomas, F, Renaud, and J-F. Guégan, eds. *Ecology and Evolution of Parasitism*, pp. 129–48. Oxford University Press, Oxford.

Darwin, C. (1872). *The Origin of Species by Means of Natural Selection, or the Preservation of Favoured Races in the Struggle for Life*. Murray, London.

Dobson, A.P. and Hudson, P.J. (1992). Regulation and stability of a free-living host-parasite system, *Trichostrongylus tenuis* in red grouse. II. Population models. *Journal of Animal Ecology*, **61**, 487–500.

Dobson, A.P. and May, R.M. (1986). Patterns of invasions by pathogens and parasites. In H.A. Mooney and J.A. Drake, eds. *Ecology of Biological Invasions of North America and Hawaii*, pp. 58–76. Springer Verlag, New York.

Drake, J.M. (2003). The paradox of the parasites: implications for biological invasions. *Proceedings of the Royal Society of London B*, **270**, S133–S135.

Drake, J.M. (2006). Heterosis, the catapult effect and establishment success of a colonizing bird. *Biology Letters*, **2**, 304–07.

Elton, C.S. (1958). *The Ecology of Invasions by Animals and Plants*. London, Methuen.

Fallon, S., Bermingham, E., and Ricklefs, R. (2005). Host specialization and geographic localization of avian malaria parasites: a regional analysis in the Lesser Antilles. *American Naturalist*, **165**, 466–80.

Fenner, F. and Ratcliffe, F.N. (1965). *Myxomatosis*. Cambridge University Press, Cambridge.

Fitze, P.S., Tschirren, B., and Richner, H. (2004). Life history and fitness consequences of ectoparasites. *Journal of Animal Ecology*, **73**, 216–26.

Holt, R.D., Barfield, M., and Gomulkiewicz, R. (2005). Theories of niche conservatism and evolution: could exotic species be potential tests? In D.F. Sax, J.J. Stachowicz, and S.D. Gaines, eds. *Species Invasions: Insights into Ecology, Evolution, and Biogeography*, pp. 259–90. Sinauer, Sunderland.

Huey, R.B., Gilchrist, G.W., and Hendry, A.P. (2005). Using invasive species to study evolution: case studies with drosophila and salmon. In D.F. Sax, J.J. Stachowicz, and S.D. Gaines, eds. *Species Invasions: Insights into Ecology, Evolution, and Biogeography*, pp. 139–164. Sinauer, Sunderland.

Jay, M. and Ascher, M.S. (1997). Seroepidemiologic studies of hantavirus infection among wild rodents in California. *Emerging Infectious Diseases*, **3**, 183–90.

Keane, R.M. and Crawley, M.J. (2002). Exotic plant invasions and the enemy release hypothesis. *Trends in Ecology and Evolution*, **17**, 164–70.

Keymer, A.E. (1981). Population dynamics of *Hymenolepis diminuta*: the influence of infective-stage density and spatial distribution. *Parasitology*, **79**, 195–207.

Lack, D. (1954). *The Natural Regulation of Animal Number*. Clarendon Press, Oxford.

Lafferty, K.D. (1993). Effects of parasitic castration on growth, reproduction and population dynamics of the marine snail *Cerithidea californica*. *Marine Ecology Progress Series*, **96**, 229–37.

Lafferty, K. (2004). Fishing for lobsters indirectly increases epidemics in sea urchins. *Ecological Applications*, **14**, 1566–73.

Lande, R., Engen, S., and Saether, B.E. (2003). *Stochastic Population Dynamics in Ecology and Conservation*. Oxford University Press, Oxford.

Lemly, A.D. and Esch, G.W. (1984). Effects of the trematode *Uvulifer ambloplitis* on juvenile bluegill sunfish, *Lepomis macrochirus*: ecological implications. *Journal of Parasitology*, **70**, 475–92.

Lyles, A.M. and Dobson, A.P. (1993). Infectious disease and intensive management: population dynamics, threatened hosts and their parasites. *Journal of Zoo and Wildlife Medicine*, **24**, 315–26.

Macarthur, R.H. and Wilson, E.O. (1967). *The Theory of Island Biogeography*. Princeton University Press, Princeton.

Mack, R.N. (1996). Biotic barriers to plant naturalization. In V.C. Moran and J.H. Hoffman, eds. *Proceedings of the IX International Symposium on Biological Control of Weeds*, pp. 39–46. University of Cape Town, Cape Town.

May, R.M. and Anderson, R.M. (1978). Regulation and stability of host-parasite population interactions. II. Destabilizing processes. *Journal of Animal Ecology*, **47**, 249–67.

Park, T. (1948). Experimental studies of interspecies competition. 1. Competition between populations of the flour beetles, *Tribolium confusum* Duval and *Tribolium castaneum* Herbst. *Ecological Monographs*, **18**, 267–307.

Peterson, A. (2003). Predicting the geography of species' invasions via ecological niche modeling. *Quarterly Review of Biology*, **78**, 419–33.

Price, T. (2008) *Speciation in Birds*. Roberts and Company, Boulder.

Ricklefs, R. and Bermingham, E. (2008). The West Indies as a laboratory of biogeography and evolution. *Philosophical Transactions of the Royal Society B*, **363**, 2393–413.

Ricklefs, R. and Cox, G. (1972). Taxon cycles in the West Indian avifauna. *American Naturalist*, **106**, 195–219.

Scott, M.E. (1987). Regulation of mouse colony abundance by *Heligmosomoides polygyrus* (Nematoda). *Parasitology*, **95**, 111–29.

Scott, M.E. and Anderson, A.M. (1984). The population dynamics of *Gyrodactylus bullatarudis* (Mongenea) on guppies (*Poecilia reticulata*). *Parasitology*, **89**, 159–94.

Smith, K.F. and Carpenter, S.M. (2006). Potential spread of introduced black rat (*Rattus rattus*) parasites to endemic deer mice (*Peromyscus maniculatus*) on the California Channel Islands. *Diversity and Distributions*, **12**, 742–48.

Soininen, J., Mcdonald, R., and Hillebrand, H. (2007). The distance decay of similarity in ecological communities. *Ecography*, **30**, 3–12.

Tabashnik, B.E. (1983). Host range evolution - the shift from native legume hosts to alfalfa by the butterfly, *Colias phiodice-eriphyle*. *Evolution*, **37**, 150–62.

Tompkins, D.M. and Begon, M. (1999). Parasites can regulate wildlife populations. *Parasitology Today*, **15**, 311–13.

Torchin, M.E. and Mitchell, C.E. (2004). Parasites, pathogens, and invasions by plants and animals. *Frontiers in Ecology and the Environment*, **2**, 183–90.

Torchin, M.E., Lafferty, K.D., and Kuris, A.M. (2001). Release from parasites as natural enemies: increased performance of a globally introduced marine crab. *Biological Invasions*, **3**, 333–45.

Torchin, M.E., Lafferty, K.D., Dobson, A.P., Mckenzie, V.J., and Kuris, A.M. (2003). Introduced species and their missing parasites. *Nature*, **421**, 628–30.

Van Riper III, C., Van Riper, M.L., Goff, M.L., and Laird, M. (1986). The epizootiology and ecological significance of malaria in Hawaiian land birds. *Ecological Monographs*, **56**, 327–44.

Warner, R.E. (1969). The role of introduced diseases in the extinction of the endemic Hawaiian avifauna. *Condor*, **70**, 101–20.

Williamson, M. (1996). *Biological Invasions*. Chapman & Hall, London.

Wolfe, L.M., Elzinga, J.A., and Biere, A. (2004). Increased susceptibility to enemies following introduction in the invasive plant *Silene latifolia*. *Ecology Letters*, **7**, 813–20.

Zietara, M.S. and Lumme, J. (2002). Speciation by host switch and adaptive radiation in a fish parasite genus *Gyrodactylus* (Monogenea, Gyrodactylidae). *Evolution*, **56**, 2445–58.

CHAPTER 15

Immune defence and invasion

Anders P. Møller and László Z. Garamszegi

15.1 Introduction

Invasion biology aims to understand the processes accounting for invasions and the subsequent consequences of such invasions (e.g. Lockwood *et al.* 2007). Successful invasion is a multi-step ecological process that requires an ability to disperse, subsequently become established and finally expand (Fig. 15.1). Each of these three component processes must be completed in order to achieve a successful invasion. Therefore, analysis of later steps in the chain of processes will require that previous steps are considered simultaneously.

This chapter deals with the effects of immune defence on successful invasions by animals. The main reason for investigating the relationship between immune defence and invasions is that no invasion takes place in isolation because invaders are accompanied by their parasites, but also by their symbionts when invading. Furthermore, the novel environment that is invaded already contains a number of other animals that each has their own parasite and symbiont fauna that may facilitate or prevent successful invasion by other taxa. Several recent studies have suggested that parasites may play an important role in invasions (e.g. Mitchell and Power 2003; Torchin *et al.* 2003). Therefore, by inference, hosts differing in level of immune defence may also differ in their invasion ability.

The immune system is the most sophisticated physiological defence system in animals. The diversity of defences produced and their intensity imply that there are costs involved in its development, maintenance, and activation. The immune system can be divided into innate and adaptive, induced defences (e.g. Abbas *et al.* 1994; Roitt *et al.* 1996). While many immune responses impair or prevent

parasites from becoming established in a host, the immune system can also show pathological responses to parasite exposure. The immune system of animals has a number of similarities to other kinds of defence systems (e.g. defences against herbivores in plants, defence against predators in many animals) suggesting that some of our arguments may have applications beyond host–parasite interactions in birds.

There are few good datasets on different components of invasion in major taxa of animals and plants. Paradis *et al.* (1998) estimated natal and breeding dispersal distances in birds based on extensive recovery or re-trap data from the UK. While this dataset provided information on dispersal, there is no information on whether these dispersers eventually reproduced. A particularly extensive and useful dataset on establishment is that based on introductions of birds to oceanic islands (Veltman *et al.* 1996; Cassey 2002), which not only provides information on success, but also on the number of attempts and the actual size of inocula during different attempts. This 'experiment' is unique because it has been going on for more than 150 years, hence qualifying as a long-term experiment. This allows for rigorous analysis of many factors associated with the invasion process, but also for assessment of the role of ecological features of different species in that process. The dataset on introduction success for birds introduced to oceanic islands also has weaknesses given that introduction success is confounded by ability to sustain stress during long-distance transport for many months. However, this dataset does provide useful information about the ability to sustain severe stress that species may also experience during long distance dispersal or migration, and then become established.

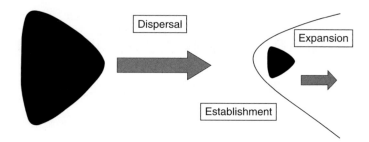

Figure 15.1 The three steps in a successful invasion: dispersal, establishment, and expansion.

Finally, range expansion can be estimated by two measures: expansion by introduced species in their new range, and range expansion following recent climate change (Hyytiä *et al.* 1983; Väisänen *et al.* 1998). These estimates reflect the degree of expansion by the actual populations that were introduced into new habitats or adapted their current distributions to the changing environments. Moreover, total range size can reflect whether species with large breeding ranges have been successful in expansion at a large scale, although this measure may be confounded by large overall population sizes and hence a large amount of standing genetic variation. These measures reflect expansions acting at different temporal and spatial scales.

Invasion success, including all three processes, can be assessed in different ways. Møller and Mousseau (2007) estimated maximum distances between island populations of breeding birds and the nearest mainland, producing a dataset that integrates dispersal, introduction success, and subsequent expansion. Moreover, a successful colonization of more than a single continent can also indicate invasion success as a whole. For example, a species that had established breeding populations both in the Palaearctic and the Nearctic can be interpreted to have the ability to invade a novel continent and then dramatically expand its range. Expansion from an initial introduction should result in occupation of a large range and potentially further expansion into other continents, if fully successful. For example, the European starling *Sturnus vulgaris* and house sparrow *Passer domesticus* have been introduced on many occasions to different continents, with subsequent extensive range expansions.

Natal dispersal is generally much longer than breeding dispersal (e.g. Paradis *et al.* 1998), making juveniles much more likely to account for invasions. Hence we predicted that the immune function of young birds would be more important than the immune function of adults as predictors of invasion success. Alternatively, young individuals constitute the majority of individuals that disperse into novel environments, become established there, and subsequently expand the breeding range, simply because they are more common. If this is the case, we should expect the fraction of young individuals in stable populations to be larger in species with a low adult survival rate, and hence invasion success to be predicted by adult survival rate.

Parasitism and host defences against parasitism may affect all three processes in a successful invasion. We review evidence suggesting that the ability of hosts to resist parasites contributes to dispersal and the successful invasion of islands and continents, and this ability of hosts to resist parasites accounts for differences in range size among taxa (and hence successful expansion). Because there is hardly any published information on immunity and invasion success (see Møller and Cassey 2004; Møller *et al.* 2004; Snoeijs *et al.* 2004 for exceptions), we present a number of novel analyses linking immunity to invasions, using extensive datasets on birds. We will argue that to gain a better understanding of the invasion process, we should really introduce all three components (dispersal ability, establishment success, and subsequent expansion) in the statistical models to analyse the effects contributing to successful establishment. The relationship between these three components of the invasion process is crucial for understanding how the invasion process works, and for the first time we present such estimates here.

15.2 Measures of immune function

Modern achievements in physiological measurements allow the characterization of immune mechanisms in a broad range of species based on standard tools. The focus on host immune defences instead of parasite loads is crucial in the context of large-scale, parasitism-mediated spatial movements for at least two reasons. First, hosts can harbour several species of parasites, of which only few are debilitating and can have an evolutionary impact on invasion. Because we generally cannot identify the parasite species that potentially drive hosts to move from an area, host investment in immune responses may more reliably reflect the effect of parasites on their hosts (see Møller *et al.* 1999). Second, invading species will likely confront novel parasite faunas, and the status of the immune system may be a better predictor of how species will cope with new parasitic challenges in the invaded area than the parasites left behind.

We use four different datasets on immune function in our analyses of invasions and immunity. The main reason for doing so is that no single measure is sufficient for capturing information on the different aspects of immunity (Tizard 1991; Abbas *et al.* 1994; Roitt *et al.* 1996; Pastoret *et al.* 1998). In addition, similarly to invasion, immune defence is a complex mechanism, consisting of several components that play different roles.

First, we used an extensive dataset on the size of three different immune defence organs, namely bursa of Fabricius, thymus, and spleen collected by a taxidermist. The bursa of Fabricius is responsible for the B-cell repertoire of the immune system, while the thymus is responsible for the T-cell repertoire (Tizard 1991; Abbas *et al.* 1994; Roitt *et al.* 1996). In contrast, the spleen is a secondary lymphoid organ used as a major site of storage of leucocytes (Tizard 1991; Abbas *et al.* 1994; Roitt *et al.* 1996). We have previously used the relative size of these immune defence organs, after adjusting for body mass, to investigate relationships between immunity and the

prevalence of avian influenza (Garamszegi and Møller 2007).

Second, we used an extensive dataset on the inflammatory response of the patagium to injection with the mitogenic phytohemagglutinin stimulating T-lymphocyte proliferation, followed by local recruitment of inflammatory cells and increased expression of major histocompatibility complex molecules at the site of injection (Goto *et al.* 1978; Abbas *et al.* 1994; Parmentier *et al.* 1998; Martin *et al.* 2006). This estimate can be obtained for both adults and nestlings, and thus can indicate the efficiency of the T-cell mediated immune response in two different life stages.

Third, we quantified natural antibodies (NAbs) and complement, whose function is to recognize and initiate the complement enzyme cascade that eventually ends in cell lysis (Carroll and Prodeus 1998). NAbs and complement provide a link between innate and acquired immune defence (Carroll and Prodeus 1998; Ochsenbein and Zinkernagel 2000). NAbs occur in immunologically naïve animals and thus do not require prior antigen exposure (Ochsenbein and Zinkernagel 2000). NAbs have been shown to provide resistance to malarial parasites (Congdon *et al.* 1969), and naturally occurring concentrations can kill bacteria *in vivo* (Ochsenbein *et al.* 1999). Accordingly, interspecific variation in NAbs relates to anti-bacterial defence in birds (Møller *et al.* 2009).

Fourth, we used estimates of the concentration of lysozyme in eggs as a fundamental component of innate antibacterial immune defence that is transferred from the mother to the egg in birds, digesting the peptoglycanes that are major components of bacterial cell walls (Tizard 1991; Braun and Fehlhaber 1996; Roitt *et al.* 1996; Pastoret *et al.* 1998; Kudo 2000).

The relationships between the components of immune defence were generally weak with most correlation coefficients being small (Table 15.1). All variables with correlations greater than 0.20 were included in the analyses presented below to ascertain that the immunological variables reported were indeed the best predictors. Therefore, they reflect partially different immunological mechanisms for different aspects of immunity (e.g.

Table 15.1   Pearson product-moment correlation coefficients between different measures of immune function. Number in parentheses are number of species. BF—bursa of Fabricius, TH—thymus, SP—spleen, NT—nestling T-cell response, AT—adult T-cell response, NNAb—nestling NAb, ANAb—adult NAb, NC—nestling complement, AC—adult complement, LY—lysozyme. Body mass was entered as an additional variable in all analyses (*–$p<0.05$, ** – $p<0.01$, *** – $p<0.001$).

	BF	TH	SP	NT	AT	NNAb	ANAb	NC	AC
TH	0.32* (51)								
SP	0.53*** (97)	−0.04 (63)							
NT	−0.04 (48)	0.67*** (40)	−0.14 (61)						
AT	−0.03 (47)	0.32* (40)	0.14 (60)	0.09 (56)					
NNAb	−0.48 (37)	0.04 (31)	−0.01 (44)	0.20 (39)	−0.10 (29)				
ANAb	0.62** (24)	0.33 (23)	0.48* (28)	−0.22 (25)	−0.44* (29)	0.59** (21)			
NC	0.07 (37)	−0.30 (31)	0.12 (44)	−0.09 (39)	0.10 (29)	−0.17 (21)	0.18 (21)		
AC	0.04 (24)	−0.42 (23)	0.03 (28)	−0.63** (25)	−0.17 (29)	0.20 (21)	0.23 (31)	−0.11 (21)	
LY	−0.32 (18)	−0.31 (17)	−0.15 (25)	−0.10 (24)	0.14 (25)	0.01 (47)	0.02 (14)	−0.14 (14)	−0.02 (14)

cell-mediated and humoral, specific and general, primary and secondary) that may be involved in defence against different parasites, or may be active during different life-stages. Although we used extensive data on various immune mechanisms, it is possible that the effects of unmeasured components of the immune system are associated with invasions. For example, genetic variation in the major histocompatibility complex (MHC) that determines parasite resistance can have consequences for dispersal and the colonization of islands (e.g. Seddon and Baverstock 1999; Sommer 2003).

15.3   Statistical and comparative analyses

We related measures of the three processes of invasions to estimates of immune defence as listed above. Because sample sizes sometimes varied considerably, hence violating an important assumption about statistical analyses, we weighted these by sample size (Møller and Nielsen 2007).

Because phenotypic traits of related taxa may be similar due to common descent rather than convergent evolution, we calculated statistically independent linear contrasts in our comparative analyses, as implemented in the software by Purvis and Rambaut (1995). We relied on composite phylogenies derived from many sources to produce the most recent phylogenetic relationships. We did not find qualitatively different results in analyses of

species-specific data and independent contrasts, and, therefore, we only report the analyses of species-specific data here.

15.4   Results

15.4.1   Tests for bias in samples

Although the introduction dataset for birds has been analysed extensively, there has been no explicit test for bias in composition relative to the variables under investigation. Here we provide such a test by comparing the immunity of species introduced with those not included from the western Palaearctic for which we had data on the variables of interest.

Species of birds that were used in introduction events did not differ significantly from excluded species in terms of ten different immune variables (Table 15.2). These results imply that the sample of species used in the subsequent analyses is representative of the entire western Palaearctic fauna with respect to immune traits.

15.4.2   Relationships between the three processes of invasions

Dispersal was not significantly predicted by introduction success of birds (after adjusting for number of releases, natal dispersal distance: Pearson $r=-0.09$, $N=24$ species, $p=0.68$; breeding dispersal distance: Pearson $r=-0.10$, $N=27$ species, $p=0.64$).

Table 15.2 Tests for difference in immune function in relation to whether bird species had or had not been involved in introductions on oceanic islands.

Variable	Wald χ^2	p	Slope±SE	N
Bursa of Fabricius	0.01	0.91	−0.095±0.808	121
Thymus	0.05	0.83	0.116±0.527	63
Spleen	0.29	0.59	10.351±0.648	162
Nestling T-cell response	0.21	0.65	0.490±1.071	83
Adult T-cell response	2.99	0.08	−1.951±1.129	85
Nestling NAbs	1.22	0.27	−1.366±1.234	51
Adult NAbs	0.06	0.81	0.260±1.070	31
Nestling complement	0.01	0.93	−0.182±9.340	51
Adult complement	0.17	0.68	1.187±2.876	31
Lysozyme concentration in eggs	0.90	0.34	−0.871±0.918	33

Likewise, introduction success was not significantly predicted by overall invasion success (distance from islands to mainland: $r=0.13$, $N=37$, $p=0.47$), nor was dispersal significantly related to distance to mainland (natal dispersal: $r=-0.003$, $N=59$, $p=0.94$; breeding dispersal: $r=0.15$, $N=58$, $p=0.28$). Therefore, these three processes of invasions can be considered statistically independent.

15.4.3   Dispersal and immunity

We related natal dispersal distance to immune responses. Natal dispersal distance was predicted by T-cell response of nestlings, with species with longer dispersal distances having stronger immune responses (Fig. 15.2; Møller *et al.* 2004). In addition, natal dispersal distance was predicted by NAbs in nestlings, with species with higher levels of natural antibodies having longer dispersal distances (weighted model (same result for unweighted model): $F_{1,30}=6.19$, $r^2=0.17$, $p=0.019$, slope±S.E.=$0.709±0.285$). This suggests that strong innate anti-bacterial defences promote dispersal. Snoeijs *et al.* (2004) showed for the great tit *Parus major* that humoral immune response was positively related to dispersal distance.

15.4.4   Establishment and immunity

Introduction success was predicted by T-cell mediated response of nestlings (Fig. 15.3a; Møller and Cassey 2004). This effect of T-cell mediated immune response was stronger for large than for small inocula (Fig. 15.3b; Møller and Cassey 2004).

15.4.5   Expansion and immunity

Maximum expansion after successful introduction was on average 560 km (S.E.=1 km, after back-transformation from log-transformed data), median 860 km, range 10–7675 km, $N=28$ species. Expansion by introduced species estimated as the change in range

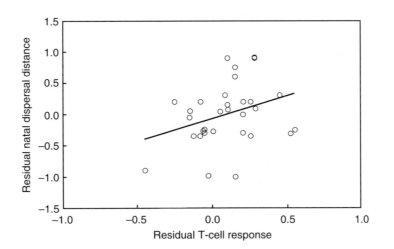

Figure 15.2   Natal dispersal distance in relation to T-cell mediated immunity of nestlings in different species of birds. Modified after Møller *et al.* (2004).

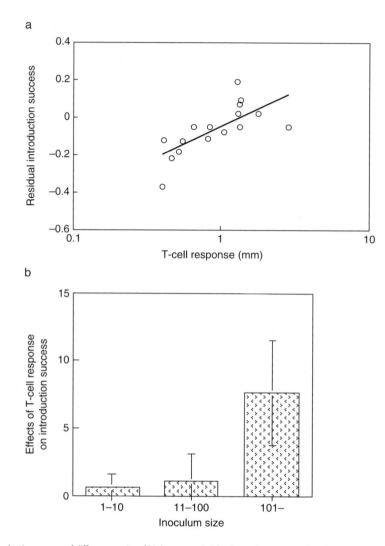

Figure 15.3 (a) Introduction success of different species of birds on oceanic islands in relation to T-cell mediated immune response of nestlings. (b) The effect of T-cell mediated immune response on introduction success depends strongly on inoculum size. Modified after Møller and Cassey (2004).

size was not significantly related to any of the immune defence variables.

Range expansion in Finland following recent climate change was estimated as the change in northernmost breeding range between 1974–79 and 1986–89 (Hyytiä et al. 1983; Väisänen et al. 1998) after inclusion of change in total breeding range as a confounding variable because species that have become common invariably are expected to expand their range limits, while species that have become rare are

expected to have contracted their range (Thomas and Lennon 1999). That change in northernmost breeding range was predicted by spleen mass, with species expanding their range having small spleens and species contracting their range having large spleens (Fig. 15.4; partial effect of spleen in a model that included change in abundance: $F_{1,70}=6.34$, $r^2=0.08$, $p=0.014$, slope±S.E.=−28.264±11.220). Thus, species that expanded their range the most had smaller spleens for their body size.

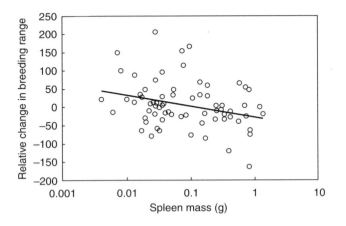

Figure 15.4 Breeding range expansion for the northern range in Finland in relation to relative size of the spleen, after adjusting for body size, in different species of birds.

Expansion as reflected by total range size was predicted by T-cell mediated response of nestlings ($F_{1,82}$=9.20, r^2=0.11, p=0.0032, slope±S.E.=0.434±0.143). Species with stronger immune responses had larger total breeding ranges.

15.4.6 Invasion and immunity

We related invasion success from the data set of Møller and Mousseau (2007) to immune responses. This measure of invasion success was not significantly related to T-cell response of nestlings ($F_{1,58}$=0.97, p=0.33), or NAbs in nestlings ($F_{1,30}$=0.68, p=0.42). However, species with long maximum distances to the nearest mainland had low adult survival rates ($F_{1,74}$=9.46, r^2=0.11, p=0.0029, slope±S.E.=−1.697±0.552) and hence a large proportion of young individuals. This second series of analyses suggests that immune responses are not related to all measures of dispersal, and that invasion success leading to establishment of isolated breeding populations on islands is unrelated to immune responses.

Another measure of invasion is whether a species has colonized more than one continent (Palaearctic and Nearctic). Species that had conquered two continents had almost twice as large total breeding ranges as species that only occurred on one continent (one continent: 15.45×10⁶ km², N=432, two continents: 28.64×10⁶ km², N=94). Species that had expanded to two continents had significantly longer distances to the mainland than species with a breed-

ing range restricted to one continent (maximum distance between islands and mainland: Wald χ^2=18.19, p<0.0001, slope±S.E.=−1.190±0.279), suggesting that two independent measures of expansion are positively correlated. Furthermore, species that had colonized two continents had higher adult survival rates than species that had only colonized a single continent (Wald χ^2=19.80, p<0.0001, slope±S.E.=4.369±0.982), suggesting that the ability to survive may have rescued initially small populations during invasions. Finally, an ability to colonize multiple continents was related to nestling complement, with species colonizing

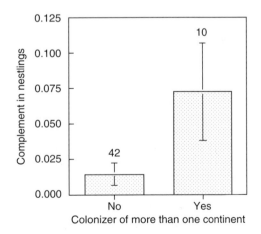

Figure 15.5 Ability to colonize multiple continents in relation to levels of complement in nestlings of different species of birds. Values are means ±SE; the number of species is indicated above the histograms.

multiple continents having higher levels of complement (Fig. 15.5; Wald $\chi^2=16.80$, $p<0.0001$, slope±S.E.$=11.374\pm2.775$). This finding suggests that the ability of young birds to defend themselves against bacterial infections was superior in species colonizing two continents compared to species restricted to a single continent.

15.5 Discussion

The main results of this analysis of the relationship between the three component processes of successful invasions and immunity in birds are presented in Fig. 15.6. More specifically, we found that a number of measures of immune function in young birds, but not in adults, predicted success. Bird species with strong T-cell mediated immune responses in young birds had long natal dispersal distances, elevated introduction success—especially at large inoculum sizes—and large total breeding ranges. In addition, bird species that have colonized both the Palaearctic and the Nearctic had higher

levels of complement in nestlings than species with breeding ranges restricted to a single continent.

Dispersal implies movement into new environments that may have parasite strains that a dispersing individual has not yet encountered. This potentially raises problems of responding rapidly and efficiently to such novel parasites to avoid risking mortality and the loss of fecundity. T-cell response of nestlings in different species of birds was positively related to natal, but not to breeding, dispersal distance (Møller *et al.* 2004). We consider the differential effect on immunity of young birds compared to adults to be significant because young birds disperse much further than adults from one breeding site to another (Paradis *et al.* 1998). This analysis was controlled for effects of habitat specialization because habitat specialists that exploit a small number of very specific habitats are more likely to encounter specialist parasites than hosts that can be found in all different kinds of habitats (Møller *et al.* 2004). In fact Møller *et al.* (2005) and Møller and Rózsa (2005) have shown that avian host species with strong T-cell responses

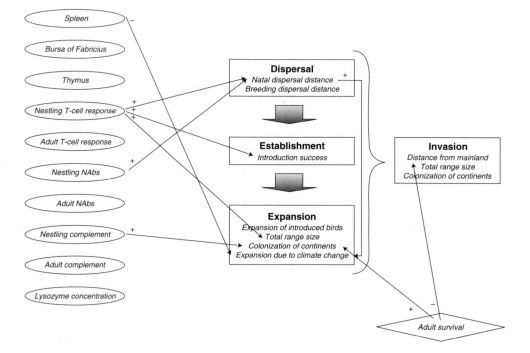

Figure 15.6 Summary of the analyses of invasion success in relation to immunity in birds. The sign of the effects reflects whether a positive or a negative relationship was found.

in nestlings are exploited by specialist parasites (fleas, chewing lice) that seem to be more virulent (Regoes *et al.* 2002; Garamszegi 2006).

Establishment success was not random with respect to immunity because the T-cell response of nestlings promoted introduction success, especially at large inoculum sizes. Drake (2003) has shown in a model that the relationship between the number of individuals introduced and the probability of establishment depends upon the relationship between virulence and the fraction of the population infected at introduction. The finding that T-cell mediated immune response did not predict introduction success for small inocula, but did so for large inocula (Møller and Cassey 2004) is in agreement with this prediction.

Subsequent expansion by introduced birds was not accounted for by any of the immunity variables. However, several other measures of expansion were significantly related to immunity. Total breeding range size was larger in species with strong T-cell response in nestlings. This finding could be due to greater exposure to a more diverse fauna of parasites in species with a large total range, or species with a large breeding range have higher levels of standing genetic variation (Møller *et al.* 2008), with greater genetic variation accounting for stronger immune responses. We also found that complement in nestlings was elevated in species that had expanded into two continents rather than just one. Again, we consider the two mechanisms just listed to be possible explanations. Finally, range expansion in northern Finland associated with recent climate change was greater in species with relatively small spleens. Because the spleen is a secondary lymphoid organ mainly used for storage of leucocytes (Tizard 1991; Abbas *et al.* 1994; Roitt *et al.* 1996), an interpretation of this finding is that bird species with relatively low levels of infection are better able to expand the breeding range northwards.

Why were some measures of immunity, but not others, related to the three variables accounting for successful invasions? *A priori* we predicted that the immune function of young birds would be more important than the immune function of adults. The simple reason is that the ability to cope with novel environments is a feature of young individuals that

generally disperse much further than older ones. Another explanation is that young individuals constitute the majority of individuals that disperse into novel environments, become established there, and subsequently expand the breeding range of a species, simply because young individuals are more common. We did not find much support for this prediction because only two analyses showed a significant effect of adult survival rate, and then only one of these in the predicted direction. Invasion estimated as the maximum distance to the nearest mainland decreased with increasing adult survival rate, as expected. In contrast, species that had managed to colonize more than one continent had a higher adult survival rate than those that only occurred on one continent. This finding is more in line with expectations from small populations being buffered by high adult survival rate, thus being partly shielded from effects of demographic stochasticity (Legendre *et al.* 1999).

The interspecific patterns that we have described here beg for investigation of the underlying mechanisms involved and clarification of causal relationships between invasions and immune traits. Which individuals are involved in invasions, and what is their parasite and immune status? With current range expansion related to climate change occurring at a remarkable pace of 1–2 km per years (Thomas and Lennon 1999; Brommer 2004), there is a real possibility of characterizing individuals involved in such expansion compared to the average individual near the current distribution border and, importantly, well within the main breeding range of a species.

In conclusion, we have shown that dispersal, establishment, and expansion are not processes that are random with respect to the immune status of the individuals of different species. However, particular selection processes involved in these three steps are partially different, as different components of the immune defence may play different roles at different levels. In general, we suggest that successful invasive species will be characterized by strong investment in immune function of the young individuals that are the primary dispersive agents colonizing new areas and expanding current distribution borders as these may change in response to altered environmental conditions.

References

Abbas, A.K., Lichtman, A.H., and Pober, J.S. (1994). *Cellular and Molecular Immunology*. Saunders, Philadelphia.

Braun, P. and Fehlhaber, K. (1996). Studies of the inhibitory effect of egg albumen on gram-positive bacteria and on *Salmonella enteritidis* strains. *Archiv für Geflügelkunde*, **60**, 203–07.

Brommer, J.E. (2004). The range margins of northern birds shift polewards. *Annales Zoologici Fennici*, **41**, 392–97.

Carroll, M.C. and Prodeus, A.P. (1998). Linkages of innate and adaptive immunity. *Current Opinions in Immunology*, **10**, 36–40.

Cassey, P. (2002). Life history and ecology influences establishment success of introduced land birds. *Biological Journal of the Linnean Society*, **76**, 465–80.

Congdon, L.L., Farmer, J.N., Longenecker, B.M., and Breitenbach, R.P. (1969). Natural and acquired antibodies to *Plasmodium lophurae* in intact and bursaless chickens. II. Immunofluorescent studies. *Journal of Parasitology*, **55**, 817–24.

Drake, J.M. (2003). The paradox of the parasites: implications for biological invasion. *Biology Letters*, **270**, S133–S135.

Garamszegi, L.Z. (2006). The evolution of virulence and host specialization in malaria parasites of primates. *Ecology Letters*, **9**, 933–40.

Garamszegi, L.Z. and Møller, A.P. (2007). Prevalence of avian influenza and host ecology. *Proceedings of the Royal Society of London B*, **274**, 2003–12.

Goto, N., Kodama, H., Okada, K., and Fujimoto, Y. (1978). Suppression of phytohaemagglutinin skin response in thymectomized chickens. *Poultry Science*, **52**, 246–50.

Hyytiä, K., Kellomäki, E., and Koistinen, I., eds. (1983). *Suomen Lintuatlas*. SLYn Lintutieto Oy, Helsinki.

Kudo, S. (2000). Enzymes responsible for the bactericidal effect in extracts of vitelline and fertilisation envelopes of rainbow trout eggs. *Zygote*, **8**, 257–65.

Legendre, S., Clobert, J., Møller, A.P., and Sorci, G. (1999). Demographic stochasticity and social mating system in the process of extinction of small populations: The case of passerines introduced to New Zealand. *American Naturalist*, **153**, 449–63.

Lockwood, J.L., Hoopes, M.F., and Marchetti, M.P. (2007). *Invasion Ecology*. Blackwell, Malden.

Martin, L.B., Han, P., Lewittes, J., Kuhlman, J. R., Klasing, K.C., and Wikelski, M. (2006). Phytohemaggutinin-induced skin swelling in birds: histological support for a classical immunecological technique. *Functional Ecology*, **20**, 290–09.

Mitchell, C.E. and Power, A.G. (2003). Release of invasive plants from fungal and viral pathogens. *Nature*, **421**, 625–27.

Møller, A.P. and Cassey, P. (2004). On the relationship between T-cell mediated immunity in bird species and the establishment success of introduced populations. *Journal of Animal Ecology*, **73**, 1035–42.

Møller, A.P. and Mousseau, T.A. (2007). Determinants of interspecific variation in population declines of birds from exposure to radiation at Chernobyl. *Journal of Applied Ecology*, **44**, 909–19.

Møller, A.P. and Nielsen, J.T. (2007). Malaria and risk of predation: a comparative study of birds. *Ecology*, **88**, 871–81.

Møller, A.P. and Rózsa, L. (2005). Parasite biodiversity and host defenses: chewing lice and immune response of their avian hosts. *Oecologia*, **142**, 169–76.

Møller, A.P., Christe, P., and Lux, E. (1999). Parasitism, host immune function, and sexual selection. *Quarterly Review of Biology*, **74**, 3–20.

Møller, A.P., Martín-Vivaldi, M., and Soler, J.J. (2004). Parasitism, host immune response and dispersal. *Journal of Evolutionary Biology*, **17**, 603–12.

Møller, A.P., Christe, P., and Garamszegi, L.Z. (2005). Coevolutionary arms races: increased host immune defense promotes specialization by avian fleas. *Journal of Evolutionary Biology*, **18**, 46–59.

Møller, A.P., Garamszegi, L.Z., and Spottiswoode, C. (2008). Genetic similarity, distribution range and sexual selection. *Journal of Evolutionary Biology*, **21**, 213–25.

Møller, A.P., Erritzøe, J., and Rózsa, L. (2010). Ectoparasites, uropygial glands and hatching success in birds. *Oecologia* (in press).

Ochsenbein, A.F. and Zinkernagel, R.M. (2000). Natural antibodies and complement link innate and acquired immunity. *Immunology Today*, **21**, 624–30.

Ochsenbein, A.F., Fehr, T., Lutz, C. *et al.* (1999). Control of early viral and bacterial distribution and disease by natural antibodies. *Science*, **286**, 2156–59.

Paradis, E., Baillie, S.R., Sutherland, W.J., and Gregory, R.D. (1998). Patterns of natal and breeding dispersal in birds. *Journal of Animal Ecology*, **67**, 518–36.

Parmentier, H.K., de Vries Reilingh, G., and Nieuwland, M.G.B. (1998). Kinetic immunohistochemical characteristic of mitogen-induced cutaneous hypersensitivity responses in chickens divergently selected for antibody responsiveness. *Veterinary Immunology and Immunopathology*, **66**, 367–-76.

Pastoret, P., Gabriel, P., Bazin, H., and Govaerts, A. (1998). *Handbook of Vertebrate Immunology*. Academic Press, San Diego.

Purvis, A. and Rambaut, A. (1995). Comparative analysis by independent contrasts (CAIC). *Computer and Applied Biosciences*, **11**, 247–51.

Regoes, R.R, Nowak, M.A., and Bonhoeffer, S. (2000). Evolution of virulence in a heterogeneous host population. *Evolution*, **54**, 64–71.

Roitt, I., Brostoff, J., and Male, D. (1996). *Immunology*. Mosby, London.

Seddon, J.M. and Baverstock, P.R. (1999). Variation on islands: major histocompatibility complex (Mhc) polymorphism in populations of the Australian bush rat. *Molecular Ecology*, **8**, 2071–79.

Snoeijs, T., Van de Casteele, T., Adriaensen, F., Matthysen, E., and Eens, M. (2004). A strong association between immune responsiveness and natal dispersal in a songbird. *Biology Letters*, **271**, S199–S201.

Sommer, S. (2003). Effects of habitat fragmentation and changes of dispersal behaviour after a recent population decline on the genetic variability of noncoding and coding DNA of a monogamous Malagasy rodent. *Molecular Ecology*, **10**, 2845–51.

Thomas, C.D. and Lennon, J.J. (1999). Birds extend their ranges northwards. *Nature*, **399**, 213.

Tizard, I. (1991). *Veterinary Immunology*. Saunders, Philadelphia.

Torchin, M.E., Lafferty, K.D., Dobson, A.P., McKenzie, V.J., and Kuris, A.M. (2003). Introduced species and their missing parasites. *Nature*, **421**, 628–30.

Väisänen, R.A., Lammi, E., and Koskimies, P. (1998). *Distribution, Number and Population Changes of Finnish Breeding Birds*. Otava, Keuruu (in Finnish).

Veltman, C.J., Nee, S., and Crawley, M.J. (1996). Correlates of introduction success in exotic New Zealand birds. *American Naturalist*, **147**, 542–57.

CHAPTER 16

Infection, immunity, and island adaptation in birds

Kevin D. Matson and Jon S. Beadell

16.1 Introduction

Islands are discrete units with clear boundaries which makes them ideal natural laboratories for the study of evolution and ecology and, in particular, facilitates the study of the geography of host–parasite interactions. One such connection between island geography and host–parasite dynamics centres on island-dwelling landbirds and the role parasite communities play in shaping their physiology. Interest in this particular interaction arose in part from accounts of the plight of the endemic birds of Hawaii; research conducted since the 1980s has clearly identified a role of infectious disease in driving these birds to the brink of extinction and in some cases beyond. It is less clear what, if any, generalizable differences exist between islands and continents in terms of host–parasite coevolution. For example, it is uncertain whether the prevailing view of islands as simplified ecosystems with reduced species richness extends to parasite communities and translates into attenuated disease risk on islands. Ongoing challenges include identifying the causes (e.g. from differential parasite exposure) and consequences (e.g. for fitness) of general differences in physiology between island and continental birds and gaining insight into the ecological and evolutionary influences shaping immune function. Meeting these challenges will ultimately require a combination of measurements of immune defences and disease risk, as well as parameters of avian life-history.

In this chapter, we first delineate its geographical and biological scope, review briefly the avian immune system, and present the immune indices most commonly associated with ornithological, ecological, and evolutionary studies. Second, we outline (in terms of interactions and evolutionary processes) the possible paths that could potentially lead to immunological differentiation and increased disease susceptibility in insular birds. Third, as a case study, we review the interaction that provoked, to a great extent, the interest in immunity and parasitism in insular taxa: the endemic Hawaiian honeycreepers and their introduced parasites. Fourth, we assess the impacts of different diseases on other insular (and continental) birds to evaluate if the increased susceptibility of some native Hawaiian birds to particular diseases is an anomaly or if the system remains a broadly relevant and informative example.

16.2 Background

16.2.1 Definitions and scope

It is important to explicitly acknowledge the interactions with which we are concerned and the scope of this chapter. In this chapter the term 'islands' generally refers to relatively small and isolated oceanic islands or archipelagos, two classic examples being the Galápagos and Hawaiian islands. In limited cases, we also reference studies of nearshore and offshore continental islands. Other isolated habitat patches (e.g. mountain peaks, ponds, forest fragments) are not explicitly considered here, but similar interactions in these insular environments can sometimes be envisioned.

Generally, we cast a fairly wide net when considering 'parasites'. Basically, when taken in tandem with immunity, we use this term to primarily refer to the micro- (e.g. viruses, bacteria, fungi, and protists) and macro- (e.g. worms) parasites that interact with and potentially harm internal tissues of the host and that are therefore under surveillance by the host's immune system. In some cases ectoparasites, like lice and ticks that breach the skin surface to feed on blood, similarly meet this host-interaction criterion.

Two other groups with immunological roles, which do not fall clearly within the bounds of parasites, also need to be taken into account when considering factors, geographical or otherwise, that shape immune systems. The first group, which is substantially broader and more encompassing than the parasite group, can be thought of as the entire immunologically-relevant molecular universe (e.g. antigenic, immunogenic, inflammatory, and toxic agents, but also self/nonself differentiation). Exposure routes to this molecular universe are varied and include ingestion, inhalation, and absorption, but exposure to nonself (or more precisely, 'inappropriate self') can also originate endogenously, for example via cancerous or otherwise damaged cells. The second group with an immunological role comprises organisms that engage in primarily commensal relationships with hosts, but interact with host immune systems nonetheless. Gut microflora communities are an ideal example. On the one hand, these bacteria directly shape immune responses; on the other hand, uncompromised immune systems are vital for controlling these populations and relegating them to their commensal status (Kelly *et al.* 2005; Magalhaes *et al.* 2007; Brisbin *et al.* 2008). This second group highlights the fine lines distinguishing parasitic relationships from commensal and mutual ones. Moreover, coevolution between hosts and typical parasites (as defined above) can similarly lead to parasites that have negligible detrimental consequences for host fitness, effectively blurring the lines between these categories of biological interaction even further.

16.2.3 Avian immune systems

Immune systems in general and the avian immune system in particular have been described by others in great detail (e.g. Glick 2000). The following is only a cursory review of the main immune subsystems and axes. Overall, immune defences segregate functionally between innate (also termed 'natural' or 'nonspecific') and acquired (also termed 'adaptive' or 'specific') immunity (Davies *et al.* 1998; Abbas and Lichtman 2001). Immunologists view innate immunity as a first line of defence: it is essentially always present, it offers immediate protection, and it can stimulate or enhance acquired immunity (Davies *et al.* 1998; Abbas and Lichtman 2001). Forms of innate immunity range from physical and chemical barriers (e.g. unbroken skin and stomach acid) to molecular and cellular mechanisms (e.g. bacteriolytic enzymes and natural killer cells), all of which function to prevent or limit the initial assault (Davies *et al.* 1998; Abbas and Lichtman 2001).

Unlike innate immunity, acquired immunity is highly specific. This high specificity is associated with other valuable qualities such as immunological memory, but adaptive immunity requires days to weeks before being maximally protective (Abbas and Lichtman 2001). Acquired immune function splits along mechanistic lines into humoral and cell-mediated defences. Since lymphocytes are the central cellular players, however, acquired immunity is more precisely subdivided by lymphocyte type: B cells, CD8+ cytotoxic T cells, and CD4+ helper T cells. B cells produce antibodies; antibodies bind to extracellular antigens and protect via several processes (e.g. opsonization, neutralization, agglutination; Davies *et al.* 1998). CD8+ cytotoxic T cells defend against intracellular threats (both exogenous and endogenous) by destroying directly any cells exhibiting foreign or inappropriate antigens (Davies *et al.* 1998). CD4+ helper T cells subtypes coordinate different immune responses. One subtype (Th1) promotes inflammation, macrophage function, and defences against intracellular threats; another subtype (Th2) promotes antibody production and defence against extracellular threats including worms (Davies *et al.* 1998).

When viewed from an evolutionary and ecological perspective rather than a strictly immunological one, host immune responses to parasites also have been compared along two axes: 'tactics' (i.e. constitutive vs. induced) and 'specificity' (i.e. non-specific vs. specific; Schmidt-Hempel and Ebert 2003). These two axes by and large reflect the innate/acquired dichotomy. That is, most innate defences are both nonspecific and constitutively present, and most acquired defences are specific and must be induced. Cleary, as Schmidt-Hempel and Ebert (2003) point out using lymphocyte populations as one example, some induced responses do have constitutive foundations that in theory could be separately quantified. In practice, however, the distinction between constitutive (or baseline) levels and induced (or response) levels is often hazy. The immune system is highly dynamic, and some of the most ever-present constitutive defences (e.g. phagocytic activity) can be induced (either positively or negatively) through concomitant short-term changes to health, immunological, or physiological status. Such dynamism among defences, regardless of where they fall along a 'tactics' axis, obfuscates evolutionary and ecological interpretations.

16.2.3 Indices of immune function

Quantification of immune function in non-traditional study species, particularly in free-living animals, presents an array of challenges, which oftentimes partially dictate what can be measured. Instead of dividing descriptions of the assays most commonly used by ecologists and evolutionary biologists along functional or mechanistic lines, we divide the assays into two groups based on study design and the allied logistical constraints. On the one hand, there are methods that are optimal for broader comparative studies because a single blood sample collected at capture can be used. On the other hand, there are other methods that require recapture and sampling at two or more time points. Because reliable recapture of wild birds is often unfeasible, this second group of methods is most germane to studies limited to one or a few study species amenable to recapture (e.g. box- or cavity-nesting species).

Immunological methods based on a single blood sample collected at capture provide a snapshot of current immune status and are often characterized as quantifying innate and constitutive defences. For an immune index to fully meet these criteria, the measured parameter should be relatively unresponsive to short-term and condition-related fluctuations. For example, one assay, which quantifies the abilities of blood plasma to agglutinate and lyse exogenous red blood cells, appears to be insensitive to the many physiological changes associated with lipopolysaccharide- (LPS-) induced endotoxic shock (Matson *et al.* 2005). This insensitivity to changes in

inflammatory status actually enhances the interpretability of haemolysis and haemagglutination data from field-collected samples: current health status need not be considered when interpreting results among individuals.

Unfortunately, most assays that can be used to analyse single blood samples collected at capture either are known to be sensitive to various stimuli that have effects on different time scales or have not been well-studied in this regard. One traditional method for indexing immune function in both clinical and nonclinical settings is white blood cell (WBC, or leucocyte) quantification using blood smears (Bounous and Stedman 2000). Clinical applications and studies in domestic and wild animals have shown WBC counts change over hours or days in animals that are sick (Hawkey *et al.* 1982; Averbeck 1992) or stressed (Gross and Siegel 1983; Vleck *et al.* 2000). In well studied species, baseline ranges corresponding to healthy individuals can be established (Bounous and Stedman 2000), and individuals outside these ranges can be identified (Hawkey *et al.* 1982; Averbeck 1992). Interpretation of WBC data among individuals without this type of background understanding is difficult. Meaningful comparisons among populations of the same species are possible (Matson 2006; Buehler *et al.* 2008), but comparisons among dissimilar species, which can differ in their baseline WBC concentrations and relative proportions, only serve to highlight the complexity of immune system evolution (Matson *et al.* 2006a) and rarely result in clear connections between, for example, physiology and life-history.

Another method, assaying the abilities of whole blood or blood plasma to kill microorganisms in vitro, has been revitalized recently by ecological immunologists (Tieleman *et al.* 2005; Matson *et al.* 2006b; Millet *et al.* 2007; Buehler *et al.* 2008a). This method has the great benefit of being readily interpreted as a result of its functional nature (i.e. better *in vitro* killing abilities equates to better *in vivo* bacterial defence capacities). Like WBC profiles however, bacteria killing is influenced by both health/immune status (e.g. systemic inflammation; Millet *et al.* 2007) and other physiological parameters (e.g. stress; Matson *et al.* 2006b; Buehler *et al.* 2008a). Moreover, it is not always obvious that more bacteria killing is better. Any immunological function must be viewed in the context of other complementary immune functions (which could make up for any deficit) and in the context of other features of life-history or immunity (self-reactivity) that might be detrimentally affected by a high response in one particular aspect of immune function.

Measuring concentrations of immunologically-relevant plasma proteins, in particular acute phase proteins, is yet another approach for characterizing immune defences using a single blood sample. The existence of assays based on protein function (e.g. the iron-binding abilities of haptoglobin (Matson 2006) or the bacteriolytic abilities of lysozyme (Millet *et al.* 2007), rather than highly specific protein structure (as with enzyme-linked immunosorbent assays or ELISAs) allows functionally similar molecules to be quantified across species without species-specific reagents. Additionally, these assays often require only minute volumes of plasma (e.g. 5–15 uL). Given the very nature of the acute phase response, however, concentrations of these proteins are especially sensitive to intra-individual changes in inflammation and health status, and these sources of variation cannot be ignored when interpreting results. The bottom line is that as the field of ecological immunology has developed, more methods that meet the logistical constraints inherent to large comparative studies have become available, but the these methods still have important limitations when it comes to interpretation.

One way to improve interpretability of immune assay data is to perform repeated measures within individuals, ideally before and after an immunological manipulation. Under this scenario, the utility of the methods described above increases, but this comes at the expense of taxonomic diversity: in the wild, some species or groups are particularly difficult, if not impossible, to resample with the requisite regularity. In addition to making repeated measures using those methods that can technically be measured at a single time point, working within repeated measures framework, either through an initial capture and a subsequent re-capture or through a brief period in captivity, makes available an additional set of assays that require at least two handling points.

Probably the most common technique requiring repeated measures is the patagial injection of phytohaemagglutinin (PHA) to induce a localized, short-term (i.e. one or two days) swelling (Tella *et al.* 2002). This swelling response, which is often said to measure cell-mediated immunity, appears to rely on both innate and acquired elements (Tella *et al.* 2002; Martin *et al.* 2006). Working on a similar time scale, other inflammatory agents like LPS are injected to induce systemic inflammation and to allow quantification of broad behavioural and physiological sickness responses (Bonneaud *et al.* 2003; Owen-Ashley *et al.* 2006). Requiring a slightly longer time course of days to weeks, primary and secondary antibody responses to vaccinations with either antigenic particles (e.g. exogenous red blood cells) or purified soluble antigens are used to measure acquired humoral immunity (Hanssen *et al.* 2005; Smits and Baos 2005; Hanssen 2006; Mendes *et al.* 2006). Assays of wound-healing capacities, which effectively integrate a suite of *in vivo* immunological processes, have also been applied with some success in comparative studies of captive mammals (Martin *et al.* 2007), even though these assays can be influenced by stress (Padgett *et al.* 1998; French *et al.* 2006).

The functional perspective from which wound healing capacities (and, to an extent, *in vitro* microbial killing abilities) can be interpreted is an important asset, particularly in light of criticisms that highlight the disconnection between some measures of immune function and actual disease resistance (Adamo 2004). Combining direct assays of disease susceptibility (e.g. via experimental infections) with assays of immune defences will provide the strongest evidence to support or refute existing hypotheses linking hosts, parasites, and island-life. Such tests of disease susceptibility, however, are often undesirable or impossible to perform when working in non-domestic vertebrate animals as the result of legal, ethical, or practical constraints. Remarkably, precisely this type of testing played a pivotal part in elucidating the severe effects of malaria (*Plasmodium relictum*) in the Hawaiian honeycreepers. With a few exceptions, however, few studies have attempted to determine explicitly what, if any, changes in immune physiology are correlated with this increased susceptibility and island-life in general (Matson 2006; Beadell *et al.* 2007).

16.3   Paths to immune function differentiation

The immunological, ecological, and evolutionary foundations of reduced resistance and increased susceptibility in insular populations are poorly understood and have only begun to be investigated (Jarvi *et al.* 2001; Lindström *et al.* 2004; Matson 2006; Beadell *et al.* 2007). In terms of evolutionary processes and ecological interactions, there are a number of possible paths that could lead to immunological differentiation and potentially to increased disease susceptibility in insular birds. Associated with the transition from continental origins to island existence, genetic differences might arise through random or nonrandom processes acting on relatively small island populations of hosts, parasites, or both. Likewise, the ecological differentiation that is associated with small isolated patches of habitat and that is characteristic of islands (compared to continents) might influence or shape host and parasite phenotypes or relax coevolutionary constraints and allow new evolutionary trajectories.

16.3.1   Genetic bases

Genetic diversity is critical for both immune function and disease resistance; however, reduced genetic diversity is often associated with island populations and their establishment (Frankham 1997). Beyond the body of work relating genetics and immunity in humans and domestic animals (e.g. artificial selection in chickens, Leshchinsky and Klasing 2001), avian ecologists have begun investigating the links among genetics of wild birds, comparative indices of immune function (as described above) and disease resistance (Hawley *et al.* 2006; Whiteman *et al.* 2006). The reductions in genetic diversity in island populations, which adds another facet to these sorts of studies (e.g. Whiteman *et al.* 2006), can come about or can be accentuated through several mechanisms.

Oceanic island populations are likely to experience extreme founder effects (or founder bottlenecks; Berry 1986; Frankham 1997, 1998). The inverse relationship between isolation (i.e. the distance between an island and its source of colonizers) and immigration rate (*sensu* MacArthur and Wilson 2001) means

that the most isolated islands are likely to be colonized by the fewest individuals and subjected to the strongest founder effects (Frankham 1997). Thus, gene frequencies between the source population and the island population will differ—rare alleles in the source population might disappear entirely, common alleles might become fixed. Founder effects can have immunological consequences; Hawley *et al.* (2006) connect such an effect to disease outbreaks in populations of the house finches (*Carpodacus mexicanus*) that were introduced to new continental locations. The potential loss of genetic diversity from founder effects could be further amplified through other genetic processes that exert a strong effect on small populations, like genetic drift and inbreeding (Frankham 1997). Indeed, like founder effects, inbreeding can have immunological consequences in wild birds (Reid *et al.* 2003, 2007). Thus, without considering any fitness costs or benefits of immune system maintenance, it is easy to envisage the loss of immunologically-relevant genetic diversity through essentially purely stochastic processes (e.g. rare colonization events by very small populations, genetic drift/bottleneck).

Attenuated parasite and pathogen pressure on islands might weaken the selective pressures that maintain immune system function. There are a number of hypotheses linking the disease ecology of different habitats and the evolution of avian immune systems. Specifically it is believed that some environments like islands (but also xeric and marine habitats) have relatively low pathogen pressures and other environments like continents (and especially humid and tropical habitats) have particularly high pathogen pressures (Piersma 1997; Mendes *et al.* 2005; Tieleman 2005; Tieleman *et al.* 2005; Matson 2006). Under slackened pathogen pressures, natural selection could lead to diminished immune function as an evolutionary response to immune system costs. While there is an on-going discussion concerning the sources, magnitudes, and most important currencies (e.g. energy, nutrients, etc.) of these immunological costs, many studies do document their existence. Among those that have be investigated are the energetic (Martin *et al.* 2002), autoimmunological (Råberg *et al.* 1998), growth (Klasing 1987), survival and reproduction (Hanssen *et al.* 2005; Hanssen 2006) costs associated with immune system development (Lee *et al.* 2008), maintenance (Buehler 2008b), and use (Bonneaud *et al.* 2003).

16.3.2 Ecological bases

Unlike the many genetic effects that reduce variability upon island establishment, the unpopulated island environments, new ecological opportunities, and empty niches available to the earliest colonizers can lead to dramatic evolutionary changes and new diversity (e.g. through processes like adaptive radiation). Insofar as immunity and physiology are traded-off, adaptation to a new island environment (e.g. dietary changes, behavioural shifts, etc.) could cause the immune system to follow a new trajectory. Thus, it is worthwhile explicitly considering how both parasite ecology and host ecology differ on islands and how these ecological differences might impact on host–parasite coevolution. These ecological differences potentially represent one aspect of an 'island syndrome', which is normally characterized by reduced interspecific interactions (e.g. predator–prey, but in this case host–parasite) in the simplified ecosystems of islands (Adler and Levins 1994; Hochberg and Møller 2001; Blumstein and Daniel 2005).

The theoretical underpinnings of island colonization processes (MacArthur and Wilson 2001) and some empirical studies (Horner-Devine *et al.* 2004; Bell *et al.* 2005) suggest that seral and climax communities of micro- and macroparasites on islands likely relate to island size and differ from equivalent communities on continents. Such disparities could translate to broad differences in overall microbe and parasite community composition (i.e. richness, abundance, diversity) but also to more narrow differences in the presence/absence of particular disease-causing organisms (Steadman *et al.* 1990; Super and Van Riper 1995).

When a volcanic island rises from the sea, the first permanent landmass is devoid of all terrestrial life. Microorganisms are undoubtedly among the first individual colonizers, given their small size and great dispersal abilities. Entire communities of microorganisms and macroparasites might be deposited on islands after hitchhiking on and inside of their associated hosts, which themselves might

or might not become permanent island residents. Like their hosts, parasites are subject to genetic consequences of colonization. For example, some genetic diversity of parasites is likely lost upon arrival through the same mechanisms affecting host genetic diversity. To an extent, these effects might be abated because parasite populations are likely larger and their generation times shorter than those of vertebrate hosts. Through ecological processes that are analogous to the loss of alleles via founder effects, specific strains of symbiotic microbes (either commensals or pathogens) or other endo- or ectoparasites might never arrive to an island environment if the initial founders of the host populations do not possess them. In fact, infected or diseased individuals might be the least likely to arrive on distant islands. The absence of suitable vectors, which are key to the transmission cycle of some diseases, would also place powerful limitations on the establishment of the vector-dependent parasite faunas.

Parasites that successfully arrive to an island must interact with a different assemblage of hosts. The diversity and composition of host communities can drive disease risk (LoGiudice *et al.* 2003). Coevolved epidemiological equilibria might be disrupted in the absence of a secondary reservoir species, which like vectors are an important element in the transmission cycle of some diseases. Epidemiological models suggest disease transmission interacts with host density (Gao and Hethcote 1992; McCallum *et al.* 2001; Bonsall and Benmayor 2005), which can be relatively higher or lower on islands (MacArthur *et al.* 1972). By affecting the frequency of interactions within and among metapopulations, spatial and temporal population structures of hosts can also shape the impacts of disease via altered virulence and transmission rates (Haraguchi and Sasaki 2000; Boots *et al.* 2004). Single small isolated islands and the small host populations that live there should be the least structured; populations on archipelagos and continents are likely more structured.

The annual cycle dynamics affecting host–parasite relationships are also different on islands compared to continents. In insular landbirds, migration is eliminated. Additionally, weather conditions and seasonality are moderated by the oceanic climates that dominate small islands. Seasonal changes in blood parasite prevalences have been found in both winter migrant and resident populations (Perez-Tris and Bensch 2005). The combination of this finding and the evidence that in the same local breeding populations geographically-restricted parasites are less prevalent than migration-dispersed parasites (Perez-Tris and Bensch 2005) offers a glimpse into the complexity of interactions between epidemiology and annual cycles (see also Hurtado 2008).

Too few data exist to fully and effectively contrast the microbial and parasite communities between islands and continents. Likewise, not much is known about the physiological and immunological differentiation of island hosts (Lindström *et al.* 2004; Matson 2006; Beadell *et al.* 2007). Ultimately more studies that simultaneously evaluate the effects of island life on both hosts and parasites are very much needed, particularly interdisciplinary ones that bring together fields like molecular genetics, veterinary medicine, and microbial ecology. Nevertheless, we predict the numerous opportunities for initial and sustained differences between island and continent parasite communities will promote dissimilar coevolutionary processes and endpoints in some hosts. Furthermore, the particularities of host–parasite coevolution on islands might lead to and help maintain idiosyncratic changes in both insular–parasite virulence and transmission and insular–host immune system architecture.

16.4 Hawaiian model of disease susceptibility

As a case study, we turn now to the interaction that generated much of the initial interest in immunity and parasitism in insular taxa: the endemic Hawaiian honeycreepers (Drepanidini) and their introduced parasites. Hawaii has become a natural laboratory for understanding complex interactions among birds, vectors, and diseases as well as the influence of these interactions on epidemiology, ecology, and adaptation. In particular, the dramatic impact of avian malaria (*Plasmodium relictum*) and avian pox (*Avipoxvirus*) on Hawaii's honeycreepers has elevated certain features of this system to become components of a model for understanding the susceptibility of island fauna to novel intro-

duced diseases. Critical components of this model include:

(1) the extreme isolation of the Hawaiian avifauna and a presumed reduction in exposure to parasites;
(2) the role of introduced vectors and non-native hosts in increasing parasite exposure; and
(3) the old age of the honeycreeper radiation and the implied role of small population sizes and inbreeding in constraining response to novel disease.

Here, we briefly review the experimental results that form the foundation of Hawaii's contribution to the island susceptibility model and then we examine components of the model more closely to evaluate their assumptions and broader applicability.

The co-occurrence in Hawaii of endemic and recently introduced non-native species in the same epidemiological environment provided a fortuitous framework for demonstrating that disease severity and mortality in endemic honeycreepers is generally higher than that observed in introduced birds of continental origin. For example, iiwi (*Vestiaria coccinea*) and Laysan finches (*Telespiza cantans*) that were experimentally challenged with malaria infection exhibited mortality rates of 90–100 per cent, while three non-native species appeared to be refractory (e.g. van Riper *et al.* 1986; Atkinson *et al.* 1995; summary in Jarvi *et al.* 2001). Similarly, a small trial in which endemic and non-native birds were exposed to a Hawaiian strain of avian pox revealed 75 per cent mortality in the Hawaii amakihi (*Hemignathus virens*) compared to 0 per cent mortality in house sparrows (*Passer domesticus*) and Japanese white-eyes (*Zosterops japonicus*) (Jarvi *et al.* 2008). Additionally, a wealth of field observations revealed declines in populations of endemic species, but not introduced species, at low elevations where mosquitoes are common (Warner 1968; van Riper *et al.* 1986). Combined, these studies suggest that the particular strains of parasite found in Hawaii are not unusually virulent. Instead, certain Hawaiian island endemic species appear to be unusually susceptible.

The Hawaiian model suggests that the drastic difference in susceptibility may be attributed in part to an immunological naivety in honeycreepers arising from reduced exposure to pathogens.

Unfortunately, we do not have any record of parasite exposure in the Hawaiian avifauna prior to the mid-1800s; however, extreme isolation combined with the sequential rise and fall of these islands likely constrained the diversity of parasites present. The Hawaiian Islands are located approximately 4,000 km from North America and are amongst the most isolated islands in the world. The largest islands, which are home to most of the honeycreepers, range in age from 4.7 Ma (Kauai) to 0–0.5 Ma (Hawaii), but these islands are only the newest links in a volcanic chain that has been developing over the last 34 million years (Price *et al.* 2002). We might speculate that the islands have received a continuous input of possible parasite colonists from seabirds and migrating birds (almost exclusively ducks and shorebirds arriving from Arctic breeding grounds) as well as from rare avian vagrants, some of which gave rise to current endemics and others of which may have simply deposited their parasites before going extinct. In addition, some parasites may have hitchhiked to Hawaii along with the chickens that were established by early Polynesians. In both cases, though, colonization of endemic forest birds by some of these parasites would have been prevented by host range restrictions and habitat segregation (e.g. forest versus aquatic). In addition, transmission of many diseases would have been impossible since mosquitoes (and presumably many other vectors) were historically absent from the islands. Without solid evidence one way or another, we are left to assume that, while Hawaiian honeycreepers were not totally free from parasites, and certainly not free from broader antigenic stimulation, exposure to many parasites was likely reduced.

The role that introduced vectors and avian hosts played in increasing parasite exposure is much better documented. In 1826, the epidemiological environment changed dramatically when larva of *Culex quinquefasciatus* were released into a stream by sailors from a visiting ship, and this vector became established soon thereafter (Warner 1968). These mosquitoes allowed for transmission of both avian malaria and avian pox, though competence for the extant strain of avian malaria may have only developed following secondary colonization of Hawaii by a South-Pacific strain of *C. quinquefasciatus*

(Fonseca *et al.* 2000, 2006). Avian pox is thought to have been established earlier than malaria (Warner 1968; Jarvi *et al.* 2008; Fleischer, unpublished data), but both diseases likely arrived with the hundreds of exotic birds that were introduced to Hawaii from continental sources, many species of which became established in the wild (Long 1981; van Riper *et al.* 1986). The link to these introduced birds is not proven, but genetic evidence suggests that the poxvirus found in Hawaiian forest birds is not closely related to the fowlpox commonly found in chickens and may be composed of several strains arising from multiple introductions (Tripathy *et al.* 2000; Jarvi *et al.* 2008). On the other hand, the strain of avian malaria in Hawaii is composed of a single mitochondrial lineage that appears to have originated in the Old World, where it exhibits an extremely broad passerine host range (Beadell *et al.* 2006). In the case of malaria, mortality in continental birds appears to be rare, except in the case of species that are removed from their native range (e.g. penguins in zoos) and exposed to novel strains of parasites (Bennett *et al.* 1993; Valkiunas 2005). Thus, in Hawaii, the presence of a large pool of competent, but unaffected, continental hosts provided a stable reservoir for disease even as endemic host populations declined.

If susceptibility to disease depends on the extent to which a parasite presents a novel challenge to a host, then we might ask, does the Hawaiian model inform the evolutionary time scale on which a parasite becomes novel? Given that the loss of genetic and cultural determinants of pathogen recognition and response is likely influenced by a host's unique history of exposure to parasites and stochastic demographic events that may randomly alter genetic composition, there is probably no characteristic time frame (Matson 2006; Beadell *et al.* 2007). Genetic data indicate that the Hawaiian honeycreepers likely diverged from a mainland ancestor no more than 5–6 Ma ago (Fleischer and McIntosh 2001) and have been resident on the islands for at least 4 Ma (based on cytochrome b diversity; Fleischer *et al.* 1998). Although they share a long period of residency in the Hawaiian Islands, the honeycreepers are not uniformly susceptible to disease. When challenged with malaria, Laysan finches and iiwi exhibit relatively high susceptibility compared to apapane (*Himatione sanguinea*) and amakihi (van Riper *et al.* 1986; Atkinson *et al.* 2000; Yorinks and Atkinson 2000; summary in Jarvi *et al.* 2001). In fact, evidence suggests that some amakihi are resistant to superinfection (Atkinson *et al.* 2001a) and that some populations of amakihi have apparently persisted at low elevations where disease is most prevalent and vectors are most abundant (Woodworth *et al.* 2005; Eggert *et al.* 2008). The divergent responses of iiwi and amakihi do not correlate with genetic diversity at neutral (Eggert *et al.* 2008) or MHC loci (Jarvi *et al.* 2004), which is relatively high in both species. Instead, the divergent responses could reflect the chance loss of rare pathogen-recognition alleles in iiwi, or other stochastic changes in regulatory components of the immune system. Beyond honeycreepers, if we consider other endemic Hawaiian taxa representing independent evolutionary trajectories in the same epidemiological environment, time of isolation similarly lacks predictive power. On the one hand, the elepaio (*Chasiempis sandwichensis*, Monarchidae; estimated time of colonization between 2–5.5 Ma; Filardi and Moyle 2005) is extremely susceptible to pox (Vanderwerf *et al.* 2006) and the Hawaiian crow or alala (*Corvus hawaiiensis*, Corvidae), another relatively old colonist (Fleischer *et al.* 2001) appears to be impaired by both pox and malaria (Massey *et al.* 1996). On the other hand, the omao (*Myadestes obscurus*, Turdidae; estimated time of colonization between 4–7 Ma; Fleischer *et al.* 2001; Miller *et al.* 2007) exhibited only transient infections when challenged with malaria and these individuals were immune when re-challenged (Atkinson *et al.* 2001b).

Reflecting on the discussion above, two major points suggest that care is warranted when using the Hawaiian model to predict the outcome of introduced diseases in other isolated island faunas. First, exposure of island taxa to parasites over their evolutionary history remains a black box. Even if we could catalogue the various parasites to which island taxa have been exposed, we still would have only a faint idea of how those have contributed to maintaining pathogen surveillance machinery and to shaping the immune response to novel pathogens. And second, the evolutionary time scale on which a parasite becomes novel can vary since genetic determinants of disease susceptibility are likely to be subject to chance events. This is particularly true in island taxa that are subject to small founder sizes and periodic bottlenecks. We might expect both of these factors to increase the variance in disease susceptibility observed among other island taxa.

16.5   Re-evaluation: examples and exceptions

Our review of avian pox and avian malaria in Hawaii indicates that the Hawaiian honeycreepers are not uniformly susceptible to these novel disease challenges and that other endemic Hawaiian taxa vary in susceptibility. Acknowledging that the evolution of disease susceptibility may be strongly influenced by stochastic processes, we now look beyond Hawaii. Below, we assess the impacts of different diseases on other insular (and continental) birds to evaluate if the increased susceptibility of some native Hawaiian birds to particular diseases is an anomaly or if the system remains a broadly relevant and informative example.

Hawaii is extremely isolated, but among island archipelagos, Samoa, Easter Island, the Cook Islands, the Pitcairns, and French Polynesia are equally or even more isolated from mainland sources of disease (Blackburn *et al.* 2004). Many of these other archipelagos are host to small, threatened, and sometimes endangered populations of endemic birds with residence times on the order of millions of years (Filardi and Moyle 2005; Cibois *et al.* 2007). If we consider the effects of even just a single type of pathogen across these and other island faunas, we observe that the outcomes are quite different. For example, a Hawaiian-like strain of malaria has been detected in endemic Marquesan warblers, but these birds do not appear to have experienced drastic declines (Beadell *et al.* 2006). A diverse collection of avian malaria parasites has also been detected in the endemic birds of American Samoa. Here, the detection of low peripheral parasitemias hints at a long coevolutionary association with endemic hosts (Jarvi *et al.* 2003) and suggests that these hosts might enjoy some level of immunity and protection against potential introductions of similar parasites (Atkinson *et al.* 2006).

Avian malaria also appears to have little consequence for endemic Berthelot's pipits (*Anthus berthelotii*) residing on the Macaronesian islands, which lie just off the north-west coast of Africa (Illera *et al.* 2008). In contrast, avian malaria may have caused up to 80 per cent mortality in the New Zealand endemic mohua (or yellowhead, *Mohoua ochrocephala*) that were transported to a wildlife park where malaria had only recently become established (Tompkins and Gleeson 2006). Most of these examples only speculate about the possible consequence of disease exposure and highlight the difficulty of quantifying disease effects in wild populations. Few systems have been studied as closely as Hawaii, but below, we examine two relatively well-studied island disease systems to determine if components of the Hawaiian model apply.

16.5.1   Mauritius

Mauritius presents a disease system that shares many components of Hawaii's system, and at least one endemic species appears to exhibit unusual susceptibility to introduced disease. Mauritius lies approximately 900 km from Madagascar and approximately 1,900 km from the coast of Africa. Like Hawaii, it is host to many migratory shorebirds and seabirds, as well as several endemic and introduced landbird species (Avibase 2009 <http://avibase.bsc-eoc.org/checklist.jsp>). Among these, the pink pigeon (*Columba mayeri*), an endangered endemic species that has been resident on the island for approximately 1.5 Ma (Johnson *et al.* 2001), has garnered significant attention. The species exhibits low genetic diversity attributable to recent recovery from a bottleneck of fewer than 20 individuals (Swinnerton *et al.* 2004). Avian pox and *Leucocytozoon* sp. (Peirce *et al.* 1997) have been implicated in earlier declines, but the best evidence for direct parasite-driven mortality comes from studies of *Trichomonas gallinae* (Swinnerton *et al.* 2005; Bunbury *et al.* 2008). This pathogen may have been introduced to Mauritius via rock doves as early as the eighteenth century (Swinnerton *et al.* 2005), and the parasite was first detected in pink pigeons in 1992 (Bunbury *et al.* 2008). Since then, studies have documented significant mortality in nestlings (Swinnerton *et al.* 2005) and an increased probability

of mortality in individuals with persistent disease (Bunbury *et al.* 2008). In addition, the pathogenicity observed in pink pigeons contrasts with that observed in non-native doves, which exhibit similar infection prevalence but little or no clinical sign of disease (Swinnerton *et al.* 2005; Bunbury *et al.* 2007).

At first glance, the pink pigeon would appear to conform to the Hawaiian model of an old endemic island species that is unusually susceptible to an introduced pathogen. Two considerations, however, warrant closer attention. First, it is unclear whether different *Trichomonas* strains may be responsible for the varying responses observed in different hosts. Second, *Trichomonas* has been documented to cause severe mortality even in continental taxa, including mourning doves in the USA (Greiner and Baxter 1974) and woodpigeons in Spain (Hofle *et al.* 2004). Therefore, although we might be inclined to apply the Hawaiian model in explaining the apparent susceptibility of the pink pigeons, we cannot rule out taxon-specific interactions between doves and *Trichomonas* strains that may drive disease epidemiology independently of island factors such as degree of inbreeding and historic isolation from disease.

16.5.2   Galápagos

The Galápagos avifauna remains largely intact (Harris 1973; Gibbs *et al.* 1999). However, because of recent declines and the threat of extinction facing at least some endemic species (Grant and Grant 1999; Grant *et al.* 2005), the Galápagos avifauna may be at risk of replaying the events in Hawaii (Wikelski *et al.* 2004). The age of at least several endemic species approaches or exceeds that of the Hawaiian honeycreepers: Darwin's finches (Geospizinae) 1.2–2.8 Ma (Grant 1994, Sato *et al.* 2001); Galápagos dove (*Zenaida galapagoensis*) > 2.0 Ma (Johnson and Clayton 2000); and Galápagos mockingbird (*Nesomimus parvulus*) 0.6–5.5 Ma (Arbogast *et al.* 2006). The Galápagos archipelago is only about 1,000 km from mainland South America, making it much less geographically isolated than the Hawaiian archipelago. As a consequence, the islands are home to a wider diversity of migrant species (Avibase 2009   <http://avibase.bsc-eoc.org/checklist.jsp>), which may have served over evolutionary time to increase exposure of endemic birds to parasites. Historically, the Galápagos supported just a single mosquito species (Hardy 1960), but *Culex quinquefasciatus* was introduced within the last century (Peck *et al.* 1998; Whiteman *et al.* 2005). Although avian malaria does not yet appear to have colonized the Galápagos (Padilla *et al.* 2004), avian pox probably became established prior to the early 1900s (Wikelski *et al.* 2004) and has now been detected in a wide range of hosts. As in Hawaii, the virus probably did not originate in domesticated fowl (Thiel *et al.* 2005). Avian pox has been implicated in the mortality of Galápagos mockingbirds (Vargas 1987; Curry and Grant 1989), but the disease is not uniformly detrimental in Darwin's finches (Kleindorfer *et al.* 2006). Since no resident or non-native species have been challenged with the virus, it is unclear if the endemic island species that are affected are unusually susceptible or if the fitness effects recorded in certain Galápagos species are attributable to strain-specific virulence.

The consequence of other putative pathogens is similarly difficult to deduce. *Trichomonas* was likely introduced to the Galápagos via introduced rock doves in the 1970s (Harmon *et al.* 1987). *Trichomonas* has since been detected in endemic Galápagos doves, however any negative consequences remain speculative (Harmon *et al.* 1987; Padilla *et al.* 2004). Blood-borne parasites in the genus *Haemoproteus* have also been detected in Galápagos doves (Padilla *et al.* 2004; Santiago-Alarcon *et al.* 2008). This parasite is generally considered to be relatively benign in wild birds (Bennett *et al.* 1993), but the high prevalences and high parasitemias found in endemic doves (relative to introduced rock doves; Padilla *et al.* 2004) might indicate that these endemic birds are unusually susceptible. Finally, although domestic fowl host a wide variety of pathogens (Gottdenker *et al.* 2005), a survey of endemic passerines living near poultry farms did not find evidence of previous exposure to these pathogens (Soos *et al.* 2008). The studies reviewed above highlight the difficulty in assessing the effects of disease in wild populations. As in Hawaii, though, the few studies that do provide solid evidence point to a mixed outcome for endemic species exposed to introduced disease.

16.6 Concluding thoughts

The negative effects of some diseases on some island host species are undisputed. But continental taxa can suffer as well from the effects of introduced or newly emergent diseases. Thus, one central issue that remains unaddressed is whether island birds are in fact more susceptible to disease than continental birds when confronted with equally novel pathogens. To qualify as 'novel' in this regard, an interaction should be sufficiently distinct from or outside of the contemporary, evolved host–parasite equilibrium. The nature of such an equilibrium, of course, depends on many factors including host phylogeny, as evidenced by West Nile Virus. Following the 1999 arrival of this virus to North America, many continental bird species played host to the virus, but certain species, including members of the corvid and owl families exhibited particularly high susceptibility (Bowen and Nemeth 2007; Gancz *et al.* 2004). Daszak *et al.* (2000) provide other examples in their summary of 30 emergent infectious diseases affecting wildlife. All but one (avian malaria in Hawaii) primarily affect continental areas, and the three diseases found in continental birds result in high mortality. One way to reconcile these and other findings is to hypothesize that novel host-parasite interactions transpire on both continents and islands but simply do so with relatively greater impacts on small island populations. An alternative hypothesis is that more parasites appear novel to island birds compared to mainland birds because of the suite of genetic, ecological, and geographic differences associated with island life, but many of these unique features of islands effectively impede the establishment of most parasite faunas. Regrettably, with so many quantifiable and stochastic variables, the outcome of any particular host-parasite interaction in terms of host fitness, immune response, and pathology, still remains unpredictable.

References

Abbas, A.K. and Lichtman, A.H. (2001). *Basic Immunology: Functions and Disorders of the Immune System*. W.B. Saunders Company, Philadelphia, PA.

Adamo, S.A. (2004). How should behavioural ecologists interpret measurements of immunity? *Animal Behaviour*, **68**, 1443–49.

Adler, G.H. and Levins, R. (1994). The island syndrome in rodent populations. *Quarterly Review of Biology*, **69**, 473–90.

Arbogast, B.S., Drovetski, S.V., Curry, R.L. *et al.* (2006). The origin and diversification of Galápagos mockingbirds. *Evolution*, **60**, 370–82.

Atkinson, C.T., Woods, K.L., Dusek, R.J., Sileo, L.S., and Iko, W.M. (1995). Wildlife disease and conservation in Hawaii: pathogenicity of avian malaria (*Plasmodium relictum*) in experimentally infected Iiwi (*Vestiaria coccinea*). *Parasitology*, **111**, S59–S69.

Atkinson, C.T., Dusek, R.J., Woods, K.L., and Iko, W.M. (2000). Pathogenecity of avian malaria in experimentally-infected Hawaii amakihi. *Journal of Wildlife Diseases*, **36**, 197–204.

Atkinson, C.T., Dusek, R.J., and Lease, J.K. (2001a). Serological responses and immunity to superinfection with avian malaria in experimentally-infected Hawaii amakihi. *Journal of Wildlife Diseases*, **37**, 20–27.

Atkinson, C.T., Lease, J.K., Drake, B.M., and Shema, N.P. (2001b). Pathogenicity, serological responses, and diagnosis of experimental and natural malarial infections in native Hawaiian thrushes. *Condor*, **103**, 209–18.

Atkinson, C.T., Utzurrum, R.C., Seamon, J.O., Savage, A.F., and LaPointe, D.A. (2006). Hematozoa of forest birds in American Samoa – evidence for a diverse, indigenous parasite fauna from the South Pacific. *Pacific Conservation Biology*, **12**, 229–37.

Averbeck, C. (1992). Haematology and blood chemistry of healthy and clinically abnormal great black-backed gulls (*Larus marinus*) and herring gulls (*Larus argentatus*). *Avian Pathology*, **21**, 215–23.

Beadell, J.S., Ishtiaq, F., Covas, R. *et al.* (2006). Global phylogeographic limits of Hawaii's avian malaria. *Proceedings of the Royal Society of London B*, **273**, 2935–44.

Beadell, J.S., Atkins, C., Cashion, E., Jonker, M., and Fleischer, R.C. (2007). Immunological change in a parasite-impoverished environment: divergent signals from four island taxa. *PLoS One*, **2**, e896.

Bell, T., Ager, D., Song, J. *et al.* (2005). Larger islands house more bacterial taxa. *Science*, **308**, 1884.

Bennett, G.F., Peirce, M.A., and Ashford, R.W. (1993). Avian haematozoa: mortality and pathogenicity. *Journal of Natural History*, **27**, 993–1001.

Berry, R.J. (1986). Genetics of insular populations of mammals, with particular reference to differentiation and founder effects in British small mammals. *Biological Journal of the Linnean Society*, **28**, 205–30.

Blackburn, T.M., Cassey, P., Duncan, R.P., Evans, K.L., and Gaston, K.J. (2004). Avian extinction and mammalian introductions on oceanic islands. *Science*, **305**, 1955–58.

Blumstein, D.T. and Daniel, J.C. (2005). The loss of anti-predator behaviour following isolation on islands. *Proceedings of the Royal Society of London B*, **272**, 1663–68.

Bonneaud, C., Mazuc, J., Gonzalez, G. *et al.* (2003). Assessing the cost of mounting an immune response. *American Naturalist*, **161**, 367–79.

Bonsall, M.B. and Benmayor, R. (2005). Multiple infections alter density dependence in host-pathogen interactions. *Journal of Animal Ecology*, **74**, 937–45.

Boots, M., Hudson, P.J., and Sasaki, A. (2004). Large shifts in pathogen virulence relate to host population structure. *Science*, **303**, 842–44.

Bowen, R.A. and Nemeth, N.M. (2007). Experimental infections with West Nile virus. *Current Opinion in Infectious Disease*, **20**, 293–97.

Bounous, D.I. and Stedman, N.L. (2000). Normal avian hematology: chicken and turkey. In B.F. Feldman, J.G. Zinkl, N.C. Jain, eds. *Schalm's Veterinary Hematology*, pp. 1147–54. Lippincott Williams & Wilkins, Philadelphia.

Brisbin, J.T., Gong, J., and Sharif, S. (2008). Interactions between commensal bacteria and the gut-associated immune system of the chicken. *Animal Health Research Reviews/Conference of Research Workers in Animal Diseases*, **9**, 101–10.

Buehler, D.M., Bhola, N., Barjaktarov, D. *et al.* (2008a). Constitutive immune function responds more slowly to handling stress than corticosterone in a shorebird. *Physiological and Biochemical Zoology*, **81**, 673–81.

Buehler, D.M., Piersma, T., Matson, K., and Tieleman, B.I. (2008b). Seasonal redistribution of immune function in a migrant shorebird: annual-cycle effects override adjustments to thermal regime. *American Naturalist*, **172**, 783–96.

Bunbury, N., Jones, C.G., Greenwood, A.G., and Bell, D.J. (2007) *Trichomonas gallinae* in Mauritian columbids: implications for an endangered endemic. *Journal of Wildlife Diseases*, **43**, 399–407.

Bunbury, N., Jones, C.G., Greenwood, A.G., and Bell, D.J. (2008). Epidemiology and conservation implications of *Trichomonas gallinae* infection in the endangered Mauritian pink pigeons. *Biological Conservation*, **141**, 153–61.

Cibois, A., Thibault, J.-C., and Pasquet, E. (2007). Uniform phenotype conceals double colonization by reed-warblers of a remote Pacific archipelago. *Journal of Biogeography*, **34**, 1150–66.

Curry, R.L. and Grant, P.R. (1989). Demography of the cooperatively breeding Galápagos mockingbird, *Nesomimus parvulus*, in a climatically variable environment. *Journal of Animal Ecology*, **58**, 441–63.

Daszak, P., Cunningham, A.A., and Hyatt, A.D. (2000). Emerging infectious diseases of wildlife - threats to biodiversity and human health. *Science*, **287**, 443–49.

Davies, D.H., Halablab, M.A., Clarke, J., Cox, F.E.G., and Young, T.W.K. (1998). *Infection and Immunity.* Taylor and Francis, London.

Eggert, L.S., Terwilliger, L.A., Woodworth, B.L., Hart, P.J., Palmer, D., and Fleischer, R.C. (2008). Genetic structure along an elevational gradient in Hawaiian honeycreepers reveals contrasting evolutionary responses to avian malaria. *BMC Evolutionary Biology*, **8**, 315.

Filardi, C.E. and Moyle, R.G. (2005). Single origin of a pan-Pacific bird group and upstream colonization of Australasia. *Nature*, **438**, 216–19.

Fleischer, R.C. and McIntosh, C.E. (2001). Molecular systematics and biogeography of the Hawaiian avifauna. *Studies in Avian Biology*, **22**, 51–60.

Fleischer, R.C., McIntosh, C.E., and Tarr, C.L. (1998). Evolution on a volcanic conveyor belt: using phylogenetic reconstructions and K-Ar-based ages of the Hawaiian Islands to estimate molecular evolutionary rates. *Molecular Ecology*, **7**, 533–45.

Fonseca, D.M., LaPointe, D.A., and Fleischer, R.C. (2000). Bottlenecks and multiple introductions: population genetics of the vector of avian malaria in Hawaii. *Molecular Ecology*, **9**, 1803–14.

Fonseca, D.M., Smith, J., Wilkerson, R., and Fleischer, R.C. (2006). Pathways of expansion and multiple introductions illustrated by large genetic differentiation among worldwide populations of the southern house mosquito. *American Journal of Tropical Medicine and Hygiene*, **74**, 284–89.

Frankham, R. (1997). Do island populations have less genetic variation than mainland populations? *Heredity*, **78**, 311–27.

Frankham, R. (1998). Inbreeding and extinction: island populations. *Conservation Biology*, **12**, 665–75.

French, S.S., Matt, K.S., and Moore, M.C. (2006). The effects of stress on wound healing in male tree lizards (*Urosaurus ornatus*). *General and Comparative Endocrinology*, **145**, 128–32.

Gancz, A.Y., Barker, I.K., Lindsay, R., Dibernardo, A., McKeever, K., and Hunter, B. (2004). West Nile virus outbreak in North American owls, Ontario, 2002. *Emerging Infectious Diseases*, **10**, 2135–42.

Gao, L.Q. and Hethcote, H.W. (1992). Disease transmission models with density-dependent demographics. *Journal of Mathematical Biology*, **30**, 717–31.

Glick, B. (2000). Immunophysiology. In G.C. Whittow, ed. *Sturkie's Avian Physiology*, 5th Edn, pp. 657–85. Academic Press, London.

Gibbs, J.P., Snell, H.L., and Causton, C.E. (1999). Effective monitoring for adaptive wildlife management: lessons from the Galápagos Islands. *Journal of Wildlife Management*, **63**, 1055–65.

Gottdenker, N.L., Walsh, T., Vargas, H. *et al.* (2005). Assessing the risks of introduced chickens and their pathogens to native birds in the Galápagos Archipelago. *Biological Conservation*, **126**, 429–39.

Grant, P.R. (1994). Population variation and hybridization: comparison of finches from two archipelagos. *Evolutionary Ecology*, **8**, 598–617.

Grant, P.R. and Grant, B.R. (1999). The rarest of Darwin's finches. *Conservation Biology*, **11**, 119–26.

Grant, P.R., Grant, B.R., Petren, K., and Keller, L.F. (2005). Extinction behind our backs: the possible fate of one of the Darwin's finch species on Isla Floreana, Galápagos. *Biological Conservation*, **122**, 499–503.

Greiner, E.C. and Baxter, W.L. (1974). A localized epizootic of *Trichomoniasis* in mourning doves. *Journal of Wildlife Diseases*, **10**, 104–06.

Gross, W.B. and Siegel, H. (1983). Evaluation of the heterophil/lymphocyte ratio as a measure of stress in chickens. *Avian Diseases*, **34**, 759–61.

Hanssen, S.A. (2006). Costs of an immune challenge and terminal investment in a long-lived bird. *Ecology*, **87**, 2440–6.

Hanssen, S.A., Hasselquist, D., Folstad, I., and Erikstad, E.K. (2005). Cost of reproduction in a long-lived bird: incubation effort reduces immune function and future reproduction. *Proceedings of the Royal Society of London B*, **272**, 1039–46.

Haraguchi, Y. and Sasaki, A. (2000). The evolution of parasite virulence and transmission rate in a spatially structured population. *Journal of Theoretical Biology*, **203**, 85–96.

Hardy, D.E. (1960). *Insects of Hawaii. Vol. 10.* University of Hawaii Press, Honolulu.

Harmon, W.M., Clark, W.A., Hawbecker, A.C., and Stafford, M. (1987). *Trichomonas gallinae* in columbiform birds from the Galápagos Islands. *Journal of Wildlife Diseases*, **23**, 492–94.

Harris, M.P. (1973). The Galápagos avifauna. *Condor*, **75**, 265–78.

Hawkey, C.M., Hart, M.G., Knight, J.A., Samour, J.H., and Jones, D.M. (1982). Haematological findings in healthy and sick African grey parrots (*Psittacus erithacus*). *Veterinary Record*, **111**, 580–82.

Hawley, D.M., Hanley, D., Dhondt, A.A., and Lovette, I.J. (2006). Molecular evidence for a founder effect in invasive house finch (*Carpodacus mexicanus*) populations experiencing an emergent disease epidemic. *Molecular Ecology*, **15**, 263–75.

Hochberg, M.E. and Møller, A.P. (2001). Insularity and adaptation in coupled victim-enemy interactions. *Journal of Evolutionary Biology*, **14**, 539–51.

Hofle, U., Gortazar, C., Ortiz, J.A., Knispel, B., and Kaleta, E.F. (2004). Outbreak of trichomoniasis in a woodpigeon

(*Columba palumbus*) wintering roost. *European Journal of Wildlife Research*, **50**, 73–77.

Horner-Devine, M.C., Lage, M., Hughes, J.B., and Bohannan, B.J.M. (2004). A taxa-area relationship for bacteria. *Nature*, **432**, 750–53.

Hurtado, P. (2008). The potential impact of disease on the migratory structure of a partially migratory passerine population. *Bulletin of Mathematical Biology*, **70**, 2264–82.

Illera, J.C., Emerson, B.C., and Richardson, D.S. (2008). Genetic characterization, distribution and prevalence of avian pox and avian malaria in the Berthelot's pipit (*Anthus berthelotii*) in Macaronesia. *Parasitology Research*, **103**, 1435–43.

Jarvi, S.I., Atkinson, C.T., and Fleischer, R.C. (2001). Immunogenetics and resistance to avian malaria in Hawaiian honeycreepers (Drepanidinae). *Studies in Avian Biology*, **22**, 254–63.

Jarvi, S.I., Farias, M.E.M., Baker, H. *et al.* (2003). Detection of avian malaria (*Plasmodium* spp.) in native land birds of American Samoa. *Conservation Genetics*, **4**, 629–37.

Jarvi, S.I., Tarr, C.L., McIntosh, C.E., Atkinson, C.T., and Fleischer, R.C. (2004). Natural selection of the major histocompatibility complex (Mhc) in Hawaiian honeycreepers (Drepanidinae). *Molecular Ecology*, **13**, 2157–68.

Jarvi, S., Triglia, D., Giannouli, A., Farias, M., Bianchi, K., and Atkinson, C.T. (2008). Diversity, origins and virulence of *Avipoxviruses* in Hawaiian forest birds. *Conservation Genetics*, **9**, 339–48.

Johnson, K.P. and Clayton, D.H. (2000). A molecular phylogeny of the dove genus *Zenaida*: mitochondrial and nuclear DNA sequence. *Condor*, **102**, 864–70.

Johnson, K.P., de Kort, S., Dinwoodey, K. *et al.* (2001). A molecular phylogeny of the dove genera *Streptopelia* and *Columba*. *Auk*, **118**, 874–87.

Kelly, D., Conway, S., and Aminov, R. (2005). Commensal gut bacteria: mechanisms of immune modulation. *Trends in Immunology*, **26**, 326–33.

Klasing, K., Laurin, D., Peng, R., and Fry, D. (1987). Immunologically mediated growth depression in chicks: influence of feed-intake, corticosterone, and interleukin-1. *Journal of Nutrition*, **117**, 1629–37.

Kleindorfer, S. and Dudaniec, R.Y. (2006). Increasing prevalence of avian poxvirus in Darwin's finches and its effect on male pairing success. *Journal of Avian Biology*, **37**, 69–76.

Lee, K.A., Wikelski, M., Robinson, W.D., Robinson, T.R., and Klasing, K.C. (2008). Constitutive immune defences correlate with life-history variables in tropical birds. *Journal of Animal Ecology*, **77**, 356–63.

Leshchinsky, T.V. and Klasing, K.C. (2001). Divergence of the inflammatory response in two types of chickens. *Developmental and Comparative Immunology*, **25**, 629–38.

Lindström, K.M., Foufopoulos, J., Pärn, H., and Wikelski, M. (2004). Immunological investments reflect parasite abundance in island populations of Darwin's finches. *Proceedings of the Royal Society of London B*, **271**, 1513–19.

LoGiudice, K., Ostfeld, R.S., Schmidt, K.A. and Keesing, F. (2003). The ecology of infectious disease: effects of host diversity and community composition on Lyme disease risk. *Proceedings of the National Academy of Sciences of the USA*, **100**, 567–71.

Long, J.L. (1981). *Introduced Birds of the World*. Universe Books, New York.

MacArthur, R.H. and Wilson, E.O. (2001). *The Theory of Island Biogeography*, 2nd Edn. Princeton University Press, Princeton.

MacArthur, R.H., Diamond, J.M., and Karr, J.R. (1972). Density compensation in island faunas. *Ecology*, **53**, 330–42.

Magalhaes, J.G., Tattoli, I., and Girardin, S.E. (2007). The intestinal epithelial barrier: how to distinguish between the microbial flora and pathogens. *Seminars in Immunology*, **19**, 106–15.

Martin, L.B., Scheuerlein, A., and Wikelski, M. (2002). Immune activity elevates energy expenditure of house sparrows: a link between direct and indirect costs? *Proceedings of the Royal Society of London B*, **270**, 153–58.

Martin, L.B., Han, P., Lewittes, J., Kuhlman, J.R., Klasing, K.C., and Wikelski, M. (2006). Phytohemagglutinin-induced skin swelling in birds: histological support for a classic immunoecological technique. *Functional Ecology*, **20**, 290–99.

Martin, L.B., Weil, Z.M., and Nelson, R.J. (2007). Immune defense and reproductive pace of life in *Peromyscus* mice. *Ecology*, **88**, 2516–28.

Massey, J.G., Graczyk, T.K., and Cranfield, M.R. (1996). Characteristics of naturally acquired *Plasmodium relictum capistranoae* infections in naïve Hawaiian crows (*Corvus hawaiiensis*) in Hawaii. *Journal of Parasitology*, **82**, 182–85.

Matson, K.D. (2006). Are there differences in immune function between continental and insular birds? *Proceedings of the Royal Society of London B*, **273**, 2267–74.

Matson, K.D., Ricklefs, R.E., and Klasing, K.C. (2005). A hemolysis-hemagglutination assay for characterizing constitutive innate humoral immunity in wild and domestic birds. *Developmental and Comparative Immunology*, **29**, 275–86.

Matson, K.D., Cohen, A.A., Klasing, K.C., Ricklefs, R.E., and Scheuerlein, A. (2006a). No simple answers for ecological immunology: relationships among immune indices at the individual level break down at the species level in waterfowl. *Proceedings of the Royal Society of London B*, **273**, 815–22.

Matson, K.D., Tieleman, B.I. and Klasing, K.C. (2006b). The bactericidal competence of blood and plasma in five species of tropical birds. *Physiological and Biochemical Zoology*, **79**, 556–64.

McCallum, H., Barlow, N., and Hone, J. (2001). How should pathogen transmission be modelled? *Trends in Ecology and Evolution*, **16**, 295–300.

Mendes, L., Piersma, T., Lecoq, M., Spaans, B., and Ricklefs, R.E. (2005). Disease-limited distributions? Contrasts in the prevalence of avian malaria in shorebird species using marine and freshwater habitats. *Oikos*, **109**, 396–404.

Mendes, L., Piersma, T., Hasselquist, D., Matson, K.D., and Ricklefs, R.E. (2006). Variation in the innate and acquired arms of the immune system among five shorebird species. *Journal of Experimental Biology*, **209**, 284–91.

Miller, M.J., Bermingham, E., and Ricklefs, R.E. (2007). Historical biogeography of the new world solitaires (*Myadestes* spp.). *Auk*, **124**, 868–85.

Millet, S., Bennett, J., Lee, K.A., Hau, M., and Klasing, K.C. (2007). Quantifying and comparing constitutive immunity across avian species. *Developmental and Comparative Immunology*, **31**, 188–201.

Owen-Ashley, N.T., Turner, M., Hahn, T.P., and Wingfield, J.C. (2006). Hormonal, behavioral, and thermoregulatory responses to bacterial lipopolysaccharide in captive and free-living white-crowned sparrows (*Zonotrichia leucophrys gambelii*). *Hormones and Behavior*, **49**, 15–29.

Padilla, L.R., Santiago-Alarcon, D., Merkel, J., Miller, R.E., and Parker, P.G. (2004). Survey for *Haemoproteus* spp., *Trichomonas gallinae*, *Chlamydophila psittaci*, and *Salmonella* spp. in Galápagos Islands Columbiformes. *Journal of Zoo and Wildlife Medicine*, **35**, 60–64.

Padgett, D.A., Marucha, P.T., and Sheridan, J.F. (1998). Restraint stress slows cutaneous wound healing in mice. *Brain, Behavior, and Immunity*, **12**, 64–73.

Peck, S.B., Heraty, J., Landry, B., and Sinclair, B.J. (1998). Introduced insect fauna of an oceanic archipelago: the Galápagos Islands, Ecuador. *American Entomologist*, **44**, 218–37.

Peirce, M.A., Greenwood, A.G., and Swinnerton, K. (1997). Pathogenicity of *Leucocytozoon marchouxi* in the pink pigeon (*Columba mayeri*) in Mauritius. *Veterinary Record*, **140**, 155–56.

Perez-Tris, J. and Bensch, S. (2005). Dispersal increases local transmission of avian malarial parasites. *Ecology Letters*, **8**, 838–45.

Piersma, T. (1997). Do global patterns of habitat use and migration strategics coevolve with relative investments in immunocompetence due to spatial variation in parasite pressure? *Oikos*, **80**, 623–31.

Price, J.P. and Clague, D.A. (2002). How old is the Hawaiian biota? Geology and phylogeny suggest recent divergence. *Proceedings of the Royal Society of London B*, **269**, 2429–35.

Råberg, L., Grahn, M., Hasselquist, D., and Svensson, E. (1998). On the adaptive signicance of stress-induced immunosuppression. *Proceedings of the Royal Society of London B*, **265**, 1637–41.

Reid, J.M., Arcese, P., and Keller, L.F. (2003). Inbreeding depresses immune response in song sparrows (*Melospiza melodia*): direct and inter-generational effects. *Proceedings of the Royal Society of London B*, **270**, 2151–57.

Reid, J.M., Arcese, P., Keller, L.F., Elliott, K.H., Sampson, L., and Hasselquist, D. (2007). Inbreeding effects on immune response in free-living song sparrows (*Melospiza melodia*). *Proceedings of the Royal Society of London B*, **274**, 697–706.

Santiago-Alarcon, D., Whiteman, N.K., Parker, P.G., Ricklefs, R.E., andValkiunas, G. (2008). Patterns of parasite abundance and distribution in island populations of Galápagos endemic birds. *Journal of Parasitology*, **94**, 584–90.

Sato, A, Tichy, H., O'hUigin, C., Grant, P.R., Grant, B.R., and Klein, J. (2001). On the origin of Darwin's finches. *Molecular Biology and Evolution*, **18**, 299–311.

Schmid-Hempel, P. and Ebert, D. (2003). On the evolutionary ecology of specific immune defence. *Trends in Ecology and Evolution*, **18**, 27–32.

Smits, J.E. and Baos, R. (2005). Evaluation of the antibody mediated immune response in nestling American kestrels (*Falco sparverius*). *Developmental and Comparative Immunology*, **29**, 161–70.

Soos, C., Padilla, L., Iglesias, A. *et al.* (2008). Comparison of pathogens in broiler and backyard chickens on the Galápagos Islands: implications for transmission to wildlife. *Auk*, **125**, 445–55.

Steadman, D.W., Greiner, E.C., and Wood, C.S. (1990). Absence of blood parasites in indigenous and introduced birds from the Cook Islands, South Pacific. *Conservation Biology*, **4**, 398–404.

Swinnerton, K.J., Groombridge, J.J., Jones, C.G., Burn, R.W., and Mungroo, Y. (2004). Inbreeding depression and founder diversity among captive and free-living populations of the endangered pink pigeon *Columba mayeri*. *Animal Conservation*, **7**, 353–64.

Swinnerton, K.J., Greenwood, A.G., Chapman, R.E., and Jones, C.G. (2005). The incidence of the parasitic disease trichomoniasis and its treatment in reintroduced and wild Pink Pigeons *Columba mayeri*. *Ibis*, **147**, 772–82.

Super, R.E. and van Riper III, C. (1995). Comparison of avian hematozoan epizootiology in two California coastal scrub communities. *Journal of Wildlife Diseases*, **31**, 447–61.

Tella, J.L, Scheuerlein, A., and Ricklefs, R.E. (2002). Is cell-mediated immunity related to the evolution of life-history strategies in birds? *Proceedings of the Royal Society of London B*, **269**, 1059–66.

Tella, J.L, Lemus, J.A., Carrete, M., and Blanco, G. (2008). The PHA test reflects acquired T-cell mediated immunocompetence in birds. *PLoS One*, **3**, e3295.

Tompkins, D.M. and Gleeson, D.M. (2006). Relationship between avian malaria distribution and an exotic invasive mosquito in New Zealand. *Journal of the Royal Society of New Zealand*, **36**, 51–62.

Thiel, T., Whiteman, N.K., Tirape, A. *et al.* (2005). Characterization of canarypox-like viruses infecting endemic birds in the Galápagos Islands. *Journal of Wildlife Diseases*, **41**, 342–53.

Tieleman, B.I. (2005). Physiological, behavioral and life history adaptations of larks along an aridity gradient: a review. In G. Bota, J. Camprodon, S. Manosa, and M. Morales, eds. *Ecology and Conservation of Steppe-land Birds*, pp. 49–67. Lynx Edicions, Barcelona.

Tieleman, B.I., Williams, J.B., Ricklefs, R.E., and Klasing, K.C. (2005). Constitutive innate immunity is a component of the pace-of-life syndrome in tropical birds. *Proceedings of the Royal Society of London B*, **272**, 1715–20.

Tripathy, D.N., Schnitzlein, W.M., Morris, P.J. *et al.* (2000). Characterization of poxviruses from forest birds in Hawaii. *Journal of Wildlife Diseases*, **36**, 225–30.

Valkiunas, G. (2005). *Avian Malaria Parasites and Other Haemosporidia*. CRC Press, Boca Raton.

VanderWerf, E.A., Burt, M.D., Rohrer, J.L., and Mosher, S.M. (2006). Distribution and prevalence of mosquito-borne diseases in Oahu 'Elepaio. *Condor*, **108**, 770–77.

van Riper III, C., van Riper, S.G., Goff, M.L., and Laird, M. (1986). The epizootiology and ecological significance of malaria in Hawaiian land birds. *Ecological Monographs*, **56**, 327–44.

Vargas, H. (1987). Frequency and effect of pox-like lesions in Galápagos mockinbirds. *Journal of Field Ornithology*, **58**, 101–02.

Vleck, C.M., Vertalino, N., Vleck, D., and Bucher, T.L. (2000). Stress, corticosterone, and heterophil to lymphocyte ratios in free-living Adelie penguins. *Condor*, **102**, 392–400.

Warner, R.E. (1968). The role of introduced diseases in the extinction of the endemic Hawaiian avifauna. *Condor*, **70**, 101–20.

Whiteman, N.K., Goodman, S.J., Sinclair, B.J. *et al.* (2005). Establishment of the avian disease vector *Culex quinquefasciatus* Say, 1823 (Diptera: Culicidae) on the Galápagos Islands, Ecuador. *Ibis*, **147**, 844–47.

Whiteman, N.K., Matson, K.D., Bollmer, J.L., and Parker, P.G. (2006). Disease ecology in the Galápagos hawk (*Buteo galapagoensis*): host genetic diversity, parasite load and natural antibodies. *Proceedings of the Royal Society of London B*, **273**, 797–804.

Wikelski, M., Foufopoulos, J., Vargas, H., and Snell, H. (2004). Galápagos birds and diseases: invasive pathogens as threats for island species. *Ecology and Society*, **9**, [online].

Woodworth, B., Atkinson, C.T., LaPointe, D.A. *et al.* (2005). Host population persistence in the face of introduced vector-borne diseases: Hawaii amakihi and avian malaria. *Proceedings of the National Academy of Sciences of the USA*, **102**, 1531–36.

Yorinks, N. and Atkinson, C.T. (2000). Effects of malaria (*Plasmodium relictum*) on activity budgets of experimentally infected juvenile apapane (*Himatione sanguinea*). *Auk*, **117**, 731–38.

PART V

Applied Biogeography

CHAPTER 17

The geography and ecology of pathogen emergence

Jan Slingenbergh, Lenny Hogerwerf, and Stéphane de la Rocque

17.1 Introduction

There is strong evidence to suggest that both the ability to detect, identify, and monitor disease agents as well as the real number of emerging infectious disease events at the animal—human interface and in the food production chain are on the increase (Woolhouse 2008). Among the driving forces, human population density and growth have been singled out as key to the global dynamics of vector-borne diseases, the growing problem of drug resistance, public threats resulting from pathogens circulating in wildlife, as well as zoonotic infections involving domestic animals (Jones et al. 2008). Apart from human population pressures and the enhanced mobility of people, climate change, food and agricultural dynamics, and the progressive encroachment of forest and game reserves are also among the often cited global factors amplifying emerging infectious disease events (Daszak et al. 2001). Still, the precise mechanism by which these socio-ecological dynamics or disruptions translate into disease emergence, be it in the human host, livestock and/or fauna, are not well understood. There is in fact no single analytical framework adequate to disentangling all the different disease emergence scenarios (Pulliam 2008). Disease emergence, defined as an increase in the incidence of a disease, implies an increase in transmission rates arising from host contact dynamics, or more profound changes implicating the parasite genetic evolution, in terms of transmission-virulence trade-offs and/or host specificity (Kapan et al. 2006).

A fundamental principle in plant pests and animal or human diseases is that increases in host abundance are associated with enhanced transmission (Knops *et al.* 1999). Epidemic spread of disease may cause an increase in spillover, causing dead-end infections in host species that normally do not become infected (Power and Mitchell 2004). Environmental dynamics such as those involving a growth in livestock numbers or an increase in poaching of wildlife may cause infection of non-habitual hosts. Secondary infections may occasionally become established in contact-hosts and when this process repeats itself again and again this may produce a progressive lengthening of the transmission chain in the new host population. Eventually this could yield the emergence of a truly novel disease agent (Antia *et al.* 2003). Yet, the term 'disease emergence' is not restricted to the process of host radiation *per se* but, at least in this section of the book, applied as a collective noun to describe disease flare-up in general, be it the result of a mere increase in incidence, a geographic range expansion, a novel vector or any other transmission adjustment, a higher pathogenicity level, or an adjustment of the host species range. The terms parasite and pathogen are used interchangeably in this section.

We explore how the ecology and evolution of emerging disease agents may be better understood when applying a geographic and ecological invasion framework. We argue that initially pathogen establishment and colonization dictates the parasite–host interactions. During this phase, the specifics of parasite transmission, infection course and host specificity evolve in such a manner that invasion is supported. During the subsequent consolidation phase pathogen and host start to coevolve and the parasite life-history is accordingly adjusted. Thus, the invading parasite ecological strategy initially reflects the life history of an *r*-type strategist and, next, switches to a more *K*-strategist profile particularly once disease emergence has reached its peak and retraction has started (see further). The basic underlying assumption here is that generalist and specialist parasites behave differently, with specialization corresponding to an adaptation of the parasite to a more predictable host environment and vice versa, generalists prevailing in more dynamic host landscapes (Southwood *et al.* 1974; Sasal *et al.* 1999). It is believed that parasites move dynamically in the *r-K* continuum, adjusting life-history features (replication, mortality, and dispersal) to the most successful ecological strategy. Naturally, both *r* and *K* type parasite subpopulations co-exist in a heterogeneous host landscape.

We explore this hypothesis taking the virulence jump and host radiation of avian influenza viruses (AIV), in particularly the recent panzootic of the highly pathogenic avian influenza (HPAI) H5N1 virus as an example. More than any other emerging disease agent, the avian influenza virus has the reputation of being a highly flexible and evolvable disease agent (Lu *et al.* 2007; Webster *et al.* 1992). Influenza viruses belong to the family Orthomyxoviridae and are classified into three types, A, B, and C based on the identity of major internal protein antigens. Influenza A and C viruses can infect multiple mammalian species, while the influenza B virus is almost exclusively a human pathogen. Influenza A viruses cause the greatest morbidity and mortality in humans. The main natural habitat of Influenza A is the reservoir of wild waterfowl, as distributed across the Holarctic. Food and companion animal populations such as horses, swine, dogs, and poultry support specialized influenza A viruses which occasionally continue to exchange genetic material with the wild bird avian gene pool, mainly through a process based on re-assortment and which entails the exchange of whole genetic segments. The influenza A viral genome counts eight RNA segments which all become frequently exchanged. This confounds the nomenclature because influenza A

viruses are classified into subtypes on the basis of antigenic analyses of the hemagluttinin (HA) and neuraminidase (NA) glycoproteins. So far, 16 HA subtypes and 9 HA subtypes have been identified.

We review the coevolution of the AIV gene pool and the natural avian hosts: the reservoir of migratory waterfowl. As a next step we explore how twentieth century agro-ecological dynamics may have contributed to creating and shaping a novel avian host ecology. We then explore how the poultry invasion by the H5N1 virus may first have evolved in China, to eventually provoke an intercontinental scale disease flare-up, with subsequent retraction and virus persistence in selected countries. We conclude with predictions regarding the outcome of the ongoing HPAI H5N1 consolidation phase and the likely virus evolution process that ensues. We argue that the HPAI H5N1 as a disease is turning into an infection of the respiratory tract of domestic ducks and also geese, whilst bringing generalized infection to terrestrial poultry in situations where the broader socio-ecological, agricultural, and geographic settings are conducive.

17.2 Invasion as a framework for disease emergence

Whenever a new host ecological vacuum presents itself parasites are challenged to exploit this resource. Hence, an emerging disease event compares to an ecological invasion process. Successful invasion implies that the parasite becomes introduced and established in the new host environment, followed through by a process of colonization and, next, consolidation (Sakai *et al.* 2001). The latter is required for the sustained occupancy by the invading agent of the novel host niche. We argue that the nature of the parasite–host interaction changes during the invasion process. Whereas during the initial parasite–host–environment interplay the emphasis is on both how amenable the host environment is to parasite invasion and how invasive the parasite is, the ecological context gradually shifts towards a less intrusive occupancy pattern, and thus a more sustainable host—parasite relationship.

Viruses radiate by expanding their host range in the widest possible sense. This may entail invading a novel host type or species within the host community, a new host subpopulation, a different host body compartment, a novel organ system, tissue, or cell type. This may extend to host defence systems and the ability of the virus to survive in the environment outside the host. A hallmark of the infections cycle for viruses parisitizing the host–environment complex is the need to invade and successfully replicate (Cuevas *et al.* 2003).

Arguably, the genetic evolvability of the concerned pathogen plays a key role in the invasion process (Pulliam 2008). Single stranded RNA viruses and other microorganisms, particularly pathogens engaging in horizontal gene transfer, are more inclined to adapt to the novel host environment. The profile of a generalist, invasive, or opportunistic agent is characterized by prolific reproduction (Southwood *et al.* 1974), which, in the context of pathogenic agents, is about replicating the number of infected hosts. Generalist type pathogens tend to facilitate disease spread, often involving a broad range of hosts (Fig. 17.1). Hence, generalists, as opposed to specialists, are the first to encroach and intrude upon a new host resource.

The host environment may, in different ways, provide the right incentives to the invading pathogen. The host may become more available to the pathogen through formation of a dense and well-connected meta-population enhancing transmission. Also the presence of abundant immuno-naïve juveniles or immuno-compromised individuals invites invasion, altering the infection process within the body. True ecological invasion is called for when a single host type is showing up as a homogeneous patch in the host community, challenging the parasite host specificity.

With a host resource becoming vulnerable to invasion the parasites selected for are those that will swiftly colonize the new host resource. Rapid geospatial spread, severe disease, and a broad host range may assist this process. Eventually, the novel host supply will become finite and the parasite ecological strategy shifts accordingly, from colonization to consolidation. Sustained occupancy entails parasite permanence in predictable, long-lived host habitats; endemic foci may emerge. The infection process will reflect the more intimate host—parasite relationship. Eventually, the parasite may turn into a single host type specialist.

As already stated, the shift in ecological strategy is reflected in the parasite population parameters and life-history. *r*-selected pathogens will act as colonizers engaging in a hectic invasion process. These 'boom-and-bust' agents require demographic resilience in order to survive the often erratic invasion pattern. Eventually, *K*-selected functionalities will become important. Parasite reproduction, mortality, and dispersal will now support an equilibrium status in a more static host environment. *K*-selected parasites cling to the carrying capacity level.

Following establishment, the invasion process of the susceptible host population by the invading agent often features rapid spread and an epidemic form of disease. The increase in transmission rate may involve virulence adjustment. Yet, given the convexity of the transmission—virulence trade-off, an evolutionary stable virulence level is not very amenable to change unless strongly challenged (Alizon and van Baalen 2005). A novel transmission pattern may yield a new disease reproductive number, *R* nought. One step further would be the adjustment of the infection process within the host. With host individuals being highly susceptible, the pathogen may turn more aggressive within the host body. This may apply to different host cells, tissues, and whole body compartments. Hence, profuse virion replication may both enhance transmission at population level and facilitate wider access to within-the-host-body niches, thus altering the course of infection. Finally, in addition to a novel pathogen transmission pattern and a novel infection process, the host range may also become adjustable.

Biodiversity acts as a barrier to ecological invasion (Kennedy *et al.* 2002; Knops *et al.* 1999). Large, homogeneous host patches are unusual in the natural landscape. However, anthropogenic congregation of humans and animals, ranging from the social life of humans to factory farming of animals, is artificially sustained by means of health protection precautions ranging from hygiene to vaccination to bio-security practices to active disease control, all intended to counter pathogen intrusions. When not properly shielded, a novel host ecological vacuum tends to become colonized by invading agents. Natural homogeneous host patches are often transient

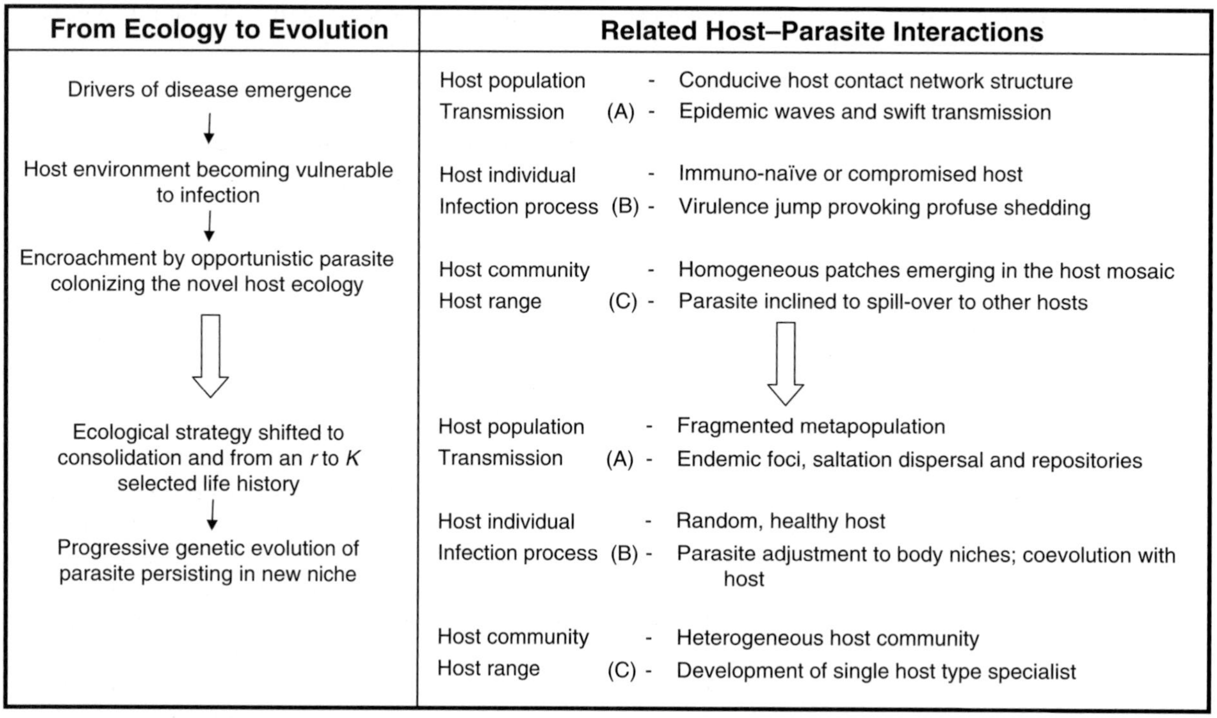

From Ecology to Evolution	Related Host–Parasite Interactions		
Drivers of disease emergence	Host population	-	Conducive host contact network structure
	Transmission	(A) -	Epidemic waves and swift transmission
↓			
Host environment becoming vulnerable to infection	Host individual	-	Immuno-naïve or compromised host
	Infection process	(B) -	Virulence jump provoking profuse shedding
↓			
Encroachment by opportunistic parasite colonizing the novel host ecology	Host community	-	Homogeneous patches emerging in the host mosaic
	Host range	(C) -	Parasite inclined to spill-over to other hosts
⇩	⇩		
Ecological strategy shifted to consolidation and from an *r* to *K* selected life history	Host population	-	Fragmented metapopulation
	Transmission	(A) -	Endemic foci, saltation dispersal and repositories
↓			
Progressive genetic evolution of parasite persisting in new niche	Host individual	-	Random, healthy host
	Infection process	(B) -	Parasite adjustment to body niches; coevolution with host
	Host community	-	Heterogeneous host community
	Host range	(C) -	Development of single host type specialist

Figure 17.1 Parasite invasion of a new host resource brings encroachment and colonization followed by a consolidation process required for sustained occupancy by, and development of, a specialist parasite. Host–parasite interactions change with this shift in parasite ecological strategy, from aggressive invader of the various host environment component parts to a more long-lived host-parasite association. Parasite genetic evolvability tends to decrease from A to B to C.

because of greater fitness obtained from a more differentiated resource exploitation pattern (Knops *et al.* 1999). Hence, pathogen specialization will be necessary to secure persistence of an invading agent.

The switch in ecological strategy can be broadly discerned from the epicurve peak and the geospatial disease dynamics; an *r*-type, invasive pathogen causes spread of disease into a wider area until the gradual, progressive retraction process kicks in. The latter paves the way for the evolution of an ever more *K*-type oriented pathogen in an increasingly well demarcated host habitat.

17.3 The coevolution of avian influenza viruses and migratory waterfowl hosts

The global bird migration pattern cannot be fully understood without recalling the profound effect of the imbalance in the geographic distribution of the

landmasses on the earth, with over 70 per cent of the total landmass found in the northern hemisphere (Orme *et al.* 2006). From south to north the landmass increases, reaching a peak at sub-Arctic latitude level. Siberia may be singled out as prominent land surface area attracting migratory wild birds. The combination of adverse climatic conditions and the burst of spring time insect life and other water bird feed resources in wetlands of the northern Palaearctic and Nearctic biogeographical areas help to explain congregations of migratory waterfowl here for spring–summer breeding (Veen *et al.* 2005). The western Siberia lowland area constitutes the world's most prominent wetland area, with approximately 10,000 small, shallow water bodies distributed in peat lands, and forms a main waterfowl summer breeding habitat (Sheng and Smith 2003).

Mallard ducks are by far the most numerous waterfowl species and also the most prominent natural AIV host, followed by other dabbling duck species (Gilbert *et al.* 2006b; Olsen *et al* 2006). Apart from waterfowl, waders and shorebirds are also among the water bird species in which AIV are found to circulate. Thus, the AIV wild bird host reservoir is characterized by great diversity. AIV circulating in their natural avian hosts randomly and continually reconstitute all eight genetic segments (Dugan *et al.* 2008). Presumably, this re-assortment mechanism assists AIV in occupying a range of water bird habitats across the northern hemisphere. The horizontal gene selection mode requires that two different AIV co-infect one and the same host cell. This, in turn, requires a good match between the life-histories of host and parasite so as to secure continual re-infection.

The peak prevalence of AIV occurs just after the summer breeding, when adults and juvenile ducks plus also other waterfowl arrive from different breeding locations (Brown *et al.* 2005; Halvorson *et al.* 1985). In Siberia, birds congregate in large numbers for moulting, in selected water bodies such as Lake Chaney, to where migratory birds arrive from both the eastern and the western Palaearctic (Brown *et al.* 2005; Halvorson *et al.* 1985). During the wing moulting period adult birds do not fly for a period of about one month. The presence of a diversity of water bird species, some of which carrying AIV, meeting up with the latest crop of immuno-naïve juvenile ducks, provides for an ideal scenario for AIV redistribution through faecal–oral transmission and virus persistence in water, faeces, and mud.

The geographic reshuffling of the AIV genes also takes place during the winter season when migratory and resident birds mix near water bodies. Particularly during mid-winter, one single AIV infected bird may kick-start the build-up of a local virus repository, at a time of the year when both the virus survival conditions outside the host are optimal and bird movements minimal.

The continual mixing of AIV across the Holarctic involves different water bird ecological settings and habitats. Available evidence suggests the mixing of all eight genetic AIV segments in most areas. Broad phylogenetic divides separate the Old and New World AIV and, also, AIV in gull species from other water birds (Olsen *et al.* 2006). Importantly, despite the mosaic of bird ecological niches, there is no

departure from the horizontal gene selection mode; AIV do not specialize in different wild bird habitats. It is therefore plausible that AIV, for a sustained association with migratory bird as hosts, derive fitness from generalist type faculties and life-history features. A horizontal genetic selection mode based on re-assortment may accommodate the variation in water bird biology, life-history, and associated eco-geographic and climate related features.

17.4   Industrial poultry and the creation of an AIV host ecological vacuum

The early days of poultry industrialization go back to the post-Second World War era when US technological advances paved the way for mass rearing of poultry in confined feeding operations, for automation of bird slaughter, and for broiler meat preservation in the form of canned products and cold storage. The first automated mass broiler slaughter plant was established during the mid-1940s. During the late 1950s and early 1960s these economies of scale started to translate into a decline in the number of poultry producers, even with a strongly increased demand for poultry meat and eggs. Improved feed conversion rates, stemming from genetic improvement and proficient poultry husbandry, made it possible to generate bulk quantities poultry protein at a cost below that of beef, mutton, or pork (Delgado *et al.* 1999).

Poultry production soon became a globally significant development, first in western countries, particularly during the 1960s and 1970s, and, next, in Latin America, the Middle East, and eastern and south-eastern Asia, mostly during the 1980s. At a global level, the growth of poultry production peaked during the 1990s. In absolute terms the number of birds continues to increase and this trend is expected to continue for several decades still (Harrison 2002).

Infectious poultry diseases started to emerge along with the scaling up of poultry production plants. Whilst many countries were able to progressively control and eliminate a growing number of the historically ubiquitous OIE notifiable livestock diseases, including Rinderpest, contagious bovine pleuropneumonia, foot and mouth disease, and brucellosis in cattle; glanders in horses and sheep; and goat pox in small ruminants, success stories in poultry disease control, such as the progressive elimination of Newcastle disease, were countered by the emergence of multiple, novel infectious diseases spreading across the globe in commercial poultry; infectious bronchitis, infectious laryngo-tracheitis, infectious bursitis and, incidentally, HPAI. Newcastle disease started to stage a comeback from the 1990s (Neuteboom and Slingenbergh 2006).

The first appearance of HPAI in poultry cannot be traced back precisely but may have occurred during the late nineteenth century. The first virus isolation was in chicken in 1959 in Scotland (Alexander 1987). HPAI H5 or H7 subtypes occasionally showed up in poultry industries worldwide, but it was not until the 1980s that evidence appeared for a non-linear increase in the cumulative number of countries where HPAI had been detected (Capua and Alexander 2004). With the increase in HPAI flare-ups continuing during the 2000s, two features emerged. First, HPAI detection sites in wild birds and poultry in the western Palearctic were often distributed near wintering sites for migratory water birds located just below the wintering line, in places where during mid-winter water bodies are partially frozen, birds congregate, and the conditions are ideal for the build-up of virus repositories (Alexander 2007). Second, HPAI tended to persist for many years in countries classified as emerging economies and where poultry industries were unable to wipe out infection and instead started to rely on routine mass vaccination. Already, prior to the major, trans-frontier HPAI H5N1 invasion, which started early 2004, Mexico, Pakistan, Iran, and China were among the countries semi-permanently infected with H5, H7, or H9 avian influenza circulating in terrestrial poultry.

17.5 The emergence of HPAI H5N1 in China

Whereas ducks, geese, and swans constitute the major avian fauna reservoir for natural or low pathogenic AIV, HPAI was until 1996 associated with terrestrial poultry species, in chicken, turkey, ostrich, or quail. Retrospective studies suggest that poultry flocks could attract infection through wild bird contact, particularly when kept in partially

confined settings and located in countries frequented by migratory birds. The mutation of the virus into a highly pathogenic form, characterized by a typical amino acid sequence at the HA cleavage site, is a gradual process and it may take several months or more of virus circulation in a terrestrial poultry flock before an HPAI virus evolves, if it does at all (Capua and Alexander 2004). The absence of HPAI in aquatic poultry species, until 1996, does not imply that domestic ducks and geese did not become AIV infected. AIV were found to circulate on a year-round basis in domestic ducks in China in the late 1970s (Duan et al. 2008), but no HPAI evolved in domestic waterfowl at the time. It is to be recalled also that the genetic makeup of domestic ducks in China, mostly Peking ducks, is hardly discernible from that of wild mallards. Traditional duck and goose husbandry systems, with small flocks scavenging around backyards, in rice paddies, and on dikes, presumably do not provide for the right setting for AIV to mutate into HPAI. H5 and H7 HPAI in commercial terrestrial poultry in China already occurred prior to the mid-1990s, as did the H9N2 virus (Li et al. 2003).

Mid-1980s domestic duck production in China started to accelerate; duck meat production volumes increased from 1985 to 2000 by over 500 per cent, up to 2 million metric tonnes (Gilbert et al. 2007). With the shift towards more industrial type production, the ducks were, in part, moved out of the wetlands and river systems. Old and new forms of poultry production became mixed up and different poultry species intermingled. Traditionally, the highest number of duck breeds is found in areas with double-cropping of rice. Domestic geese are more widely distributed across the single rice crop areas, extending to the north-eastern and extreme western parts of the country (Fig. 17.2). Production intensification brought duck and geese plants nearer to the urban centres. The broiler industry had already increased in size prior to the explosive growth of duck and goose meat production, and was already located in areas of high human density and in the proximity of coastal harbours importing soybeans (Gerber et al. 2005). The standing populations of poultry flocks in China in 2007 amounted to about 700 million ducks, 350 million domestic geese,

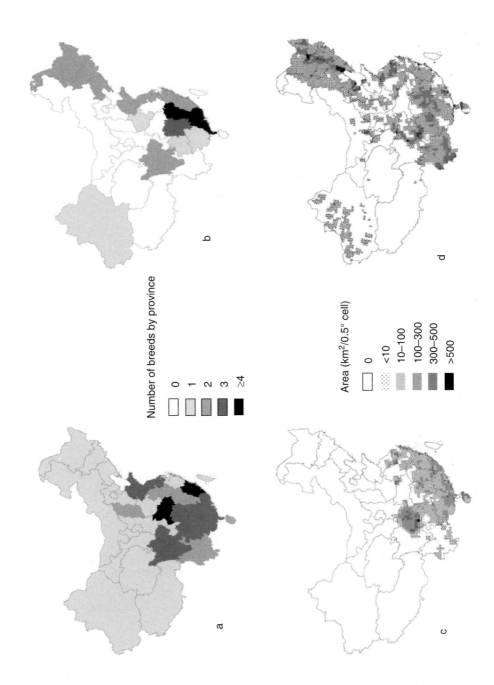

Figure 17.2   Distribution of endemic duck and goose breed number by province (a and b; data from Xu and Chen 2003) and matching distribution of respectively double-cropping and single rice crop areas per 0.5 °grid cell (c and d; modified from Frolking et al. 2002) in China.

and 4.6 billion head of chicken (FAOSTAT 2007). China constitutes the most important global hub for traditional and industrial as well as terrestrial and aquatic poultry.

17.6   The panzootic spread of HPAI H5N1

The 1996 HPAI H5N1 virus circulating in geese was highly pathogenic to chicken, as became evident with the appearance of HPAI H5N1 in Hong Kong SAR, in 1997, eventually prompting local authorities to eliminate all poultry flocks as 18 people became infected, of which 6 persons died (Claas *et al.* 1998). Systematic surveillance in aquatic poultry in southern China during the end of the 1990s revealed 21 different H5N1 genotypes, all sampled from healthy looking domestic ducks encountered in live bird markets. Still, the pathogenicity level of H5N1 inoculated ducks, chicken, and also mice progressively increased from 1999 to 2002 (Chen *et al.* 2004). With it, the H5N1 prevalence in poultry increased, particularly following the emergence of genotype Z in 2002 (Li *et al.* 2004).

In ducks, the virus increasingly manifested itself as a respiratory infection (Sturm-Ramirez *et al.* 2005). With genotype Z pathogenicity levels being much higher in chicken than in ducks, and with chicken being the more prevalent poultry species-with the broiler cycle taking only five weeks as opposed to the four months fattening required for meat ducks-the better transmissibility in chicken brought the progressive geographic expansion of H5N1 infection, much facilitated by chicken transport and trade in live bird markets (Sims 2007). The first genotype Z human infection in China dates back to 2003 (Guan *et al.* 2004). Late 2003 marked the onset of a sudden, subcontinental scatter of outbreaks across countries in eastern and southeastern Asia. Commercial poultry chains have presumably been involved in this rapid dissemination of H5N1 infection. Retrospective studies suggest that China may be singled out as the main H5N1 supply and dispersal source (Chen *et al.* 2006b). Indonesia became infected through a single introduction event, probably dating back to August 2003. Phylogenetic analysis suggests that this virus had originated from the Hunan Province in China. Multiple viruses showing up in northern Viet Nam

could be traced back to China. Korea and Japan also detected H5N1 viruses that had presumably arrived from China (Chen *et al.* 2006b). Several H5N1 epizootic waves hit eastern and southeastern Asia during 2004 and 2005 (Gilbert *et al.* 2008). With it, several clades and subclades started to evolve (Duan *et al.* 2008). Whilst wild bird H5N1 infections in Hong Kong SAR and elsewhere had occasionally been encountered in dead or clinically diseased birds, there was as yet no evidence for H5N1 circulation in healthy wild birds, until the first detection of H5N1 in six apparently healthy wild ducks sampled in China's Poyang Lake area, January to March 2005 (Chen *et al.* 2006b). The Poyang Lake flood plains form important rice production areas and provide an ideal interface for wild and domestic ducks. Further support for the circulation of H5N1 in wild birds, already discovered during the winter of 2004–2005, showed up in May 2005 in the remote Qinghai Lake area in western China where thousands of bar-headed geese had reportedly died of H5N1 (Chen *et al.* 2006a; Liu *et al.* 2005; Wang *et al.* 2008). An indication that wild birds could be vectoring H5N1 across larger distances emerged with the subsequent detection of a Qinghai-like virus in wild water birds in July 2005 in Russia (Lipatov 2009). A cluster of both poultry and wild bird infections became established between Omsk and Novosibirsk, around the Lake Chaney area. Summer 2005 marked the start of a trail of poultry infections running westwards, mainly involving poultry plants. Early autumn 2005 marked the arrival of H5N1, presumably by wild birds, in the Black Sea basin (Gilbert *et al.* 2006b). Evidence for H5N1 infections in both wild birds and poultry initially concerned the Danube delta, until mid-winter, when a series of cold weather spells propagated H5N1 vectored mainly by mute swans across western Europe (Kilpatrick *et al.* 2006). Countries in the Middle East and Africa also became infected, possibly due to poultry trade and traffic as well as wild birds (Ducatez 2006). The H5N1 panzootic peaked during the first half of 2006, when the cumulative number of countries where H5N1 had been detected surpassed 60. The number of countries with human infections reached a total of 15 countries in 2008 (WHO 2009).

17.8 The persistence of HPAI H5N1

It is possible that the annual flare-up of H5N1 during mainly the autumn and winter seasons in poultry, wild birds, as well as in humans, has to be explained differently for the eastern and western Palaearctic. With rice—duck agriculture playing a role in countries in southeastern Asia, the monsoon-driven rice harvest during early autumn enables ducks to feed on leftover rice grains in post-harvest rice paddies during autumn and early winter (Gilbert *et al.* 2006a). Ducks are slaughtered at the time of the Chinese Lunar New Year, late January to early February in the Gregorian calendar. In China, H5N1 prevalence in ducks and geese fluctuated at around 2 to 3 per cent during the winter of 2004–2005, and 6 to 9 per cent during the winter of 2006–2007, as suggested by samples taken from apparently healthy birds in live bird markets in the southern Provinces Fujian, Guangdong, Guangxi, Guiyang, Hunan, and Yunan (Smith *et al.* 2006). Also, in Thailand, H5N1 had become retracted to wetland areas with a high density of free-ranging ducks (Gilbert *et al.* 2008).

The production of meat ducks in rice paddies after the autumn harvest requires the advanced, synchronized hatching of eggs and raising of ducklings during the monsoon rains. At this time of the year it is difficult to keep ducks in rice paddies except for temporary ducklings feeding on weeds and insects (Hossain *et al.* 2005). Keeping significant numbers of adult duck layers requires a second rice harvest, preferably during the late dry season, so as to feed the ducks on leftover rice grains on a year-round basis. Double rice-cropping areas are confined to river deltas and plains, and it is here where the highest duck densities are found and H5N1 virus persists (Gilbert *et al.* 2008). The geographic distribution of H5N1 outbreaks in poultry in Viet Nam became increasingly confined to the Red River Basin and the Mekong Delta. Since 2005, Thailand has introduced strict sanitary restrictions for ducks kept in the central plains of the country and, as a result, H5N1 virus circulation has progressively declined.

In Indonesia, Bangladesh, and southern China, H5N1 persistence was also found to be associated with rice—duck agriculture, mostly in rather densely populated wetland areas. In Egypt, H5N1 mainly persist in the duck-rich Nile delta. High land pressure and irrigated cropping in Asia entails a mosaic of complex farming systems. In places with free-ranging duck rotating in small rice plots H5N1 may readily persist and, also, spread to village chicken, live bird markets, and commercial broiler plants. Whereas traditional domestic waterfowl production facilitates H5N1 virus persistence, the chicken trade and live bird markets are believed to provide for the H5N1 propagule, dispersing H5N1 virus beyond the rice-duck areas.

17.9 Conclusions

The progressive, global colonization by the H5N1 virus and the subsequent first signs of consolidation broadly match the shift from generalist to specialist type invading agent depicted in Fig. 17.1. Following the initial detection of an HPAI type H5N1 virus in domestic geese in Guangdong, in 1996, the virus became firmly established in the rice–duck agriculture areas of southern China. With the emergence of the aggressive genotype Z in 2002 the virus started to disperse more rapidly and to infect a broader avian and also mammalian host range (Webster *et al.* 2006). Commercial chicken and live bird markets formed the main medium for rapid, epizootic virus spread at a subcontinental scale. Thus, the H5N1 virus entered into contact with novel supplies of domestic ducks in the wetlands of the Mekong countries. Migratory waterfowl increasingly showed infection with the pathogenic H5N1 virus. Some waterfowl species, particularly mallards (Keawcharoen *et al.* 2008) but also mute swans, were found to acquire immunity (Kalthoff *et al.* 2008). Birds resistant to disease presumably vectored virus into wider geographic areas.

The Eurasian H5N1 invasion reached its peak in 2006. Virus retraction into lower laying moisture areas subsequently revealed the persistence of a novel pathogen increasingly causing respiratory infection in ducks and geese. The virus also continues to be secreted with faeces and is capable of sustaining a faecal–oral cycle in wild water birds, complete with virus survival outside the host body in water and ice. The virus continues to spread, especially during the winter, in commercial broilers and associated live bird markets. Hence, temperature tolerance is critically important, to virion survival in and outside the body, in poultry markets or in wild bird wintering habitat.

Available disease ecological data support the notion that virus persistence in poultry in Asia has become dependent on domestic waterfowl. Phylogenetic analysis also shows that the number of co-circulating clades and subclades is highest in wetlands supporting rice—duck agriculture. Wild birds assist in vectoring the H5N1 virus over long distances, both latitudinally, from Asia to Europe and vice versa in not more than six months, and also longitudinally, linking the southern China-based H5N1 gene pool in domestic ducks and the western Pacific flyways of migratory water birds. It is conceivable that H5N1 diversity will continue to increase and this may, eventually, assume Holarctic proportions (Duan *et al.* 2007).

The formation of H5N1 repositories outside the host in and around water bodies during mid-winter will arguably assist the further distribution of H5N1 across the northern Palaearctic breeding areas, creating the risk of occasional recurrence of infection in poultry in Europe. In the eastern Palaearctic, the overlap of migratory waterfowl flyways and domestic duck and geese distributions will also secure continued H5N1 exchanges in the future. Even with the virus circulation in China currently brought in check by means of mass vaccination, occasional virus dispersal to neighbouring eastern Asia countries remains likely.

Bangladesh, Egypt, Indonesia, and Vietnam are enzootically infected with H5N1 viruses turning into local specialists of the prevailing poultry landscapes here. Invariably, virus persistence is enhanced by high duck densities kept in irrigated crop areas. Virus dissemination is facilitated by broiler production and marketing. Hence, the local H5N1 niches are progressively shaped by the complex of socio-ecological drivers, including agriculture, land use, poultry trade, wild birds, climate, and geography.

References

Alexander, D.J. (1987). Avian influenza - historical aspects. In B.C. Easterday, ed. *Proceedings of the Second International Symposium on Avian Influenza, 1986*, pp. 4–13. University of Wisconsin, Madison.

Alexander, D.J. (2007). Summary of avian influenza activity in Europe, Asia, Africa, and Australasia, 2002-2006. *Avian Diseases*, **51**, 161–66.

Alizon, S. and van Baalen, M. (2005). Emergence of a convex trade-off between transmission and virulence. *American Naturalist*, **165**, E155–67.

Antia, R., Regoes, R.R., Koella, J.C., and Bergstrom, C.T. (2003). The role of evolution in the emergence of infectious diseases. *Nature*, **426**, 658–61.

Brown, I., Gaidet, N., Guberti, V., Marangon, S., and Olsen, B. (2005). *Mission to Russia to Assess the Avian Influenza Situation in Wildlife and the National Measures Being Taken to Minimize the Risk of International Spread*. Office International des Epizooties. Available from: http://www.oie.int/downld/Missions/2005/ReportRussia2005Final2.pdf

Capua, I. and Alexander, D.J. (2004). Avian influenza: recent developments. *Avian Pathology*, **33**, 393–404.

Chen, H., Deng, G., Li, Z. *et al.* (2004). The evolution of H5N1 influenza viruses in ducks in southern China. *Proceedings of the National Academy of Sciences of the USA*, **101**, 10452–27.

Chen, H., Li, Y., Li, Z. *et al.* (2006a). Properties and dissemination of H5N1 viruses isolated during an influenza outbreak in migratory waterfowl in western China. *Journal of Virology*, **80**, 5976–83.

Chen, H., Smith, G., Li, K. *et al.* (2006b). Establishment of multiple sublineages of H5N1 influenza virus in Asia: implications for pandemic control. *Proceedings of the National Academy of Sciences of the USA*, **103**, 2845–50.

Claas, E.C., Osterhaus, A.D., van Beek, R. *et al.* (1998). Human influenza A H5N1 virus related to a highly pathogenic avian influenza virus. *Lancet*, **351**, 472–77.

Cuevas, J.M., Moya, A., and Elena, S.F. (2003). Evolution of RNA virus in spatially structured heterogeneous environments. *Journal of Evolutionary Biology*, **16**, 456–66.

Daszak, P., Cunningham, A.A., and Hyatt, A.D. (2001). Anthropogenic environmental change and the emergence of infectious diseases in wildlife. *Acta Tropica*, **78**, 103–16.

Delgado, C., Rosegrant, M., Steinfeld, H., Ehui, S., and Courbois, C. (1999). *Livestock to 2020: The Next Food Revolution. Food, Agriculture and the Environment Discussion. Paper 28*. International Food Policy Research Institute, Washington, DC (USA); FAO, Rome (Italy) and International Livestock Research Institute, Nairobi (Kenya)

Duan, L., Bahl, J., Smith, G.J.D. *et al.* (2008). The development and genetic diversity of H5N1 influenza virus in China, 1996-2006. *Virology*, **380**, 243–54.

Duan, L., Campitelli, L., Fan, X.H. *et al.* (2007). Characterization of low-pathogenic H5 subtype influenza viruses from Eurasia: implications for the origin of highly pathogenic H5N1 viruses. *Journal of Virology*, **81**, 7529–39.

Ducatez, M.F., Olinger, C.M., Owoade, A.A. *et al.* (2006). Avian flu: multiple introductions of H5N1 in Nigeria. *Nature*, **442**, 37.

Dugan, V.G., Chen, R., Spiro, D.J. *et al.* (2008). The evolutionary genetics and emergence of avian influenza viruses in wild birds. *PLoS Pathogens*, **4**, e1000076.

Harrison, P. (2002). *World Agriculture: Towards 2015/2030*. Summary Report. Food and Agriculture Organization of the United Nations, Economic and Social Department. Available from: http://www.fao.org/docrep/004/Y3557E/Y3557E00.HTM

FAOSTAT (2007). *FAOSTAT: FAO Statistical Databases*. Food and Agriculture Organization of the United Nations (http://faostat.fao.org).

Frolking, S., Qiu, J., Boles, S. *et al.* (2002). Combining remote sensing and ground census data to develop new maps of the distribution of rice agriculture in China. *Global Biogeochemical Cycles*, **16**, 1091.

Gerber, P., Chilonda, P., Franceschini, G., and Menzi, H. (2005). Geographical determinants and environmental implications of livestock production intensification in Asia. *Bioresource Technology*, **96**, 263–76.

Gilbert, M., Chaitaweesub, P., Parakamawongsa, T. *et al.* (2006a). Free-grazing ducks and highly pathogenic avian influenza, Thailand. *Emerging Infectious Diseases*, **12**, 227–34.

Gilbert, M., Xiao, X., Domenech, J., Lubroth, J., Martin, V., and Slingenbergh, J. (2006b). Anatidae migration in the western Palearctic and spread of highly pathogenic avian influenza H5N1 virus. *Emerging Infectious Diseases*, **12**, 1650–56.

Gilbert, M., Xiao, X., Chaitaweesub, P. *et al.* (2007). Avian influenza, domestic ducks and rice agriculture in Thailand. *Agriculture, Ecosystems and Environment*, **119**, 409–15.

Gilbert, M., Xiao, X., Pfeiffer, D.U. *et al.* (2008). Mapping H5N1 highly pathogenic avian influenza risk in Southeast Asia. *Proceedings of the National Academy of Sciences of the USA*, **105**, 4769–74.

Guan, Y., Poon, L.L.M., Cheung, C.Y. *et al.* (2004). H5N1 influenza: a protean pandemic threat. *Proceedings of the National Academy of Sciences of the USA*, **101**, 8156–61.

Halvorson, D.A., Kelleher, C.J., and Senne, D.A. (1985). Epizootiology of avian influenza: effect of season on incidence in sentinel ducks and domestic turkeys in Minnesota. *Applied and Environmental Microbiology*, **49**, 914–19.

Hijmans, R.J., Cameron, S.E., Parra, J.L., Jones, P.G., and Jarvis, A. (2005). Very high resolution interpolated climate surfaces for global land areas. *International Journal of Climatology*, **25**, 1965–78.

Hossain, S.T., Sugimoto, H., Ahmed, G.J.U., and Islam, Md.R. (2005). Effect of integrated rice-duck farming on rice yield, farm productivity, and rice-provisioning ability of farmers. *Asian Journal of Agriculture and Development*, **2**, 79–86.

Jones, K.E., Patel, N.G., Levy, M.A. *et al.* (2008). Global trends in emerging infectious diseases. *Nature*, **451**, 990–93.

Kalthoff, D., Breithaupt, A., Teifke, J.P. *et al.* (2008). Highly pathogenic avian influenza virus (H5N1) in experimentally infected adult mute swans. *Emerging Infectious Diseases*, **14**, 1267–70.

Kapan, D.D., Bennett, S.N., Ellis, B.N. *et al.* (2006). Avian influenza (H5N1) and the evolutionary and social ecology of infectious disease emergence. *EcoHealth*, **3**, 187–94.

Keawcharoen, J., van Riel, D., van Amerongen, G. *et al.* (2008). Wild ducks as long-distance vectors of highly pathogenic avian influenza virus (H5N1). *Emerging Infectious Diseases*, **14**, 600–07.

Kennedy, T.A., Naeem, S., Howe, K.M., Knops, J.M.H., Tilman, D., and Reich, P. (2002). Biodiversity as a barrier to ecological invasion. *Nature*, **417**, 636–38.

Kilpatrick, A.M., Chmura, A.A., Gibbons, D.W., Fleischer, R.C., Marra, P.P., and Daszak, P. (2006). Predicting the global spread of H5N1 avian influenza. *Proceedings of the National Academy of Sciences of the USA*, **103**, 19368–73.

Knops, J.M.H., Tilman, D., Haddad, N.M. *et al.* (1999). Effects of plant species richness on invasion dynamics, disease outbreaks, insect abundances and diversity. *Ecology Letters*, **2**, 286–93.

Li, K.S., Guan, Y., Wang, J. *et al.* (2004). Genesis of a highly pathogenic and potentially pandemic H5N1 influenza virus in eastern Asia. *Nature*, **430**, 209–13.

Li, K.S., Xu, K.M., Peiris, J.S.M. *et al.* (2003). Characterization of H9 subtype influenza viruses from the ducks of southern China: a candidate for the next influenza pandemic in humans? *Journal of Virology*, **77**, 6988–94.

Lipatov, A.S., Evseenko, V.A., Yen, H-L. *et al.* (2007). Influenza (H5N1) viruses in poultry, Russian Federation, 2005–2006. *Emerging Infectious Diseases*, **13**, 539–46.

Liu, J., Xiao, H., Lei, F. *et al.* (2005). Highly pathogenic H5N1 influenza virus infection in migratory birds. *Science*, **309**, 1206.

Lu, G., Rowley, T., Garten, R., and Donis, R.O. (2007). FluGenome: a web tool for genotyping influenza A virus. *Nucleic Acids Research*, **35**, W275–W279.

Neuteboom, O. and Slingenbergh, J. (2006). The development of disease free areas across Europe. *Journal of Food Agriculture and Environment*, **4**, 23–30.

Olsen, B., Munster, V.J., Wallensten, A., Waldenstrom, J., Osterhaus, A.D.M.E. and Fouchier, R.A.M. (2006). Global patterns of influenza A virus in wild birds. *Science*, **312**, 384-8.

Orme, C.D.L., Davies, R.G., Olson, V.A. *et al.* (2006). Global patterns of geographic range size in birds. *PLoS Biology*, **4**, e208.

Power, A.G. and Mitchell, C.E. (2004). Pathogen spillover in disease epidemics. *American Naturalist*, **164**, S79–89.

Pulliam, J.R. (2008). Viral host jumps: moving toward a predictive framework. *EcoHealth*, **5**, 80–91.

Sakai, A.K., Allendorf, F.W., Holt, J.S. *et al.* (2001). The population biology of invasive species. *Annual Review of Ecology and Systematics*, **32**, 305–32.

Sasal, P., Trouvé, S., Muller-Graf, C., and Morand, S. (1999). Specificity and host predictability: a comparative analysis among monogenean parasites of fish. *Journal of Animal Ecology*, **68**, 437–44.

Sheng, Y., Smith, L.C, MacDonald, G.M. *et al.* (2003). A high-resolution GIS-based inventory of the west Siberian peat carbon pool. *Global Biogeochemical Cycles*, **18**, GB3004.

Sims, L.D. (2007). Lessons learned from Asian H5N1 outbreak control. *Avian Diseases*, **51**, 174–81.

Smith, G.J D., Fan, X. H., Wang, J. *et al.* (2006). Emergence and predominance of an H5N1 influenza variant in China. *Proceedings of the National Academy of Sciences of the USA*, **103**, 16936–41.

Southwood, T.R E., May, R.M., Hassell, M.P., and Conway, G.R. (1974). Ecological strategies and population parameters. *American Naturalist*, **108**, 791–804.

Sturm-Ramirez, K.M., Hulse-Post, D.J., Govorkova, E.A. *et al.* (2005). Are ducks contributing to the endemicity of highly pathogenic H5N1 influenza virus in Asia? *Journal of Virology*, **79**, 11269–79.

Veen, J., Yurlov, A.K., Delany, S.N., Mihantiev, A.I., Selivanova, M.A., and Boere, G.C. (2005). *An Atlas on Movements of Southwest Siberian Waterbirds*. Wetlands International, Wageningen, The Netherlands. Available from: http://global.wetlands.org

Wang, G., Zhan, D., Li, L. *et al.* (2008). H5N1 avian influenza re-emergence of Lake Qinghai: phylogenetic and antigenic analyses of the newly isolated viruses and roles of migratory birds in virus circulation. *Journal of General Virology*, **89**, 697–702.

Webster, R.G., Bean, W.J., Gorman, O.T., Chambers, T.M., and Kawaoka, Y. (1992). Evolution and ecology of influenza A viruses. *Microbiology and Molecular Biology Reviews*, **56**, 152–79.

Webster, R.G., Peiris, M., Chen, H., and Guan, Y. (2006). H5N1 outbreaks and enzootic influenza. *Emerging Infectious Diseases*, **12**, 3–8.

Woolhouse, M.E.J. (2008). Epidemiology: emerging diseases go global. *Nature*, **451**, 898–99.

WHO. (2009). Confirmed human cases of avian influenza A (H5N1). Epidemic and pandemic alert and response. [Cited 2009 Jan 28]. Available from: http://www.who.int/csr/disease/avian_influenza/country/en/

Xu, G. F. and Chen, K. W. (2003). *Photograph Album of China Indigenous Poultry Breeds*. China Agricultural Press, Beijing.

CHAPTER 18

When geography of health meets health ecology

Vincent Herbreteau

18.1 Introduction

Human deteriorations of the ecosystems have long been recognized as influencing health status by favouring the emergence of diseases as well as risk conditions for chronic diseases (Chivian and Berstein 2004; Patz *et al.* 2004). Furthermore, recent sudden awareness of the drastic and irreversible environmental changes, surpassing former predictions, has raised a serious anxiety regarding the consequences of emerging diseases worldwide. The swine flu pandemic that emerged in April 2009 with a new strain of influenza A virus subtype H1N1 (Shetty 2009) was a spectacular example of rapid spread in an interconnected world, and was furthermore a real health threat for people who could follow a near real-time report of suspected or confirmed cases with modern information technologies. A few years earlier, the SARS (Severe Acute Respiratory Syndrome) epidemic in 2002, then avian influenza epidemics in 2003, both emerging from South-east Asia, had also been spectacular disease outbreaks. With high fatality rates (contrary to the H1N1 pandemic), they generated a real 'psychosis' for people travelling in the epidemic areas. They caused a considerable reduction in tourism and economic activities, which consequently affected South-east Asia. Despite the actual risk of an influenza pandemic, the word 'psychosis' can be used if we consider the profile of human cases that were highly exposed to the viruses, and the magnitude of the epidemics that caused 774 deaths for the

SARS-associated coronavirus and 256 deaths for the H5N1 virus, according to the World Health Organization (WHO) data as of April 2009 (http://www.who.int). These figures are far from the about 880,000 people killed by malaria in 2006, or the 1.9 to 2.4 million people who died after HIV infection in 2007 (also reported from the WHO in April 2009). Epidemics are now reported to the world and examples of rapid transmission through space and time have awakened the consciousness of possible sudden exposure to unexpected pathogens.

On the one hand, understanding health patterns, and especially the dynamic of diseases, requires an ecological approach to assess the interaction of humans with their environment. On the other hand, a geographical approach is also essential in a world facing increasing movements of people, transport of goods, animal, and agricultural products, that contribute to the transmission of pathogens worldwide. It is also an answer to understanding health patterns when observing similar pathologies in different places, being now boosted by increasing surveillance and reporting of disease incidences.

In this chapter, I propose to describe the methodologies used in both health geography and health ecology and see how these approaches complement each other. Then, I attempt to review the global human-induced pressures on the environment and their impact on health. I illustrate these patterns with major illnesses in the light of ecology and geography.

18.2   From microsatellites to satellites: the contribution of geography to health ecology

Health ecology considers individuals or populations in their environment, working on the assumption that health status is conditioned by the physical, biological, and social environments. Ecosystems are governed by a dynamic equilibrium between these environments. Any perturbation can induce cascading effects on the ecology of living organisms. In this frame, humans, also causing stress on ecosystems, can suffer from ecological imbalance threatening health. This approach takes into account every factor working on individual conditions, from the contact with parasites, pathogens, or pollutants in the physical and biological environments, to sanitary or working conditions in the social environment. Health ecology differs from traditional medical approaches that focus on symptoms of individual cases, and from epidemiology that focuses on the characterization of specific diseases. Epidemiological studies usually target the identification of the causes and mechanisms of the emergence of chronic or infectious diseases, while health ecology tackles the general processes of individuals and population interactions with the environment that can result in the emergence of pathologies. It takes into consideration the complexity of ecosystems and interactions between living organisms and usually proposes a conceptual approach to disease, where observed disease occurrences are used to build or validate models.

A fundamental ecological approach of health patterns is to consider the ecological niche of pathogens and vectors in the case of infectious diseases. The concept of niche was first described as a spatial unit by Grinnell (1917). It represents the space where conditions are suitable for a given species to live, in terms of food, refuge, and breeding site availability (Grinnell 1917). Another definition, presenting a niche as a functional unit, was proposed by Elton in 1927 and refined the spatial unit by considering the ecological position of the organism in the community or ecosystem (Elton 2001). Gause (1934) also put forward another definition

with the 'competitive exclusion principle': two species cannot co-exist at the same locality if they have identical ecological requirements (Gause 1934; Hardin 1960). Finally, Hutchinson (1957) proposed considering a niche as an abstract multidimensional space, where each dimension represents the range of the environmental conditions (abiotic or biotic factors) that are required by the species. In health ecology, populations of pathogens are studied through their niche, and the dynamics of diseases through the overlaps in space and time of the niche of pathogens, with the niches of possible animal vectors and the introduction of humans. This approach requires a precise knowledge of species distribution and is enhanced by information of the pathogen and/or vector ecology that could allow their environmental niche modelling. Indeed, different organisms can have similar niches while one species can occupy different niches. Also, when organisms are introduced into a new environment, they can potentially invade an occupied niche and repress or reduce native organisms (Matthews 2004). This approach has been used to describe the invasion and emergence of zoonotic diseases.

Therefore, health ecology is a difficult challenge requiring researchers to assess the status and dynamics of ecosystems. Multidisciplinary approaches help to gather suitable information, including climate changes studies, land cover studies (using remote sensing techniques on earth observation satellites images), without neglecting primary inventories of faunal and floral diversities. Advances in genetics have also contributed to a better identification of living organisms and assessment of biodiversity. Regarding the transmission of vector-borne diseases, genetics help not just to find the limits between species and describe the niche of animals and parasites, but also to discover their evolutionary history and possible coevolution with phylogeny (and co-phylogeny) and explain their actual distribution with phylogeography (Despommier et al. 2007).

Health geography proposes a complementary approach to health ecology for studying infectious or non-infectious diseases and also health and healthcare systems. It considers the interconnections between socio-economic and ecological components that define the patterns of disease, the health status, or the needs for healthcare. It puts forward a spatial analysis of health from individuals to populations, at different scales, from a local to a regional and global point of view. Mapping techniques have been progressively integrated with epidemiological studies, first as tools to display information, as suggested by the legendary example of John Snow during the nineteenth century, then as new ways to explain disease patterns in space and time. The first steps in health geography were illustrated by Edmund Cooper with the cholera epidemic in London in 1854 (Brody *et al.* 2000). His maps showing the location of cholera deaths together with water pumps were later attributed to John Snow, who used them to support the hypothesis of a water contamination rather than an air contamination. Maps have been widely used in epidemiology to display incidence or prevalence of disease, and remain the major 'geographical' basic application of health studies. Moreover, health geography proposes techniques for spatial analysis to foresee ecological niche patterns at different scales and confront ecological observations in remote spaces. The geography of health has developed with the increasing power of computers and Geographic Information Systems (GIS) that allow us nowadays to deal with large datasets and integrate as many ecological variables as available to explain health patterns in space and time. Local observations and records are computed together to view environment and health from a global perspective (Gatrell 2002).

The geographical approach can be proposed as a synthetic tool to zoom out from the largest scale, each individual living organism, to the smallest, the globe, or vice versa. In other words, we suggest this metaphor of the magnification: from microsatellites, used as molecular markers in the field of genetics, to satellites used as ecological observers in the field of remote sensing.

Considering the relevance of such a multidisciplinary integrated approach of health study, we propose in this review to consider the global drivers of health patterns from a geographical and an ecological point of view.

18.3 From geography to ecology: an integrated approach of major health patterns

18.3.1 uman pressure, spatial inequalities, and implications for ecosystems and human health

*Global population increase: the main pressure
on ecosystems*

The earth is currently being degraded at an increasing pace, under the pressure of a growing population gaining ground on wild biotopes and threatening ecosystem balance. This global tendency has been recorded through history, since the world population was estimated at about 1 million in year 10,000 BC, 170 million in year 1, and is now over 6.5 billion (US Census Bureau 2008a).

Nevertheless, the growth rate dramatically increased during the nineteenth (1 billion people in 1804) and twentieth centuries (2 billion in 1927 and 6 billion in 1999), with the growth peaking in 1962 and 1963 at 2.20 per cent per annum (that represents a doubling time of 35 years) (US Census Bureau 2008a). This exponential growth can be compared in its overpowering acceleration to the spread of epidemics or pandemics, with a latency phase preceding an outbreak. We can also see that population growth has rarely encountered a noticeable slowdown except in the middle of the fourteenth century, with one of the deadliest pandemics, the Black Death (Morens *et al.* 2008). Since 1963, the world growth rate has been decreasing and is currently about 1.14 per cent, representing a doubling time of 61 years. The world population is expected to stabilize at over 10 billion after 2200 if current growth continues and projections are validated (United Nations 1999). Medical advances associated with a better agricultural productivity may explain the recent rapid population growth.

However, this global tendency masks regional and local differences in demographic dynamics revealing a world divide in population and health. The world's population growth occurs mostly in the poorest countries. During the second half of the twentieth century, the proportion of population living in developing countries has raised from 68 per cent to more than 80 per cent (United Nations 1999). The highest population natural increases (without regard for migration, i.e. the difference between the birth rate and death rate) are recorded in Western Africa, Central Africa, Eastern Africa, and the Middle East, as well as in Madagascar, Afghanistan, French Guiana, and Guatemala, with growth rates ranging from 2.4 per cent to 3.6 per cent, according to the Population Reference Bureau (2008) (Fig. 18.1a). Africa is growing faster than any other region and is projected to represent a fifth of the world's population in 2050, whereas it was less than a tenth in 1950. Unlike this trend, the Russian Federation and Eastern European countries have a negative growth rate. This spatial divide at a world scale also masks local divides between cities and countryside. Since 2008, more than one half of the world's population lives in urban areas (having more than 2,000 inhabitants), whereas it was less than a third in 1950 (US Census Bureau 2008b). However, most of the urban population lives in small cities or villages. Figure 18.1b reveals a spatial fracture between urbanized and rural countries. The most urbanized countries (over 75 per cent of the population living in urban areas) are located in the Americas, Oceania, and Europe, while the most rural countries (less than 25 per cent) are in Africa and Asia.

These spatial heterogeneities in the population dynamics imply local difficulties for the high-growth areas in dealing with the consecutive need for food, infrastructure, and services. In the poorest countries, this results in a concentration of poverty and induces sanitation and health problems. It is finally interesting to notice a significant negative correlation between the life expectancy and the natural population growth for the world countries (-0.56 with $p<0.05$), without getting into the complex relations between the populations dynamics and health patterns, but considered from a spatial point of view, that some regions are again distinguished by a very low life expectancy. Figure 18.1a illustrates that the lowest life expectancy is encountered in Central and Southern Africa (the lowest in Swaziland, at 33 years old), as well as Afghanistan, which also presents a less urbanized populations with high natural growth.

These population dynamics are also considered a major driving pressure on ecosystems, with possible return effects on health. A major implication of population growth is the need for agricultural products and land. The most visible consequence is the conversion of forests to agricultural lands (about 13 million hectares per year, according to FAO in 2008) that has risen as the main environmental issue since the twentieth century.

Indeed, the natural population growth is significantly correlated to the annual rate of forest cover change between 1990 and 2000 (-0.35 with $p<0.05$; Table 18.1 and between 2000 and 2005; -0.35 with $p<0.05$; Table 18.1), when compared to the forestry statistics from the Global Forest Resources Assessment 2005 (FAO 2006). The annual rate of forest change shows a latitudinal pattern, with the highest net loss of forests in equatorial and tropical countries (mainly Central Africa and South America) and an extension of forest areas in temperate countries, with China recently conducting large-scale afforestation activities.

On top of human-induced deforestation, forests can also be affected by natural disasters (or occasionally from human origin): fire, drought, storms, floods, and pests (animals or diseases). This global pressure of humans on ecosystems has direct and indirect negative consequences on human health, which can be first related to the introduction of people in wild biotopes.

Increasing proximity with wildlife: a threat for a rapid disease outbreak
By moving deeper into dense forests humans have been exposed to virulent pathogens. The first human infections of Ebola hemorrhagic fever were attributed to the handling of wild animals, chimpanzees, gorillas, monkeys, and antelopes in tropical African ecosystems (Gonzalez *et al.* 2005). Recent cases of Sylvatic Yellow Fever in Brazil (23 confirmed cases, 13 of whom died, from December 2007 to January 2008) have also shown the risk of exposure to harmful pathogens in wild environments (World Health Organization 2008). In all cases, the victims became sick after a stay in forests where they were bitten by infected mosquitoes ('jungle' yellow fever is a disease of monkeys, spread with mosquitoes). The number of yellow fever epidemics has risen during the last 20 years, affecting more countries in Western and Central Africa and South America (World Health Organization 2001). Yellow fever is enzootic in the Amazonian forest and

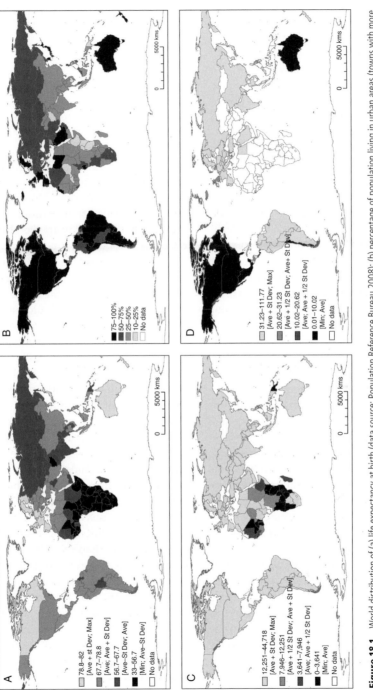

Figure 18.1 World distribution of (a) life expectancy at birth (data source: Population Reference Bureau 2008); (b) percentage of population living in urban areas (towns with more than 2,000 inhabitants; (data source: Population Reference Bureau 2008); (c) incidence of malaria in 2004 for 100,000 inhabitants (data source: GIDEON database 2009); (d) incidence of H1N1 on June 11th 2009 for 1,000,000 inhabitants (data source: WHO 2009).

Table 18.1 Summary of statistically significant correlations between the rate of forest cover changes and demographic indicators or disease incidences (correlation coefficient with $p<0.05$).

	Rate of forest cover change between	
	1990 and 2000	2000 and 2005
Demographic indicators (Source: Population Reference Bureau 2008)		
Rate of natural increase	−0.35	−0.35
Percentage of population living in urban areas	0.33	0.27
Life expectancy at birth	0.35	0.29
Diseases prevalences		
Malaria incidence in 2004 (GIDEON database)	−0.28	−0.24
Dengue incidence in 2007 (GIDEON database)	−0.35	−0.30

human cases are regularly reported in the neighbouring countries (Brazil, French Guiana, Suriname, Guyana, Venezuela, Colombia, Ecuador, Peru, Bolivia and even in the South, in Paraguay and Argentina). Infections have occurred following migrations of population in enzootic areas, illustrating the increasing proximity between humans and mosquitoes.

Forest-related activities, including logging, mining, hunting, and increasingly recreational activities expose people to unexpected pathogens for which they are not immunized. These activities are also connected, while logging or mining requires the construction of roads this later favours hunting and other recreational practices. These have been associated with cases of yellow fever, but also malaria and leishmaniasis (Chivian 2003; Patz *et al*. 2004). In these cases, humans are 'active' for transmission as they make contact with pathogens by entering wild biotopes. Therefore, the profiles of human cases correspond to people involved with activities in forests.

Further infections can occur when the circulation of the pathogens moves from wild to anthropized environments, typically brought out of forests by an animal vector. Humans are relatively passive in the transmission as they can be affected in close proximity to their living place. Consequently, people from any age or sex are likely to be affected. The development of livestock and the extent of grazing areas close to forests have been shown to attract pathogens out of the forest and cause epidemics. This was first illustrated in 1957 with the

identification of the Kyasanur virus (Flavivirus) from a forested region in the Shimoga district of Karnataka, south-west India (Pattnaik 2006). Since then, the Kyasanur forest disease has been increasingly but exclusively reported from this region, probably as a result of the population growth and development of cattle (Gould and Solomon 2008). The Kyasanur virus is transmitted by ticks (*Haemaphysalis* spp.), involves mammals as reservoirs, and can accidentally infect humans. Deforestation and the introduction of humans and cattle in proximity to wild habitats have helped the transmission of Kyasanur virus. Ticks are brought out of the forests by wild mammals and then feed on cattle or humans in these fragmented habitats that are now composed of dense forest alternating with agricultural lands.

Since 2000, increasing cases have been detected from Karnataka (Pattnaik 2006), and serological surveys have revealed the existence of another major silent focus, in the Nicobar and Andaman islands in India (Padbidri *et al*. 2002; Pattnaik 2006). Genetically close viruses causing similar tick-borne viral hemorrhagic fevers have been described in other countries. The Nanjianyin virus, isolated in Yunnan (China) during human serological investigations in 1989, is now described as a variant of the Kyasanur virus (Wang *et al*. 2009). Another variant is the Alkhurma virus that was first reported in 1992 in the Arabian Peninsula (Charrel *et al*. 2005) and has caused increasing human cases (Pattnaik 2006; Gould and Solomon, 2008). These variants have emerged in foci presenting very different biotopes than those of Karnataka, higher in latitude in China with a colder climate, and under a dry and hot climate in the Arabian Peninsula.

Another example of the transmission of pathogens following the expansion of human settlements on forested areas is the spread of the Lyme disease. This infection is the most common tick-borne disease in North America and Europe, increasingly affecting the United States, with a very high prevalence (9.1 cases per 100,000 persons in 2007) (Centers for Disease Control and Prevention 2009). Lyme disease, caused by bacteria from the genus *Borrelia*, is transmitted to humans through the bites of ticks (mostly *Ixodes scapularis* in the United States). Tick larvae can get infected while feeding on forest rodents that are the main reservoir of the disease. Adult ticks take the final bloodmeal on larger vertebrates, such as deer in the United States. The circulation of these rodents and deer at the border or out of their natural habitat contributes to the spread of ticks close to humans that can then be accidentally bitten.

Wild animals and ticks have also been responsible for the recent transmission of the Crimean-Congo hemorrhagic fever (CCHF) virus to domestic and human populations in Turkey. CCHF has emerged since 2002, with seasonal epidemics (Yilmaz *et al.* 2009). Here again, increasing pullulations of wild animals and a closer proximity between domestic animals and humans may explain the emergence.

Beyond the global driving force induced by the increase of the human population on earth, several global changes (long-term) affecting ecosystems have been observed and are suspected to have negative effects on human health.

18.3.2 Global changes, alteration of ecosystems, and expanding sources of infection

Deforestation, habitat fragmentation: consequences on the environmental niches of animal vectors and contacts with humans
The consequences for human health of the damage caused by anthropogenic actions on ecosystems are difficult to assess. However, at a worldwide scale, we can already observe some links between these ecological changes and some health patterns.

For instance, there is a significant correlation between the incidence of malaria in 2004 (source: GIDEON database, http://www.GIDEONonline.com) and the annual rate of forest cover change between 1990 and 2000 (−0.28 with $p<0.05$, for 147 countries; Table 18.1) and between 2000 and 2005 (−0.24 with $p<0.05$), using the forestry statistics from the Global Forest Resources Assessment 2005 (FAO 2006). The countries that have the highest rate of forest loss during these two periods are also the countries where the incidence of malaria was highest in 2004. This relationship should be interpreted carefully with regards to the complexity of ecological and societal factors involved in the transmission of malaria. Malaria occurs in equatorial and tropical countries where forests are denser, and we might have expected a geographical link between forest extent and the incidence of malaria. Nevertheless there is no statistically significant correlation between the surface of forested areas or the annual changes in surface and the incidence of malaria. The role of deforestation in the emergence of malaria has been explained by the consecutive increase of ecological niches for the *Anopheles* mosquito vectors of malaria, and higher numbers of humans getting into these uninhabited areas for contact (Patz *et al.* 2004). Changes in ecological niches have been interpreted in different ways. First, it could be the increase of water reservoirs in space and time, when the soil is compacted and used for agriculture or constructions, that allows mosquitoes to breed. Also, in the Amazon, deforestation has decreased the acidity of soils and water reservoirs for *Anopheles darlingi*, the main vector of malaria, which has consequently proliferated (Chivian 2003; Vittor *et al.* 2006). In West Africa, the loss of savannah and riverine forests has contributed to expand the *Anopheles* population that preferentially breeds in arid areas (Chivian 2003).

At a worldwide scale, the map showing the incidence of malaria in 2004 (Fig. 18.1c) emphasizes the very high rates of malaria in African tropical countries. These countries are also characterized by important deforestation. However, such important forest changes are observed in other countries, like South America and South-east Asia, without such high malaria incidence. Demographic variables (such as natural increase, the percentage of population living in urban areas, or life expectancy at birth) help us to understand the specificities of these African countries with high rates of malaria: their population is mostly rural, increases rapidly, and still has a low life expectancy (these three variables correlate significantly with the malaria incidence; Table 18.1), and that reflects the lack of healthcare and difficulties in living conditions. A multiple regression between the incidence of malaria in 2004 (as a dependent variable) and these three demographic variables and the annual rate of forest extent changes between 2000 and 2005 (pairwise deletion, i.e. a minimum of 150 pairs) shows that life expectancy is the best predictor. This example illustrates how a geographical approach can complete an environmental study to explain health patterns with further social variables.

Deforestation usually occurs within small patches, creating either isolated clearings in large forests or isolated small forests surrounded by agricultural fields. This is described as forest fragmentation, which leads to an increase in the border contact between humans and vector-dense areas and consequently the risk of transmission of vector-borne diseases. Wild animals living in the remaining forests have their ecological niche altered and experience changes in their density with the possible extinction or invasion of species.

Biodiversity changes: consequences on the diversity and density of vectors and pathogens

Cutting down trees radically modifies every component of the ecosystem starting with microclimate variables (local temperature, air humidity), then water and soil reserves, soil composition, soil humidity, and later the biodiversity and structure of plant and animal communities. The anthropization of the ecosystems damages habitats and affects plants and animals species, which can finally disappear. Species diversity generally decreases from rural to peri-urban to then urban ecosystems, as was calculated for mosquitoes in the Peru (Johnson *et al.* 2008). The inexorable loss of biodiversity may have different effects on the pathogenicity of infectious vector-borne agents. An intuitive understanding could be that a decrease in the diversity of animals leads to a decrease in the diversity of pathogens and consequently a lower risk of infection.

This idea may be illustrated by the greater threat of infectious diseases in tropical areas compared with regions in higher latitudes that present a general lower biodiversity (Chivian 2003). However, the diversity of animal hosts has been proved to have a diluting effect on the pathogenicity of some infectious agents. In the case of Lyme disease, an increase in the diversity of hosts has the effect of lowering the nymphal infection prevalence of ticks (LoGiudice *et al.* 2003). In epidemic areas, human encroachment, leading to a decrease in biodiversity, has contributed to higher densities of rodent population and the concentration of the bacteria responsible for Lyme disease in a few host species. The dilution effect applies to diseases transmitted by vectors (typically mosquitoes or ticks) that can feed and get infected on different hosts.

This dilution effect also concerns the density of populations and depends on the size of habitats. A local absence of deer may increase tick feeding on rodents and create a higher risk of Lyme disease transmission (Perkins *et al.* 2006). Therefore, pushing deer away from villages may have adverse effects on the risk of transmission.

From a geographical point of view, these observations can be comprehended at different scales: 1—on a global scale, a greater diversity of hosts and vectors in the tropics could be associated to a higher risk of infection; 2—on a regional scale, a decrease of biodiversity has a concentration effect on pathogens in fewer hosts for a higher pathogenicity; and 3—on a local scale, the absence of a host may also increase the prevalence in alternative hosts and induces the emergence of tick-borne or other vector-borne disease hotspots. Further observations could be proposed on the diversity of parasites and pathogens themselves regarding biodiversity and global changes.

The decrease of animal species diversity should be considered within the frame of the equilibrium between species, the species competing for food, shelter, or some species being predators of others. The extinction of one species is a break in the ecological chain that will result in major changes in the diversity and densities of remaining species. Animal species are moved and balanced following changes in their biotopes. A radical action, such as cutting down forests, makes associated wild animal species disappear, and offers a new empty niche to opportunistic animals. A few species may rapidly invade this space, as can be observed with rodent invasions in agricultural fields gained on forests. These species may also bring pathogens and cause new disease outbreaks. The lower diversity of species in peri-urban and urban environments also explains some invasions of species that cause high rates of pathogen transmission in urban areas, as has been described for *Culex* species in the United States (Kutz *et al.* 2003, Johnson *et al.* 2008).

Agricultural practices and resource management: consequences for pathogen dynamics

One of the major impacts of agriculture on the environment is the hydrological change brought about by the use of water to grow plants. Irrigation and drainage require the construction of canals, reservoirs, or dams that artificially maintain water during long periods within these infrastructures as well as in the irrigated fields. Watering systems also contribute to a smaller extent to the presence of puddles on compacted soils.

Water can be direct vector of parasites and pathogens, rapidly contaminate different areas, and create large epidemics. Several diseases are associated with water, mainly from bacterial origin (leptospirosis, cholera, botulism, dysentery, legionellosis, typhoid fever, cryptosporidiosis, colibacillosis, i.e. infection with *Escherichia coli*, etc.) and parasitic origin (schistosomiasis also known as bilharzia, dracunculiasis, echinococcosis, amebiasis, enterobiasis, etc.) but also from viral origin (poliomyelitis, i.e. polio, hepatitis A and E, gastroenteritis, etc). All these diseases may be amplified as soon as agriculture favours conditions to maintain water. Water-related diseases can be classified in three groups. First, water-borne diseases result from the consumption of water contaminated by human or animal ejections with bacteria, parasites, or viruses (in the case of cholera, hepatitis A and E). Second, water-based diseases are caused by parasites or pathogens having part of their life-cycle in water and another part in an intermediate animal vector, such as molluscs or snails (in the case of schistosomiasis and dracunculiasis) and mammals (case of leptospirosis). Third, vector-borne diseases can be linked with water without any direct connection,

but using vectors breeding or living in or near water, in particularly mosquitoes, midges, and flies. Mosquitoes, which represent the most extensive family of vector insects, can transmit malaria, filariasis, dengue fever, yellow fever, Japanese encephalitis, West Nile virus, Rift Valley fever, and chikungunya fever. For vector-borne diseases, hydrological changes affect the ecology of the vectors. On the one hand, some animals, like mammals, can respond radically to the presence of water, for example deciding whether they are attracted or not and migrate or not. On the other hand, mosquitoes or snails may find more places for breeding and their population may increase in space and density consecutively. Indeed malaria, which is the most important parasitic infectious disease in terms of prevalence, and schistosomiasis, which is the second most important parasitic disease, are particularly associated with hydrological conditions. Malaria is caused by a protozoan parasite of the genus *Plasmodium* transmitted by mosquitoes that breed in fresh or occasionally brackish water. Schistosomiasis is a chronic parasitic disease that uses freshwater snails as an intermediate host and humans as a definitive host. Leonardo *et al.* (2005) used a GIS and the analysis of satellite images with remote sensing to demonstrate the importance of the proximity to water to the incidence of malaria and schistosomiasis in the Philippines. Their study is an example of the potential of these technologies in health geography to assess prevalences and help decision-making on intervention and treatment.

Beyond the consequences of individual agricultural practices on human health, water management projects have had serious effects on ecosystems and as a result on human health. They have developed throughout the world as an answer to the increasing population and need for food during the twentieth century. They allow farmers to produce twice or three times more than the non-irrigated fields that are limited to a single harvest per year in tropical countries. At the end of the twentieth century, there were over 45,000 large dams (impoundments over 15 m high or storing at least 3 million m^3 of water) in 140 countries and over 800,000 small dams (World Commission on Dams 2000; Keiser *et al.* 2005). Since the 1980s, irrigated fields have accounted for more than half of the increase in food production (World Commission on Dams 2000). Nevertheless, these water management projects have also had considerable consequences for ecosystems, by modifying the natural flow of streams and rivers, by drying fields with drainage, or maintaining water with artificial flooding (especially for rice production). They have had direct adverse consequences on human health, by causing several noticeable epidemics worldwide. Schistosomiasis has been the major threat following the construction of dams. It has emerged since the building of the first large dams, Aswan High Dam in Egypt, Kariba Dam in Zambia, and Mozambique and Akosombo Dam in Ghana, all operating in the 1960s (Malek 1975; World Commission on Dams 2000). Rift Valley Fever, a viral zoonosis transmitted by mosquitoes (typically *Aedes* or *Culex* genera), has also spread after the building of the Aswan and Kariba dams (World Commission on Dams 2000). However, in a comprehensive study of the effects of irrigation and large dams on the incidence of malaria, Keiser *et al.* (2005) noted that research assessing health impacts must be used with care, as surveys are generally conducted after the construction of the dams and they deal with confounding factors. This study listed from the literature the surveys of malaria in relation to dam construction, but failed to relevantly quantify a global health impact due to the heterogeneity of data. It also noted that the impact of small dams on malaria is underestimated because of the lack of studies, even though their total shoreline is greater when compared with large dams.

Agricultural practices create local hydrological conditions that disrupt the continuum of the biotope in space and time. The alternation of different crops and the introduction of new species are also radical ecological changes in a biotope, since their production requires different use of water, fertilizer, or pesticides. The spread of maize production throughout the world has caused such changes. Maize has been introduced in tropical countries alternately with rice, and does not require the flooding of paddies during its growth. The use of water to grow crops (such as flooding rice fields, or watering crops) and the alternation of different crop species, condition not only the presence and absence of animal species but also their ecology. Animals living in agricultural

fields may adapt their reproduction to the presence of food and possibilities of shelter. Rodents are one such very reactive animal, moving to different fields according to the stage of crops and increasing in density during harvest. Cavia *et al.* (2005) showed the effect of corn and wheat harvest on the *Akodon azarae* population in Argentina and their ability to maximize fitness in periodically disturbed habitats. By conditioning the structure of animal populations, these agricultural practices also indirectly determine the transmission of vector-borne diseases.

Over-exploitation of natural resources for agriculture has already shown dramatic effects beyond the loss of vegetal cover and biodiversity. Agricultural fields are exposed to wind and water erosion, which is increased by agricultural practices. Irrigation also accelerates soil salination leading to unfertile lands. The loss of soil and increasing desertification (in low-rainfall areas) forces farmers to move and find more fertile agricultural spaces: this issue has also emerged as a major challenge with regard to the lack of agricultural space, famine, poverty, and health. The United Nations Convention to Combat Desertification (UNCCD) was adopted in 1994, and has been signed by 191 countries (as at September 2005). It estimates that 250 million people are affected by desertification, and about one billion people are at risk in over one hundred countries, which are also among the world's poorest.

The conversion of forests to agricultural lands has also led to dramatic soil and water pollution. A spectacular example is the pollution of the Amazon with mercury, formally found in rainforest soils, and increasingly accumulating in fishes (Passos and Mergler 2008; Patz *et al.* 2004). Agricultural practices, and especially intensive farming, also generate pollution by nitrates, pesticides, and heavy metals (lead, mercury, etc.), that cause poisoning in humans. However, health systems still fail to detect the extent of the effect of these pollutions on human health, due to the technical difficulty and high cost of investigation for such poisoning in bodies. Pesticides may cause about 10,000 deaths out of 2 million poisonings each year, according to the United Nations, and three quarters occur in developing countries with inappropriate protective measures (Quijano *et al.* 1993;

Horrigan *et al.* 2002). Pesticides also have long-term effects through increasing the risk of cancer and disrupting the immune, nervous, reproductive, and endocrine systems (Horrigan *et al.* 2002). Moreover, growth enhancers used in factory farms are found in manure and pollute ground and surface water. They may cause breast and testicular cancers in humans (Horrigan *et al.* 2002; Soto *et al.* 2004). The use of antibiotics for animal growth may also be responsible for increasing antibiotic resistance in humans and jeopardize the effectiveness of similar antibiotics in human medicine (Horrigan *et al.* 2002). Here again, the effects on humans are difficult to assess but may be dramatic if confirmed when we consider the increasing use in developing countries. Another means of pollution from intensive agriculture is air transmission, which has local effects. Substances may be released in the air from the confinement buildings or while spreading manure (Horrigan *et al.* 2002).

Some indirect impacts of agriculture on human health are linked to changes in livestock management. The intensification of livestock farming with animal husbandry has been pointed to as contributing to the development of human diseases (Chivian 2003). Animals are raised in confinement at high stocking density, to reduce space and costs, and become incubators of pathogens and parasites that can be harmful not only to their health but also to other animals and humans.

Disruption of the ecological chains in livestock management has raised serious concerns with the emergence of the bovine spongiform encephalopathy (also known as mad cow disease), first detected in 1986 in the United Kingdom (Richt and Hall 2008). Hypotheses about its origin include bad farming practices, with the administration of animal contaminated proteins in meal. Richt and Hall (2008) have recently shown that an atypical form of bovine spongiform encephalopathy had the same type of prion protein gene mutation as found in a human patient affected by the Creutzfeldt-Jakob disease, which potentially explains cattle to human transmission when eating contaminated meat.

Illnesses caused by food consumption are preponderant in public health, with more than 200 diseases described, caused by viruses, bacteria, parasites, toxins, metals, and prions (Mead *et al.*

1999). Common food-borne infections are caused by the bacteria *Campylobacter* and *Salmonella* and with a lower extent but higher mortality *Escherichia coli* and *Listeria* (Horrigan *et al.* 2002).

Transmission of pathogens from livestock to humans can occur directly, being favoured by unhealthy working conditions in farms and slaughters. One example is *Streptococcus suis* serotype 2, an enzootic bacteria affecting pigs and present in countries with extensive pig farming (Tang *et al.* 2006). Infections in humans can be associated with meningitis (about 80 per cent of all cases), septicaemia, endocarditis, and deafness (Sriskandan and Slater 2006; Tang *et al.* 2006). Human cases have been increasingly reported from Asia, with an important outbreak in 2005 in China, which caused 38 deaths out of 204 human cases (Tang *et al.* 2006). This emergence follows growth in livestock production, and probably better surveillance of infectious diseases after the H5N1 epidemics.

Livestock can act as reservoir of parasites and pathogens, which can be amplified with the density of animals. Livestock can serve as intermediate host in a complex life-cycle and present a great risk for human health because of their proximity to villages or cities. This is illustrated by Japanese encephalitis epidemics, whose responsible virus is transmitted by mosquitoes (belonging to the *Culex tritaeniorhynchus* and *Culex vishnui* groups) and also circulates in water birds and pigs (Gould and Solomon 2008). The virus reproduces and is amplified in pigs, through which mosquitoes get infected. Japanese encephalitis is widespread over South-east Asia and Australasia, where it has expanded with the development of agriculture. The annual human incidence is very high but fluctuating (between 30,000 and 50,000 cases), with about 10,000–15,000 fatal cases [C1] (Erlanger *et al.* 2009). The proximity of pig farms with irrigated rice fields creates favourable conditions for the disease emergence. The rapid population increase in South and South-east Asia, associated with the development of irrigated rice farming and pig rearing, explains the emergence and spread of Japanese encephalitis in these regions.

Other viruses have benefited from the concentration of animals in husbandry, and mutate before spilling over to other animal vectors or humans.

A major example for animal and human health is the case of influenza virus, with the frequent emergence of new strains. Intensive poultry rearing in Asia has been shown to be a major incubator of new strains of viruses causing several outbreaks (Mayer 2000). Avian influenza viruses infecting poultry and a great variety of birds are divided into two groups: subtypes H5 and H7 potentially causing the highly pathogenic avian influenza (HPAI) and the other subtypes causing the low-pathogenic avian influenza (LPAI) (Alexander 2007). Avian influenza is an ancient disease that has caused epidemics through history with a catastrophic pandemic in 1918 with the Spanish flu (Webster *et al.* 2006). However, infections have been expanding worldwide in poultry since the 1990s with the highly pathogenic H5N1 strain that has affected over 433 persons, mostly in Asia, and has killed 262 (60 per cent) from 2003 to 2009 (WHO 2009). Alexander (2007) suggests the increasing densities in poultry production, the possible changes in wild bird movements, but also higher surveillance and diagnostic capabilities, as explanations for such emergences.

Urbanization: the consequences of inadequate planning on health situation

The rapid population increase and global migration from rural to urban centres (with over one half of the total population living in urban areas, see Subsection 18.3.1) raises major health issues (Sutherst 2004). These changes have mostly occurred in developing countries, and the health consequences can be particularly dramatic in the poorest countries as the urbanization is usually not controlled and not planned. Possible consequences relevant to public health, including inadequate housing, deteriorated water, sewage, and waste management, offer favourable conditions for the emergence of water-borne, mosquito-borne, and rodent-borne diseases (Gubler 1998).

Rapid urbanization poses sanitary problems when not accompanied by infrastructures to ensure environmental health. In tropical and equatorial countries with heavy rainfall, inadequate drainage of water along roads and pavements favours the transmission of water-borne diseases. These diseases rapidly spread with insufficient sanitation and poor water treatment systems when pollution on the surface water percolates into drinking water wells. Together with inadequate waste collection and treatment it also offers habitats for different animal vectors, such as mosquito larvae that develop in standing water or used tires and containers, before the adults spread and potentially transmit dengue fever, chikungunya, West Nile fever, yellow fever, and lymphatic filariasis (Gubler 1998; Sutherst 2004; Estallo *et al.* 2008), other arthropods, such as the ticks that are potential vectors of Lyme disease (LoGiudice *et al.* 2003), small mammals, and especially rodents that look for a refuge and food in human garbage, and that are potential hosts of fleas and reservoirs of plague, leptospirosis and hantaviral, and arenaviral infections (Jittapalapong *et al.* 2009).

As soon as animal vectors and reservoirs find a suitable habitat (including shelter, a place to reproduce and protect themselves from predators and food), their population may increase rapidly in space and density, and may constitute a threat to human health. The management of waste water and garbage is a determinant of health in urban areas.

Dengue fever is the most important arboviral disease with about 2.5 billion people at risk in over 100 countries worldwide (Guha-Sapir and Schimmer 2005). Dengue fever, which has essentially caused epidemics in urban areas, is increasingly reported from peri-urban and rural areas, following a spread of its vectors (*Aedes* mosquitoes) in association with environmental changes (Chareonsook *et al.* 1999; Guha-Sapir and Schimmer 2005; Ellis and Wilcox 2009). Chareonsook *et al.* (1999) stated that Thailand, regularly affected by dengue epidemics, recorded higher dengue incidence in rural (102.2 per 100,000) than urban areas (95.4 per 100,000) in 1997. Indeed, we found a statistically significant negative correlation between dengue incidence per country in 2007 and the variable measuring the rate of forest changes, (−0.35 with forest cover change between 1990 and 2000, and -0.30 with forest cover change between 2000 and 2005; Table 18.1). This reflects again the link between population pressure, the consequences for ecosystems, and the global negative impact on human health.

Global climate changes

Evidence of a global climate change has only risen recently in the public consciousness and its irreversibility arouses great concern when considering the possible negative impacts on health. The Intergovernmental Panel on Climate Change (2007) has recognized the responsibility of human activities in most climate change, with an increase in the atmospheric concentration of greenhouse gases and aerosols and land surface changes. Causes include deforestation (reduction of the absorption of carbon dioxide and emission of gases with fires), urbanization (increase of the albedo), and agricultural practices (such as irrigation modifying the hydrological cycles, or livestock production causing large emissions of methane and nitrous oxides) (Chivian 2003; IPCC 2007; Bates *et al.* 2008). The increase of troposphere carbon dioxide (CO_2) from 280 parts per million by volume (ppmbv) 420,000 years ago to 370 ppmbv today has largely contributed to the global increase of temperatures, referred to as global warming (Chivian 2003). Some global trends in climate change are observed: oceans absorb almost all the energy from global warming; temperatures are increasing faster at higher latitudes; heat waves increase in intensity and extent with more elevated night temperatures; fresh water resources decrease; and weather extremes (especially responsible for droughts and floods) increase in intensity, higher winds accelerate and make the climate more unstable (Relman *et al.* 2008). Solomon *et al.* (2009) outlined that the CO_2 emissions responsible for climate change cause a long term increase of temperatures that could not be significantly reduced 1,000 years after the emissions stopped. Even if drastic actions are taken to reduce the human impact on CO_2 emissions, the patterns of temperatures and rainfall will evolve in a negative way for earth health and human health. Several researchers have tried to assess the consequences of climate changes on the incidence of diseases.

First, the role played by oceans as heat sinks may have direct consequences on the diseases linked to ocean temperatures. Lobitz *et al.* (2000) showed the relationship between cholera epidemics and climatic variations in Bangladesh by using remote sensing techniques. Cholera, caused by the bacterium *Vibrio cholerae*, frequently affects Bangladesh

and India and occasionally some other developing countries. It attaches preferentially to zooplankton, and especially copepods, to spread (Lobitz *et al.* 2000). The authors compared the 1992–1995 cholera epidemics to measurements of the Sea Surface Temperatures (SST) that present a similar yearly cycle. A global increase of the SST may lead to a higher risk of cholera transmission in this area.

Second, global warming, with rising temperatures and more variable precipitations, may have diverse consequences on human health. Heat waves may directly increase human mortality, as happened in Europe in 2003 when over 30,000 people died (Kosatsky 2005). Variation in rainfall may affect the supply of fresh water and the production of agricultural food that could be critical for the poorest populations. Increased atmospheric and surface temperatures and their impact on hydrological cycles necessarily disrupt the ecology of infectious diseases. They modify the distribution of the animal vectors and reservoirs with an expansion of their range, especially to higher elevations and higher temperatures (Relman *et al.* 2008). Indeed, climatic variables such as temperature, humidity, and rainfall significantly influence the development and survival of mosquitoes (Estallo *et al.* 2008). Mosquito-borne diseases that nowadays affect tropical areas (such as malaria or dengue fever) may consequently expand to higher latitudes. This was illustrated by the chikungunya epidemics transmitted by *Aedes albopictus* in northern Italy after an accidental introduction by traveller coming from India (Beltrame *et al.* 2007). The chikungunya virus first emerged in Africa before outbreaks were recorded in Central Asia, South-east Asia, and other parts of the World (Chevillon *et al.* 2008). It is transmitted by two major species of mosquitoes, *Aedes albopictus* and *Aedes aegypti*, which are also vectors of dengue virus (Chevillon *et al.* 2008). *Aedes albopictus* was first recorded in 1990 in Italy and seems to have quickly spread across the country since then (Beltrame *et al.* 2007). This expansion has already shown health consequences and highlights the threat of having these two species spread, with regards to chikungunya and dengue fever, but also yellow fever and Rift Valley fever for *Aedes aegypti* (Gould and Higgs 2009).

Third, extreme weather events linked to the El Niño Southern Oscillation (ENSO) have been associated with incidences of malaria, dengue, and Rift Valley fever (Chivian 2003). ENSO generates climate variability with warm (El Niño) and cool (La Niña) phases of surface water temperatures that cause heat waves and droughts in Africa and Asia, as well as heavy rain and floods in South America (Chivian 2003). These events occur irregularly. Different studies have shown an increase in the incidence of dengue fever with ENSO (Hales *et al.* 1996; Keating 2001). However, Guha-Sapir and Schimmer (2005) interestingly have doubts about relating the incidence of dengue epidemics with climatic changes, as regards the complexity of relations and the scarcity of data and models that have been used to make such correlations.

Climate changes were also proved to affect the distribution and the density of rodent populations. Engelthaler *et al.* (1999) relate that several hantaviral epidemics (Hantavirus Pulmonary Syndrome (HPS) outbreak in the Four Corners region in 1993, HPS outbreak in western Paraguay in 1995 and 1996) were associated with an increase in rodent populations following increasing rainfall and El Niño events. They argue that higher precipitations contributed to abundant food resources and a consequent increase in the density of rodents.

There is no linear effect between climate change and disease patterns, since climate impacts on different factors involved in the transmission of diseases. Predicting future climatic conditions and assessing their consequences to human health is also challenging with regard to the complexity of these relations and the difficulty in measuring the consequences of human activities on global climate changes. This may explain why past scenarios are regularly and radically revisited.

18.3.3 Mobility, species introduction, and environmental contaminations of pathogens

International trade, professional travel, and tourism are globally increasing following economic development and the decreasing time needed to reach remote locations. These movements have contributed to the spread of pathogens and parasites with eventually their animal vectors moved with people or products. Mobility poses a complex challenge in

the assessment of health patterns since pathogens can be spread around the world within a few days or hours, and can rapidly infect people far from the ecological area where they have been maintained and have evolved for years. Historically, the importance of transport in the emergence of epidemics has been well documented. One of the deadliest pandemics, the Black Plague also known as the Black Death, probably originated in China in the 1330s before progressing to central Asia and Europe along caravan and shipping routes (Morens *et al.* 2008). It killed about 34 million humans in Europe and 16 million humans in Asia and was scarcely confined, owing to lack of knowledge about transmission and medical care (Morens *et al.* 2008). However, a precursory quarantine (40 days) was applied to ships on arrival at port to limit the spread (Morens *et al.* 2008).

Regarding health, the question of mobility is first related to the voluntary, but most often accidental, spread of invasive alien species, which is now recognized as one of the greatest threat to ecology, economy, and health, with irreversible consequences (Matthews 2004). Species are carried with the growing transport of goods and people worldwide, and a few have become invasive. Invasive species belong to all major taxonomic groups, including viruses, fungi, algae, mosses, ferns, higher plants, invertebrates, fish, amphibians, reptiles, birds, and mammals (Matthews 2004). Some examples of alien species introduction to an ecosystem are spectacular. Among the worst, the Nile perch (*Lates niloticus*) introduced in 1954 to Lake Victoria has contributed to the extinction of more than 200 endemic fish species (Lowe *et al.* 2001). Water hyacinth (*Eichhornia crassipes*) has spread to over 50 countries, invading lakes and rivers, and with a very fast doubling time (about 12 days) has damaged fresh water ecosystems by limiting sunlight and oxygen (Lowe *et al.* 2001). Rodents are also worthy examples of invasive species. Rodents, which represent 40 per cent of mammalian species, are present worldwide on all continents other than Antarctica and inhabit most ecosystems (Wilson and Reeder 2005; Carleton 1984). They can breed rapidly, eat a large variety of food, and can adapt to fast environmental changes (Carleton 1984). When introduced into a new ecosystem, it is hardly surprising to see their population increasing rapidly and spreading before reaching a new equilibrium in the biotope.

Introduced and invasive species also bring into the ecosystem their parasites and pathogens, and these may possibly cause epidemics in humans. This was oddly illustrated with the occurrence of malaria in the neighbourhood of several international airports in Geneva, New York, and Paris, brought to these non-endemic areas with infected mosquitoes (Mayer 2000). Also, the long history of trading in the Middle East may explain the presence of two variants of flavivirus, the Kyasanur virus in India and the Alkhurma virus in the Arabian Peninsula. Exchanges of sheep or camels have probably favoured the transmission of infected ticks (Gould and Solomon 2008). So far, only a few cases of Alkhurma haemorrhagic fever have been reported in only two provinces of Saudi Arabia, Makkah and Najran, since its first isolation in 1994. Nevertheless, at the beginning of 2009, four cases were reported from the Najran province, raising awareness of this disease which has shown high lethality (25 to 30 per cent). As the ecology of the Alkhurma virus, also transmitted by ticks, remains unknown, comparisons with the other variants within the family of Flaviridae is informative, with regards to the risk of outbreak and preventative actions that could be undertaken. Further studies are needed to confirm the hypothesis of an imported infection through ticks carried on animals exchanged in the region.

Invasive species also impact vector-borne diseases through regulating the biodiversity of indigenous species and their pathogens. In addition, if transported pathogens find suitable living conditions in the place where they are introduced, they can maintain themselves, infect other animal vectors and may cause epidemics.

Rapid mobility can also allow a direct (human to human) transmission of pathogens: people travelling serving as vectors of diseases (Mayer 2000). This was recently highlighted by the worldwide diffusion of severe acute respiratory syndrome (SARS) that first appeared in southern China in November 2002 (Matthews 2004). The respiratory disease, caused by a coronavirus, rapidly spread to Asia, Europe, North and South America, causing more than 8,000 cases and over 774 deaths (WHO,

data as of April 2009, www.who.int). The illness is believed to have originated in animals traded in the region's markets (Field 2009). The virus, which originated from factory farms and animal trade, was transmitted through the air by travellers remaining infective during and after their trip. Another spectacular example is the H1N1 pandemic that started in April 2009 and rapidly spread from Mexico to countries directly connected, especially the United States, Canada, Chile and Australia. Figure 18.1d, displaying the incidence of H1N1 (for 1 million inhabitants) on 11 June 11 2009 (the day when the WHO declared phase 6 of the pandemic, after a rapid increase in incidence and worldwide distribution), shows these countries as severely affected, while the poorest countries, in Africa, remain economically isolated and unaffected. A comparison with the map showing the level of urbanization (Fig. 18.1b) also illustrates the higher transmission linked to higher population densities. Nevertheless, we note that unaffected countries may also have insufficient resources to detect the virus in the population.

Environmental contamination of infectious disease agents is a major health risk. A potential threat for human health with regards to yellow fever is the occurrence of an urban cycle if the flavivirus is imported with mosquitoes (*Haemagogus* and *Sabethes* spp.) from wild then rural areas and transmitted to the urban vector *Aedes aegypti*. The high densities of populations and *Aedes* mosquitoes in cities could then favour large epidemics and would have dramatic consequences if we consider the high lethality (20 to 50 per cent).

This problem of mobility and the threat of the potential introduction of pathogens, parasites, and/or their vectors has been now realized by most of countries following recent epidemics and pandemics (SARS, H5N1 avian influenza, and H1N1 swine flu), which may now be better prepared to curb transmission.

18.3.4 Inequalities in health offer and access to healthcare

Access to healthcare and the quality of the health systems are also determinants of the health status of populations. On one side, social inequalities result in different vulnerabilities to diseases, exacerbated by inadequate living conditions, inadequate water and waste management, hard working conditions, poverty, and the limited recourse to health services. Populations may not have sufficient resources to access healthcare (when social security is not provided, or when transport is needed to reach suitable structures: for example, specialists doctors who are unevenly distributed on the territories). Van Donk (2006) assigned poverty, unemployment, lack of secure income, and income inequality to vulnerability to HIV infection. This vulnerability is increased with difficulties in accessing healthcare and affordable prevention (Van Donk 2006). Furthermore, socio-economic disturbances such as wars, economic crises, and unemployment worsen inequalities. During the war in Nicaragua between 1983 and 1987, malaria epidemics increased in war zones where populations migrated and could not benefit from an adequate health system, while they decreased in non-war zones (Garfield 1989; Sutherst 2004).

On the other hand, health depends on the resources invested by the society in response to global change (Sutherst 2004). For instance, rapid urbanization poses sanitary problems when not accompanied by the construction of public health infrastructures with the difficulty of responding to heterogeneous needs in space and time. The use of health structures differs from rural to urban areas (Guha-Sapir and Schimmer 2005), and with regard to social status.

This problem of allocation of health resources has also been exemplified by water management projects. The construction of large infrastructures implies the resettlement of downstream communities and the development of health services and structures that condition health status. Smaller water management projects, such as drainage and irrigation, may also be accompanied by local migrations of populations. The World Commission on Dams (2000) reported different cases of such resettlements that have caused food shortage and famine in Africa and Asia. In Zimbabwe, the construction of the Kariba Dam between 1955 and 59 constrained about 57,000 people to move to lower-fertility areas and was responsible for malnutrition, as well as schistosomiasis epidemics (2000).

Similar emergences of schistosomiasis, malaria, or adverse effects on public health through resettlement have followed different dam constructions over the world. An increase in poverty may also be accompanied by the emergence of non-infectious diseases. The building of the Akosombo Dam in Ghana in the 1960s and the failures of the resettlement programme prompted migrations, especially to Côte d'Ivoire, driven by poverty and increasing infections by HIV, leading to a disproportionate prevalence (Sauvé *et al.* 2002). The World Commission on Dams (2000) concluded that there was a general lack of health impact assessment in the design of dams and infrastructures, as was already reported by Malek (1975) regarding the Aswan High Dam. These examples illustrate again the multiple consequences of man-made changes on health and how ecological and geographical (i.e. social) factors can impact upon human health. A geographical approach to health is necessary to forecast health issues for resettled populations and plan action based on former cases.

Therefore, assessment of public health needs (where and how to allocate health facilities) constitutes a fundamental task in urban planning that is unfortunately often guided by political decisions, instead of a health geography approach that would consider social, economic, and health conditions with a spatial perspective.

18.4 Conclusions

Health ecology has played a great role in the study of diseases in linking emergences or pathologies with environmental changes that have particularly accelerated since the industrialization era. In a complementary approach, health geography has contributed to a spatial understanding of local and global dynamics by multiplying scales and broadening the analysis of causative factors. Illnesses are considered globally in the geographical environment beyond traditional approaches limited to individuals. Indeed, human health status is observed within a society undergoing rapid transformations and drastically impacting on an unbalanced environment. Understanding the causes and consequences of disease is of course a very difficult task when considering the complexity of both human-induced changes and disease patterns, but some general driving forces or local mechanisms have been described and tackled in this chapter.

The risk of an epidemic or pandemic, that could be the consequence of ecological changes as well as increasing mobility and exchanges, remains a major threat for populations and an essential concern for public health. SARS and the avian influenza epidemics have shown such spectacular worldwide spread that they have positively alerted health administrations in several countries to organize surveillance and inform rapidly on human cases and on the possible emergence of diseases. The avian influenza epidemics that have been recorded in South-east Asia from 2003 may have also demonstrated that hiding an outbreak (through fear of economic impact) could have the worst economic consequences when other countries or populations lose confidence in the local health administrations and public health situation. Since then, information on epidemics is increasingly reported in a shorter time and publicly made available in journals or instantly through the Internet, in breaking news, or newsletters. Rapid reporting has also benefited from technological developments in geographic information systems with the possible management of large datasets with remote but controlled access allowing the addition and consultation of records. In 2009, the H1N1 epidemic was exemplary in the transmission of information with near real-time records and a web-based mapping of the disease incidence per country.

Other priorities are the vaccination of populations exposed to pathogens and the awareness of the risk of epidemics and general recommendations in such situations. The three epidemics cited above (SARS, H5N1, and H1N1) may have changed the global perception of the real health threat of the emergence of a highly virulent strain of H5N1 directly transmitted to humans. As mentioned in the introduction, SARS and H5N1 created a kind of psychosis for Western populations, while H1N1 may have reassured populations that a pandemic can occur with a limited number of cases and deaths.

An interdisciplinary approach to health, based on ecology and geography, appears fundamental and urgent, when noting the considerable impact on health of human-induced local and global environmental changes and facing the health threat of rapid emergence of diseases worldwide. This is a challenge on a degraded planet, since the ecological interactions involved in health patterns are complex and still remain elusive for scientists.

References

Alexander, D.J. (2007). An overview of the epidemiology of avian influenza. *Vaccine*, **25**, 5637–44.

Bates, B.C., Kundzewicz, Z.W., Wu, S., and Palutikof, J.P. (eds). (2008). *Climate Change and Water. Technical Paper of the Intergovernmental Panel on Climate Change*. IPCC Secretariat, Geneva.

Beltrame, A., Angheben, A., Bisoffi, Z. *et al.* (2007). Imported chikungunya infection, Italy. *Emerging Infectious Diseases*, **13**, 1264–65.

Brody, H., Rip, M.R., Vinten-Johansen, P., Paneth, N., and Rachman, S. (2000). Map-making and myth-making in Broad Street: the London cholera epidemic, 1854. *Lancet*, **356**, 64–68.

Carleton, M.D. (1984). Introduction to rodents, In Anderson S. and Jones Jr. J.K. (eds). *Orders and Families of Recent Mammals of the World*, pp. 255–65. John Wiley and Sons, New York.

Cavia, R., Villafañe, I.E.G., Cittadino, E.A., Bilenca, D.N., and Miño, M.H. (2005). Effects of cereal harvest on abundance and spatial distribution of the rodent *Akodon azarae* in central Argentina. *Agriculture, Ecosystems & Environment*, **107**, 95–99.

Centers for Disease Control and Prevention. (2009). *Lyme Disease*. [Online] (Updated 2 June 2009) (http://www.cdc.gov/ncidod/dvbid/LYME/index.htm).

Chareonsook, O., Foy, H.M., Teeraratkul, A., and Silarug, N. (1999). Changing epidemiology of dengue hemorrhagic fever in Thailand. *Epidemiology and Infection*, **122**, 161–66.

Charrel, R.N., Zaki, A.M., Fakeeh, M. *et al.* (2005). Low diversity of Alkhurma hemorrhagic fever virus, Saudi Arabia, 1994–1999. *Emerging Infectious Diseases*, **11**, 683–88.

Chevillon, C., Briant, L., Renaud, F., and Devaux, C. (2008). The Chikungunya threat: an ecological and evolutionary perspective. *Trends in Microbiology*, **16**, 80–88.

Chivian, E. (ed.). (2003). *Biodiversity: Its Importance to Human Health*. Interim executive summary. (http://chge.med.harvard.edu/publications/documents/Biodiversity_v2_screen.pdf).

Chivian, E. and Bernstein, A.S. (2004) Embedded in nature: human health and biodiversity. *Environmental Health Perspectives*, **112**, A12–A13.

Despommier, D., Ellis, B.R., and Wilcox, B.A. (2007). The role of ecotones in emerging infectious diseases. *Eco-Health*, **3**, 281–89.

Ellis, B.R. and Wilcox, B.A. (2009). The ecological dimensions of vector-borne disease research and control. *Cardenos de Saúde Pública*, **25**, S155–67.

Elton, C.S. (2001). *Animal Ecology*. University of Chicago Press, Chicago.

Engelthaler, D., Mosley, D., Cheek J.E. *et al.* (1999). Climatic and environmental patterns associated with hantavirus pulmonary syndrome, Four corners region, United States. *Emerging Infectious Diseases*, **5**, 87–94.

Erlanger, T.E., Weiss, S., Keiser, J., Utzinger, J. and Wiedenmayer, K. (2009). Past, present, and future of Japanese encephalitis. *Emerging Infectious Diseases*, **15**, 1–7.

Estallo, E.L., Lamfri, M.A., Scavuzzo, C.M. *et al.* (2008). Models for predicting *Aedes aegypti* larval indices based on satellite images and climatic variables. *Journal of the American Mosquito Control Association*, **24**, 368–76.

FAO (Food and Agriculture organization of the United Nations). (2006). *Global Forest Resources Assessment 2005, Main Report. Progress Towards Sustainable Forest Management*. FAO Forestry Paper 147, Rome.

Field, H.E. (2009). Bats and emerging zoonoses: henipaviruses and SARS. *Zoonoses and Public Health*, **56**, 278–84.

Garfield, R.M. (1989). War-related changes in health and health services in Nicaragua. *Social Science and Medicine*, **28**, 669–76.

Gatrell, A.C. (2002).*Geography of Health: An Introduction*, Blackwell Publishers, Oxford.

Gause, G.F. (1934). *The Struggle for Existence*. Williams & Wilkins, Baltimore.

Gonzalez, J.P., Herbreteau, V., Morvan, J., and Leroy, E., (2005). Ebola virus circulation in Africa: a balance between clinical expression and epidemiological silence. *Bulletin de la Société de Pathologie Exotique*, **98**, 210–17.

Gould, E.A. and Higgs, S. (2009). Impact of climate change and other factors on emerging arbovirus diseases. *Transactions of the Royal Society of Tropical Medicine and Hygiene*, **103**, 109–21.

Gould, E.A. and Solomon, T. (2008). Pathogenic flaviviruses. *Lancet*, **371**, 500–9.

Grinnell, J. (1917). The niche-relationships of the California Thrasher. *Auk*, **34**, 427–33.

Gubler, D.J. (1998). Resurgent vector-borne diseases as a global health problem. *Emerging Infectious Diseases*, **4**, 442–50.

Guha-Sapir, D. and Schimmer, B. (2005). Dengue fever: new paradigms for a changing epidemiology. *Emerging Themes in Epidemiology*, **2**, 1–10.

Hales S., Weinstein, P., and Woodward, A. (1996). Dengue fever epidemics in the South Pacific: driven by El Niño Southern Oscillation? *Lancet*, **348**, 1664–65.

Horrigan, L., Lawrence, R.S., and Walker, P. (2002). How sustainable agriculture can address the environmental and human health harms of industrial agriculture. *Environmental Health Perspectives*, **110**, 445–56.

IPCC (Intergovernmental Panel on Climate Change). (2007). *Climate Change 2007: the Physical Science Basis (Summary for Policy Makers)*. IPCC, Switzerland.

Jittapalapong, S., Herbreteau, V., Hugot, J.P. *et al.* 2009. Relationship of parasites and pathogens diversity to rodents in Thailand. *Kasetsart Journal*, **43**, 106–17.

Johnson, M.F., Gomez, A., and Pinedo-Vasquez, M. (2008). Land use and mosquito diversity in the Peruvian Amazon. *Journal of Medical Entomology*, **45**, 1023–30.

Hardin, G. (1960). The competitive exclusion principle. *Science*, **131**, 1292–97.

Hutchinson, G.E. (1957). Concluding remarks. *Cold Spring Harbor Symposia on Quantitative Biology*, **22**, 415–27.

Keating, J. (2001). An investigation into the cyclical incidence of dengue fever. *Social Science and Medicine*, **53**, 1587–97.

Keiser, J., De Castro, M.C., Maltese, M.F. *et al.* (2005). Effect of irrigation and large dams on the burden of malaria on a global and regional scale. *American Journal of Tropical Medicine and Hygiene*, **72**, 392–406.

Kosatsky, T. (2005). The 2003 European heat waves. *Eurosurveillance*, **10**, 552. Available at: http://www.eurosurveillance.org/.

Kutz, F.W., Wade, T.G., and Pagac, B.B. (2003). A geospatial study of the potential of two exotic species of mosquitoes to impact the epidemiology of West Nile virus in Maryland. *Journal of the American Mosquito Control Association*, **19**, 190–98.

Leonardo, L.R., Rivera, P.T., Crisostomo, B.A. *et al.* (2005). A study of the environmental determinants of malaria and schistosomiasis in the Philippines using Remote Sensing and Geographic Information Systems. *Parasitologia*, **47**, 105–14.

Lobitz, B., Beck, L., Huq, B., Fuchs, G., Faruque, A.S.G., and Colwell, R. (2000). Climate and infectious disease: Use of remote sensing for detection of *Vibrio cholerae* by indirect measurement. *Proceedings of the National Academy of Sciences of the USA*, **97**, 1438–43.

LoGiudice, K., Ostfeld, R.S., Schmidt, K.A., and Keesing, F. (2003). The ecology of infectious disease: Effects of host diversity and community composition on Lyme disease risk. *Proceedings of the National Academy of Sciences of the USA*, **100**, 567–71.

Lowe, S., Browne, M., and Boudjelas, S. (2001). *100 of the World's Worst Invasive Alien Species - A Selection from the Global Invasive Species Database*. Invasive Species Specialist Group, Auckland, New Zealand.

Malek, E.A. (1975). Effect of the Aswan High Dam on prevalence of schistosomiasis in Egypt. *Tropical and Geographical Medicine*, **27**, 359–64.

Matthews, S. (2004). *Tropical Asia Invaded: The Growing Danger of Invasive Alien Species*. Global Invasive Species Programme, Cape Town.

Mayer, D.J. (2000). Geography, ecology and emerging diseases. *Social Science and Medicine*, **50**, 937–52.

Mead, P.S., Slutsker, L., Dietz, V. *et al.* (1999). Food-related illness and death in the United States. *Emerging Infectious Diseases*, **5**, 607–25.

Morens, D.M., Folkers, G.K., and Fauci, A.S. (2008). Emerging infections: a perpetual challenge. *Lancet Infectious Diseases*, **8**, 710–19.

Padbidri, V.S., Wairagkar, N.S., Joshi, G.D. *et al.* (2002). A serological survey of arboviral diseases among the human population of the Andaman and Nicobar islands India. *Southeast Asian Journal of Tropical Medicine and Public Health*, **33**, 794–801.

Passos, C.J.S. and Mergler, D. (2008). Human mercury exposure and adverse health effects in the Amazon: a review. *Cardenos de Saúde Pública*, **24**, S503–20.

Pattnaik, P. (2006). Kyasanur forest disease: an epidemiological view in India. *Reviews in Medical Virology*, **16**, 151–65.

Patz, J.A., Daszak, P., Tabor, G.M. *et al.* (2004). Unhealthy landscapes: policy recommendations on land use change and infectious disease emergence. *Environmental Health Perspectives*, **112**, 1092–98.

Perkins, S.E., Cattadori, I.M., Tagliapietra, V., Rizzoli, A.P., and Hudson, P.J. (2006). Localized deer absence leads to tick amplification. *Ecology*, **87**, 1981–86.

Quijano, R., Panganiban, L., and Cortes-Maramba, N. (1993). Time to blow the whistle: dangers of toxic chemicals. *World Health*, **46**, 26–27.

Relman, D.A., Hamburg, M.A., Choffnes, E.R., and Mack, A. (2008). *Global Climate Change and Extreme Weather Events: Understanding the Contributions to Infectious Disease Emergence: Workshop Summary*. National Academies Press, Washington.

Richt, J.A. and Hall, S.M. (2008). BSE case associated with prion protein gene mutation. *PLoS Pathogens*, **4**, e1000156.

Sauvé, N., Dzokoto, A., Opare, B. *et al.* (2002). The price of development: HIV infection in a semiurban community of Ghana. *Journal of Acquired Immune Deficiency Syndrome*, **20**, 402–08.

Shetty, P. (2009). Preparation for a pandemic: influenza A H1N1. *Lancet Infectious Diseases*, **9**, 339–40.

Solomon, S., Plattner, G.K., Knutti, R., and Friedlingstein, P. (2009). Irreversible climate change due to carbon dioxide emissions. *Proceedings of the National Academy of Sciences of the USA*, **106**, 1704–09.

Soto, A.M., Calabro, J.M., Prechtl, N.V. *et al.* (2004). Androgenic and estrogenic activity in water bodies receiving cattle feedlot effluent in eastern Nebraska, USA. *Environmental Health Perspectives*, **112**, 346–52.

Sriskandan, S. and Slater, J.D. (2006). Invasive disease and toxic shock due to zoonotic *Streptococcus suis*: an emerging infection in the East? *PLoS Medicine*, **3**, e187, 0595–97.

Sutherst, R.W. (2004). Global change and human vulnerability to vector-borne diseases. *Clinical Microbiology Review*, **17**, 136–73.

Tang, J., Wang, C., Feng, Y. *et al.* (2006). Streptococcal toxic shock syndrome caused by *Streptococcus suis* serotype 2. *PLoS Medicine*, **3**, e151, 0668–76.

United Nations Secretariat, Population Division. (1999). *The World at Six Billion* (e-book) United Nations, New York (http://www.un.org/esa/population/publications/sixbillion/sixbillion.htm).

U.S. Census Bureau. (2008a). *Historical Estimates of World Population* (http://www.census.gov/ipc/www/idb/worldpop.html).

U.S. Census Bureau. (2008b). *Total Midyear Population for the World: 1950-2050* (http://www.census.gov/ipc/www/idb/worldpop.html).

Van Donk, M. (2006). "Positive" urban futures in sub-Saharan Africa: HIV/AIDS and the need for ABC (A Broader Conceptualization). *Environment and Urbanization*, **18**, 155–75.

Vittor, A.Y., Gilman, R.H., Tielsch, J. *et al.* (2006). The effect of deforestation on the human-biting rate of *Anopheles darlingi*, the primary vector of *falciparum* malaria in the Peruvian Amazon. *American Journal of Tropical Medicine and Hygiene*, **74**, 3–11.

Wang, J., Zhang, H., Fu, S. *et al.* (2009). Isolation of Kyasanur Forest disease virus from febrile patient, Yunnan, China. *Emerging Infectious Diseases*, **15**, 326–28.

Webster, R.G., Peiris, M., Chen, H., and Guan, Y. (2006). H5N1 outbreaks and enzootic influenza. *Emerging Infectious Diseases*, **5**, 3–8.

Wilson, D.E. and Reeder, D.M. (eds). (2005). *Mammal Species of the World*, 3rd Edn. Johns Hopkins University Press, Washington.

World Commission on Dams. (2000). *Dams and Development: A New Framework for Decision-Making* (e-book). Earthscan Publications, London (http://www.dams.org/report/contents.htm).

World Health Organization. (2001). *Fact Sheet No 100: Yellow Fever* (http://www.who.int/mediacentre/factsheets/fs100/en/).

World Health Organization. (2008). *Yellow Fever in Brazil* (http://www.who.int/csr/don/2008_02_07/en/index.html).

World Health Organization. (2009). *Confirmed Human Cases of Avian Influenza A (H5N1)* (http://www.who.int/csr/disease/avian_influenza/country/en/).

Yilmaz, G.R., Buzgan, T., Irmak, H. *et al.* (2009). The epidemiology of Crimean-Congo hemorrhagic fever in Turkey, 2002–2007. *International Journal of Infectious Diseases*, **13**, 380–86.

Conclusion

Serge Morand and Boris R. Krasnov

The contributors of this book have explored various facets of host–parasite interactions in various spatial and temporal contexts, with major references to ecology and evolution. In this final conclusion, we will present six general perspectives emerging from the results, opinions, and ideas introduced throughout chapters of this volume.

The historical biogeography of host–parasite relationships helps us to understand present and future ecology of both hosts and parasites

Eric Hoberg and Daniel Brooks illustrated how the biogeography of hosts and their parasites should be magnified by history, which can be revealed by comparisons of phylogenetic trees of hosts, parasites, and areas. Moreover, following the suggestion of Katharina Dittmar, the inclusion of fossil records—often neglected in historical biogeography—may give a more accurate scenario for the temporal scales of macroevolutionary events. Fossil records can be used as calibration points for both hosts and parasites and/or to estimate ancestral distribution areas. Eric Hoberg and Daniel Brooks also emphasized that the observed phylogeographic mosaics of interactions are ephemeral in space and time in the microevolutionary sense, but that these mosaics emerge from a deeper macroevolutionary and historical landscape. An important message, then, is that historical biogeography is not only useful for understanding the present, as the study of the past is also important to explore the future.

Macroecology identifies global rules of interactions and explores the underlying processes

Global patterns of parasite diversity have received less attention than those concerning free-living species, but new advances were presented for terrestrial environment by Frédéric Bordes and coauthors and for marine environment by Klaus Rohde. As suggested by Klaus Rohde, there is a need for autoecological studies of some key species from a variety of habitats to elucidate mechanisms that may explain diversity patterns. However, the necessity of exploring these mechanisms may come up against the problem of our poor knowledge on valid parasitological data from areas that are still largely unexplored.

The lack of geographical patterns in parasites, such as the latitudinal gradient in parasites of mammals as shown by Frédéric Bordes and coauthors, does not mean that the search for ecological laws is in vain. Optimism comes from Boris Krasnov and Robert Poulin, who showed species-specific stability in abundance and host specificity of several parasite taxa. Geographical stability may represent parasite species attributes and parasite responses to their external environment. Boris Krasnov and Robert Poulin advocated that fundamental rules in parasite ecology exist and that parasite populations may be governed by some common principles. The existence of common principles is further illustrated by Robert Poulin and Boris Krasnov, who showed the role of distance in investigations of large-scale biogeographical patterns. Hence, distance decay relationships provide simple null patterns, and these patterns should be incorporated as underlying phenomena in other biogeographical studies. However, there is a need to improve the analyses of biogeography of host–parasite relationships by incorporating epidemiological data and phylogenetic information.

*Microevolution: from phylogeographic mosaics
to landscape evolutionary epidemiology*

Nadir Alvarez and co-authors emphasized that evolutionary processes usually take place in a dynamic geographical context and consequently that patterns of genetic variation are highly structured in space and time. Reciprocally, studies of geography of co-interacting species need to explore how selection and adaptation act. Starting from the initial definition of landscape epidemiology, which was proposed as an integrative approach to understand the spatial spread of disease agents, Julie Dieter and co-authors reviewed the challenges remaining to be addressed. Molecular epidemiological information should include the distribution of genetic variation associated with host susceptibility and/or parasite virulence. The outcomes of genetic–environmental interactions should be taken into account for a better understanding of evolutionary processes.

*Immuno-ecology as an integrative part of studies
of emerging infectious diseases, invasion biology,
and conservation biology*

The recent emergence of immuno-ecology as a new scientific field is based on the integration of methods from behavioural ecology, physiology, and immunology. Immune functions are put in the framework of evolutionary ecology. Here, several contributions showed how immune functions may vary geographically among host species or populations.

Using islands as models in the study of the geography of host–parasite interactions, Kevin Matson and Jon Beadell showed the possible paths that have led to immunological differentiation and increased disease susceptibility in insular birds. However, Kevin Matson and Jon Beadell considered that the outcome of any particular host–parasite interaction in terms of host fitness, immune response, and pathology, remains highly unpredictable. This is not the sentiment of Serge Morand and co-authors, who showed that immune functions may structure parasite communities and/or may respond to parasite diversity, and suggested that immune responses are proximal mechanisms of parasite ecological organization. Anders Møller and László Garamszegi showed that hosts differing in their level of immune defence may also differ in

their invasion ability, and argued that the three components of the invasion processes (dispersal ability, establishment success, and subsequent expansion) are dependent on the immune status of invaders. Pascale Perrin and co-authors emphasized that pathogens have imposed strong selection on genetics, behavioural ecology, and potentially cultural and social structures of humans as is the case in any other animal species. They emphasized the crucial importance of large-scale investigations of geography of human parasites in relation to geography of human defences.

Methods and models are central, but progress is needed

Almost all contributions have emphasized the central role of statistical models for exploring the past, present, and future of host–parasite interactions. As underlined by Eric Hoberg and Daniel Brooks, discovery-based or *a posteriori* methods are powerful tools for exploring evolutionary and biogeographic history of the associated clades. Caroline Nieberding and co-authors proposed a co-phylogenetic program that integrates ecological traits of the interacting organisms and their geographical variation in order to identify the processes that may determine observed co-phylogenetic patterns. This program could help as in predicting the evolutionary outcome of host–parasite interactions.

As emphasized by Nadir Alvarez and co-authors, the future of comparative phylogeography is also to incorporate niche-based modelling. By using both actual and paleoecological data, these modelling approaches will help to infer past co-occurrences in space and time and parallel dispersal pathways. Eric Waltari and Sarah Perkins showed how ecological niche modelling can be applied to important questions regarding parasites, such as prediction of current or potentially invasive parasites and emerging diseases. Ecological niche modelling is becoming an important and increasingly useful tool in studies of host–parasite interactions.

Macroecological studies will gain much from using global gap analysis to identify and quantify biases in geographic sampling for parasites. Mariah Hopkins and Charles Nunn showed that gap analyses give spatially-explicit statistical corrections for sampling biases in host–parasite biogeography and help to better understand how patterns of sampling effort impact our knowledge of host–parasite biogeography.

Health ecology: a new integrative discipline?

In a world of global changes, parasites and pathogens are considered a serious threat, as reviewed by Vincent Herbreteau. Epidemiological environment is consistently disturbed by major changes in land use, migration, and population pressure, all of which act on host–parasite co-adaptation and in the appearance or diffusion of new infectious diseases. In this context, Katharina Dittmar advocated combining past and extant parasitological and climate distributions in order to get better long term predictive power regarding the parasitological consequences of climate change.

As humanity continues to homogenize the Earth's biota, there is an urgent need to predict evolutionary outcomes of bioinvasions. Mark Torchin and co-authors showed that the phenomenon of enemy release may play an important role in evolutionary responses in contemporary human-mediated invasions. They argued that biogeography may influence the extent to which introduced species escape from natural parasites.

Global changes are accompanied by the spread of emerging infectious diseases. Jan Slingenbergh and Stéphane de la Rocque explored how pathogen establishment and colonization may dictate the future of parasite–host interactions, particularly virulence evolution and host range extension, using the recent panzootic of a highly pathogenic avian influenza (HPAI) H5N1 virus as an example. They emphasized that veterinary and medicine should incorporate the concepts of evolution and ecology to improve risk analyses and control methods.

Finally, following Vincent Herbreteau, we strongly encourage an interdisciplinary approach to health issues, based on ecology and geography; a challenging perspective since the ecological–social–environment interactions involved in health patterns are very complicated.

Index